The Invention of Surgery

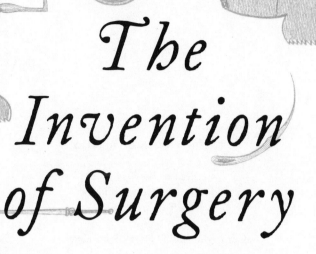

The
Invention
of Surgery

A History of Modern Medicine:
From the Renaissance to
the Implant Revolution

DAVID SCHNEIDER, MD

PEGASUS BOOKS

NEW YORK LONDON

THE INVENTION OF SURGERY

Pegasus Books, Ltd.
148 W 37th Street, 13th Floor
New York, NY 10018

Copyright © 2020 David Schneider, MD.

First Pegasus Books cloth edition March 2020

Interior design by Maria Fernandez

ISBN: 978-1-64313-316-4

10 9 8 7 6 5 4 3 2 1

Printed in the United States of America

Distributed by W. W. Norton & Company, Inc.

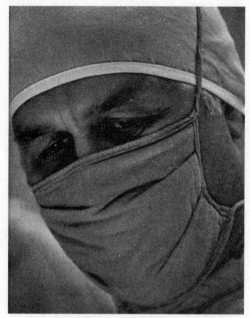

For my father, J. E. Schneider, DVM

1932–2007

Contents

Preface

"Dr. Schneider, this is Karen Lambert and I'm calling from Belize. You fixed my shoulder a few years ago and I'm dealing with an awful emergency." The phone line crackled and gapped, credibly confirming the Central American origin. "My husband and I are on an eco-tour vacation, and two days ago at a zip line park, his harness broke and he fell twenty feet. His elbow dislocated and broken bones were sticking out of his arm."

Karen went on to explain that her husband, Mark, had been shuttled to a small hospital in a nearby town, but that she had not been allowed to see him since he was admitted forty-eight hours before. The local doctor had reduced the elbow (aligning the joint and broken bones) but had not operated. Frantic, she pleaded for help to extricate her husband from the unsophisticated infirmary and convey him to the United States.

My team and I sprang into action, and working with a local air ambulance company, facilitated transport to Denver the next day on a private jet staffed with nurses. An ambulance met them at Denver International Airport, and after rushing him to my level one trauma center, we prepared for emergency surgery at 2:00 A.M.

All day long I had prepared for the worst. I was worried about a life-threatening infection, and knew for certain that loss of his arm was a distinct possibility. At best, I was hoping that we could minimize lifelong disability and hope for reasonable function. When I met with Mark and Karen in the pre-op holding area, they looked understandably exhausted and emotionally empty. Mark was lying on a gurney in a white surgery smock; Karen was still wearing the logoed travel company T-shirt, khaki shorts, and adventure sandals. Just as I was about to give them "the big talk" about doing our best to save his arm, Mark looked up at me with travel-weary eyes and informed me:

"I want to be able to play softball this summer and I don't want scars."

Stunned, I tried recalibrating, sensing the need to educate Mark about the grave consequences he was facing. I couldn't break through. He equated Karen's successful shoulder reconstruction and stabilization to his present situation. I mentioned the 80 percent death rate associated with "open fractures" just one hundred years ago, and tried to convey the complex nature of the ligaments, tendons, and muscles that held his elbow together, and the challenge of closing gnarly traumatic lacerations, but to no avail. Although I liked Mark's optimism, I was concerned he wasn't grasping the potential for serious complications, and the certainty that his arm would never be the same.

By dint of miracle, his surgery went extremely well. He didn't die, he didn't lose his arm, and he (somehow) ended up with superb function and had no disability. In fact, his scars were unnoticeable.

At his final appointment (following a softball game), we recalled his travails and assessed his final outcome. I tried one last time to discuss how close he came to losing his arm and how just a short time ago he would have almost certainly perished from this injury. Mark is an aerospace engineer, so common here in Boulder, Colorado, and despite his great intelligence, has no perspective about modern surgery. In fact, almost no one does, including surgeons. So when I reviewed his progress and compared it to how it would have gone just seventy-five years ago, Mark was shocked that there was no such thing as plates and screws, or even antibiotics, in the years before World War II.

Not that long ago, no one believed in germs. Although the first anesthetic drugs were discovered in the mid–19th century, surgery was still extremely dangerous until a small group of physicians and scientists were able to prove that the minuscule organisms that invisibly inhabit our world are the cause of infections. This knowledge triggered a revolution in medicine and surgery, and the first triumph was convincing surgeons to wash their hands before surgery.

An agonizing interval of seventy years passed from the acceptance of the germ theory to the development of antibiotics. During that period, surgery slowly developed, but to our modern eyes was highly limited in scope and efficacy. But a simultaneous series of inventions, like the development of polymers and transistors, modern alloys and antibiotics, and the undergirding establishment of private health insurance and Medicare, made modern surgery what it is.

Implant surgery, such as joint replacement, cardiac stent placement, lens surgery, and neurosurgical shunts, only became possible about fifty years ago. Implants, now numbering in the millions per year worldwide, were unthinkable a century ago, but this modern marriage of science, art, hubris, imagination, madness, bravery, and patience is nothing short of an *implant revolution*.

There are many encyclopedias of surgery and compendia of surgeons' biographies. There are a few books written in recent decades that truly bring to life some of the renegades and pioneers who helped make our world modern. What is missing is a narrative that interconnects those lives, weaves together their tales, and explains "how we got to now."

In this book, therefore, I set out to tell the story about the invention of surgery. In modern historiography, it has become au courant to presuppose that there are really no "lone geniuses" and almost no "eureka" moments. That is simply not true in the topic of surgery. There are many virtuosi who saw further in their underrated genius, challenged the status quo, and improved the lot of mankind more than in any other field. Here are their stories.

Introduction

Life is short, and Art long; the crisis fleeting; experience perilous, and decision difficult. The physician must not only be prepared to do what is right himself, but also to make the patient, the attendants, and externals cooperate.

—Hippocrates, *Aphorisms*, Section 1

The fact is that he whose purpose is to know anything better than the multitude do must far surpass all others both as regards his nature and his early training.

—Galen, *On the Natural Faculties*[1]

As the junior resident on the hand surgery service, I spend more time tending to the patients on the hospital floor and in the emergency room, and less time in the operating room. This summer has been hectic, with multiple "replants" (the reattachment of fingers after trauma suffered at factories, lumber mills, and backyard fireworks mishaps). Patients get airlifted or ambulanced to our trauma center from all over our region in hopes of saving their hands.

Two days ago, a young Amish boy suffered the loss of three fingers in a barnyard accident. Gabriel is five years old, but speaks no English—typical for a child here in central Pennsylvania, growing up in a cloistered community retaining the simplicity of a bygone era. In fact, almost no one in his family can communicate well with us. I have treated some Amish and Old Order Mennonite patients who speak modern English effortlessly, but some sects of Amish barely break away from their "Low German" dialect.

My job this morning is to change the leeches on Gabriel's fingers. You read that right. It sounds positively medieval, but there is a role for leeches in modern medicine. Once the hand surgeon has completed the daunting challenge of reattaching fingers, which includes the tasks of realigning and stabilizing the bones, stitching together the tendons, and sewing the nerves and blood vessels with microscopic suture, he must monitor the blood flow within the arteries and veins to see if the finger will thrive. Leeches are used for their bizarre ability to secrete hirudin, a natural anti-coagulant from their salivary glands that facilitates hematophagy, the ingestion of blood. Attaching a medicinal leech to a finger decongests the digit, thus increasing the chance of survival. The leech swells with feeding, and once it is fully engorged, it must be replaced with a ravenous collaborator to continue the digit-saving bacchanal.

As I walk into Gabriel's room, I am greeted with a blast of furnace-hot air laced with the essence of barnyard manure. In an effort to accentuate his fingers' vasodilation (expansion of the blood vessels), we keep patients' rooms at 95°F. Inside the room are more than twenty people, all of whom are Amish; the men with characteristic Abe Lincoln beards, black wool trousers, suspenders, and white shirts, and women with bonnets and flowing navy-blue dresses to their ankles. I am reminded that most Amish bathe once a week, and the combination of heavy wool dark clothing, blistering hot and muggy Pennsylvania summers, and farm animal occupations make this room reek, even to me, the son of a large-animal veterinarian.

I have brought a jar of fresh leeches, skinny and dark wormlike creatures. I lean over the stoic Gabriel, his hand in a massive dressing three times the size of a boxing glove. As I undo the layers of white cotton dressing, my community of witnesses leans closer; I seem to be the only one sweating in the oven of room 765. With the removal of the final loose layer of gauze we are all staring at three huge leeches, each attached to a finger. They are crimson and india ink–black, immobile, and drunk with blood. They look ready to explode. I begin to tug on the first parasite, and it won't budge. A wave of anticipation pulses through the throng, and now twenty faces are within feet of mine, and a mixture of pig, horse, and cow manure wafts pungently toward me with essences of molasses, scrapple (bacon remnant), and chow-chow (pickle relish) mixed in. I could vomit.

With a little more force, I am finally able to pluck the little vampire off a finger and the assembly groans in appreciation, "Yaaaa." I repeat this two more times, with the fingers oozing at the attachment sites. One by one I then reach into the little jar and pick out a slimy creature and drag it onto a finger. With a little wriggle, the slug-like animal positions itself on the finger and a firm linkage is established. Gabriel has remained motionless the entire time, and he and I make eye contact again. We have no words in common beyond a simple greeting, but we do share at this moment thousands of years of medical tradition—the art of bloodletting. Although bleeding a patient is no longer practiced in America, there are still places in the world where bloodletting occurs in a fashion that goes back 2,500 years to the very beginning of medicine. My medical forefathers couldn't have dreamed of refabricating fingers to a hand, even one hundred years ago. But they would have been enchanted by the notion of leeches sucking "bad blood."

On the left bank of the River Seine, in Paris's labyrinthine Latin Quarter, are situated dozens of buildings associated with the Sorbonne, including the Université Paris 5 René Descartes. Located on the rue de l'École de Médecine, the university's greatest building is a 17th-century colonnaded structure that houses an enthralling museum of medicine and a library. Inside the building, at the end of the lobby, stands a life-size stone sculpture of a veiled woman who gently lifts a shroud away from her face and upper body, revealing her placid countenance and exposed breasts. The sculpture is titled *La Nature se dévoilant à la science*, or *Nature is revealed through science*.

In this place of great learning, this monument captures the very essence of the scientific program of the Renaissance and Scientific Revolution, wherein mankind removed the opaque veil from the distinctive beauty of nature. Centuries had passed since the philosophical and artistic revolutions of ancient Greece, but as medieval darkness gave way to the light of learning, a rekindling of an enlightened curiosity took hold across Europe. The 15th century was a time of exploration, innovation, and reinvention of communication via new technology—much like our current time.[2] Figures like Leonardo da Vinci, Christopher Columbus, and Johannes Gutenberg upended the status quo, much like Steve Jobs, Elon Musk, Jack Dorsey, and Mark Zuckerberg have done over the last decades—and not without controversy.

The Renaissance is a convenient starting point to trace the origins of modernity in medicine, in part because so little had changed from the time of Hippocrates to the 15th century. Even as the Western world was awakening from a great, thousand-year sleep, it was still mostly pointless to consult with a physician, and likely, more dangerous to be under the care of even the wisest doctor. As is beautifully detailed by David Wootton in *Bad Medicine*, a patient inflicted with almost any malady, *in any era before 1865*, would have been better served by suffering alone, away from the "care" of a physician.

Therefore, the two towering figures of Western medicine, Hippocrates and Galen, had actually done very little to improve the lot of men and women under their philosophical care for almost two thousand years. And certainly, they had contributed *nothing* to the practice of surgery. Nonetheless, it is critical to understand that these fathers of medicine—even though they were merely pulling so many levers behind the curtain like the Great Oz—influenced every Western physician over the last two thousand years, and so their theories matter.

My undertaking in this work is to explore the metamorphosis of the understanding of the way the body works, how disease happens, and the near-miraculous ways 21st-century surgeons can resuscitate, reconstruct, and even reimagine human beings. I will spend little time examining ancient Asian medicine, or the oral traditions of healers in primitive societies. While there may have been surprising perceptions among antiquity's shamans, dead-end, unlinked intellectual insights are not the focus of this work. The foundational breakthroughs that led to the invention of surgery—from the invention of science itself to the discovery of cells, germs, modern materials, and outcomes research—is the thrust of this book.

Stephen Greenblatt, in his enchanting book *The Swerve*, relates the story of the near-mythical poem, "On the Nature of Things," by the Epicurean poet, Lucretius. Lost to antiquity, the poem was remembered for its insights and artistry, but no one in the Middle Ages had ever read it. All that remained were stories about its greatness, similar to the legends of the Colossus of Rhodes or the Hanging Gardens of Babylon. After disappearing for 1,500 years, it was discovered in 1417 by an Italian scribe and book hunter, Poggio Bracciolini, in a southern Germany monastery.

Poggio concealed himself for three weeks in the monastery and copied its 7,400 Latin lines from ancient papyrus, returning to Rome with his treasure. Within a few decades, Gutenberg invented the printing press, and soon copies of Lucretius's poem would be printed and distributed around the Western world. The discovery of "On the Nature of Things" helped make the world modern, turning away "from a preoccupation with angels and demons and immaterial causes and to focus instead on things in this world; to understand that humans are made of the same stuff as everything else and are part of the natural order; to conduct experiments without fearing that one is infringing on God's jealously guarded secrets . . . to legitimate the pursuit of pleasure and the avoidance of pain . . . to find the mortal world is enough."[3] While it is challenging to pinpoint all the causes of the Renaissance, surely this poem, with its groundbreaking claims, helped the world "swerve" toward modernity.

What made the poem so radical? To start with, Lucretius claims that everything is made of invisible particles. He further postulates that these particles are eternal (which would be a foundational claim of Antoine Lavoisier, one of the fathers of chemistry). Harvard University philosopher George Santayana has called this "the greatest thought that mankind has ever hit upon."[4] In addition, our poet tells us that humans are not unique, we are in a primitive battle for survival, there is no afterlife, religions are cruel, and the highest goal of life is the enhancement of pleasure and the reduction of pain. Radical, indeed. When these assertions were resurrected at the end of the Middle Ages, one can see why they were so iconoclastic. As Gustave Flaubert has said, "Just when the gods had ceased to be, and the Christ had not yet come, there was a unique moment in history, between Cicero and Marcus Aurelius, when man stood alone."[5] These ponderings would help transform astrology to astronomy, alchemy to chemistry, and, eventually, Aristotelian cosmology to Newtonian Physics.

Hippocrates's life spans the triad of great philosophers—born ten years after Socrates, most of Plato's life, and overlapping Aristotle by fourteen years. Not just a physician, Hippocrates was a renowned author, a pillar of the culture, a patriot of Greece, and a moralist. The "Hippocratic corpus," the sixty pieces of writing that are attributed to him and his followers (though by some estimates, almost half are falsely attributed), contains works that were probably written over a century or two. John Block concluded, "Hippocrates first gave the physician an independent standing, separating him from the cosmological speculator. Hippocrates confined the medical man to medicine."[6] All early "healers" were natural philosophers, and Aristotle said that it was the task of these philosophers to look into the principles of health and disease. This started with an obsession with the "correct regimen" and proper diet. "How to find the diet that would maintain the body in health and free it from disease was a problem that invited speculation about the constituents of body and of food, as well as about the structure, the functions and the activities of the body and its parts."[7]

The ancient truth-seekers mulled over the function of the body without any knowledge of cells, germs, genes, cancer, even bodily organs; is it any wonder that disease was a complete enigma? If primitive man in every corner of the world was transfixed with the starry sky, found meaning in the mutable moon, contemplated the traversing of the sun, and considered the pulsing of the tides and respirations of the winds, how much more significance would be achieved by turning inward to our bodies and examining the motions, ebbs and flows of a pounding heart, of breathing, even urination and defecation?

Siddhartha Mukherjee, in *The Emperor of All Maladies*, says the ancient Greeks were "preoccupied with fluid mechanics—with waterwheels, pistons, valves, chambers, and sluices—a revolution in hydraulic science originating with irrigation and canal-digging and culminating with Archimedes discovering his eponymous principle of buoyancy in his bathtub. This preoccupation with hydraulics also flowed into Greek medicine and pathology. To explain illness—all illness—Hippocrates fashioned an elaborate doctrine based on fluids and volumes, which he freely applied to pneumonia, boils, dysentery, and hemorrhoids."[8]

Hippocrates, and later, his disciple Galen, would explain the inner workings and dysfunction of the body with the concept of the Four Humors (liquids). Thinking like a hydraulic engineer, Hippocrates theorized that our bodily vessel is a container of blood, phlegm, black bile, and yellow bile. "In the process of digestion, food and drink are turned into the bodily juices, the humors," writes Owsei Temkin, ". . . they are the nourishment of the body, i.e., of its tissues, which consequently owe their existence to the humors. The [Aristotelian] elements of fire, earth, and water do not exist as such in the body; they are represented by yellow bile, black bile, and phlegm, respectively."[9] Air, Aristotle's fourth element, is the pneuma of the Stoics (the vital spirit or creative force of a person), and is the vehicle of vital and psychic functions.

To understand Hippocrates's 4th-century B.C.E. mindset, remember that English physician William Harvey's breakthrough experimentation of blood circulation was still almost two thousand years away. The ancients had no concept of circular blood flow, which we trace starting at the heart, coursing through the aorta and subsequent smaller vessels—all the way to the narrowest blood vessels, the capillaries—with a gradual reversal to the thin-walled, low-pressure veins that form tributaries, like ever-widening rivers on their way to the ocean, to the massive vena cava that empties back into the heart. If you are reading this book you probably understand that blood does not simply "dump" into your muscles, like a container of meat being splashed with blood. Instead, your muscles are thoroughly perfused with tiny blood vessels, too small to see with the naked eye. There is no reservoir in our body where all the juices (Hippocrates's simplified bile, blood, and phlegm) collect together. Why would he conjecture this way?

It seems that Aristotle was the first to scientifically dissect an animal, and perhaps it was his pupil, Diocles who was the first to dissect a human.[10] In the ancient world, human dissection was permitted until being outlawed by the Romans. The Hippocratic physicians would have been allowed to dissect the dead, but this was before embalming and refrigeration, and would have demanded fairly quick action before putrefying flesh made investigation too repulsive. Presented with a recently deceased person or animal, an ancient physician would have likely made cuts into the abdomen, finding smelly bowels filled with half-digested food and large blood vessels containing congealed blood, dark purple in color. In the abdominal cavity, surrounding the bowels, there was abdominal fluid, like warm apple juice. Handling the organs, the kidneys, liver, and spleen would be crimson and full of gelatinous, molten blood. The gallbladder, anchored below the liver, would have been large and pear-shaped. Slicing into it, pea-sized gallstones would have tumbled out with yellowish fluid oozing over the examiner's hands. In the thoracic cavity, home to the lungs and heart, pulmonic fluid would be discovered surrounding the lungs and filling the lobes with frothy liquid, like watery tea in a sea sponge. The trachea and bronchial airways would almost certainly have had a mucous residue so common in a dying man. In conclusion, we have the four humors: blood, yellow bile, black bile, and phlegm. Without understanding organ function, the early anatomist would deliberate over these fluids, seeking a unifying "theory of everything" to explain the workings of the most interesting system in the universe. There must have been a singular moment (upon a particular corpse), when Hippocrates formulated and simplified his four humors theory; would there ever be a philosophical offering more contemplated and recited than this?

The individual humors, when dominant in a human being, contributed to the personality and behavior. Each one of the four personality types, based upon one of the humors, is familiar to our ears. In Greek, black bile is melancholia, upon which our word for a depressed, "melancholic" person is based. A calm, "cool-headed" person had an overabundance of phlegm, and was therefore "phlegmatic." An irritable, crotchety, or "bilious" person has too much yellow bile and is "choleric." If a patient had a predominance of blood, and was spirited or intemperate, he would have been called "sanguine," from the Latin word for "blood."

And here is the very important observation about the dominance of Hippocratic theory all the way to the Scientific Revolution. Even the savants of the Renaissance, who were forced to contemplate the function of the body in a world without science, were powerless to resist the allure of Hippocratic musings. Because the philosophical foundation was a fraud, medicine was ineffectual, even lethal. The Hippocratics provided much explanation for why the therapies worked: it never occurred to them they did not. [11] If Hippocrates is the Father of Medicine, it is a dubious paternity; we can't identify any success associated with his (or his followers') theories.

The most logical intervention for a sanguine person would have been to decrease the volume of blood. If the patient was "hot-headed," or if a disease was causing redness and heat (we would describe them as having a "fever"), the Hippocratic physician would bleed the patient. This was classically performed by cutting a vein (venesection), but later by cupping (suctioning the skin with a cup) or applying leeches. "Bloodletting" was therefore the ancient art of trying to achieve a balance in humors, and explains why so many patients were bled (often, to death). Consider all the times you have been sick with a fever. That fever, a bodily increase in temperature, is a systemic reaction to a bacterial or viral attack that is now easy to explain in scientific terms. Had you lived a mere five generations ago, you likely would have been bled bedside by your community physician.

The Roman Empire began in 31 B.C.E., with the consolidation of Greece and Hellenistic Egypt under one ruler, Augustus Caesar. Augustus ruled until 14 B.C.E., and Rome was the center of a powerful, peaceful kingdom for two hundred years. Greek city-states assimilated under Roman rule, and in turn, the Early Empire embraced classical Greek culture.

Into this period of relative peace and order was born the other great physician of antiquity, Galen (130–200 C.E.). Like Hippocrates, Galen was from east of the Aegean Sea, and was born in Pergamum in Asia Minor (present day Bergama, Turkey). Like Hippocrates's island of birth (Kos), Pergamum was home to a sanctuary of the healing god Asclepius. Galen's training started at home, extended to Smyrna and Corinth, and ended at Alexandria. Owsei Temkin has written: "The founding of Alexandria was an important event in the history of ancient scholarship, science, and medicine. From the 3rd century B.C.E. until its conquest by the Arabs in 642 C.E., Alexandria was the foremost center of medical study and especially of anatomy." As will become plain in this book, there has always been a "center of scientific and medical learning" in the world. "For a time, it seems, anatomy could be studied on human bodies, until Roman law put an end to such study and confined anatomy to animal dissection." [12]

Galen returned to Pergamum, flush with his Alexandrian education, where he became physician to the gladiators. An early "sports medicine physician," it has become clear that Galen probably performed no human dissection during his career, but he was exposed to deep anatomy during his surgical treatment of gladiatorial injuries. Galen was later summoned to Rome by the emperor Marcus Aurelius, and it seems that he spent his last forty years there, writing, teaching, and attending to the emperor.

Galen was not just an influential physician. He was a philosopher, a dazzling and highly industrious author who wrote in a cultivated Greek style, a scientist and skilled dissector (albeit, of monkeys and pigs). Highly prolific, "his preserved works alone would fill about a dozen volumes of approximately one thousand pages each."[13] If Aristotle was the first to perform animal dissection, and the first to postulate that the organs of the body had individual function, it was Galen who raised animal dissection and vivisection (dissecting on live animals) to another level. The major revolution in anatomical learning had occurred in Alexandria, led by Herophilus and Erasistratus, who were both contemporaries of Epicurus, in the 3rd century B.C.E. Alexander the Great had just founded his city at the time of their birth; it was a frontier city on the Mediterranean near the mouth of the Nile, surrounded by barbarians. It is possible that dissection (and even, shockingly, vivisection) was performed on convicted criminals in that city. Steven Johnson has described the "hummingbird effect, an innovation, or cluster of innovations, in one field [that] ends up triggering changes that seem to belong to a different domain altogether . . . sometimes change comes about through the actions of political leaders or inventors . . ."[14] The young city of Alexandria, as a Hellenistic outpost, was the ideal laboratory for the Greek natural philosophers, with a tradition (handed down from Alexander) of assimilating local customs and leaders and inculcating international students. For almost one thousand years it was the greatest city of learning in the world and had the largest library (of papyrus scrolls). The marriage of ancient Egyptian scholarship, Greek philosophical insight and empiricism, and contributions of conquered Persian and Indian peoples made Alexandria the ideal city for Galen to complete his studies.

Galen's great work *On Anatomical Procedures* is a wonder. It was his last major work, and was based on a lifetime of anatomical investigation. It has been said that Galen, though not the founder of the science of anatomy, was its first important witness, and this work is his pièce de résistance. Like most anatomy books, there is an abundance of information on bones, muscles, blood vessels, and organs. However, much of the writing is imbued with Hippocratic humoral physiology; it is laughably wrong when examined today, but it was the authoritative work until Vesalius's *De humani corporis fabrica* (*On the Fabric of the Human Body*) was published in 1543. So highly regarded was Galen that Vesalius, as we shall see, had to tiptoe around criticizing the master and gently sow the first seeds of doubts about his authority.

Galen became a true pioneer when he performed anatomic experimentation. Tragically, it involved vivisection, but instead of conjecture about the imbalance of humors, Galen became the first to uncover organ function. "By tying and untying the ureters Galen proves the flow of urine from the kidneys to the bladder; he severs the spinal cord at different levels and describes the ensuing loss of motion and sensibility; he ligates the recurrent nerves [that lead from the brain to the vocal cords] and he has discovered and notes the subsequent loss of voice."[15] This 2nd century natural philosopher upended centuries of Aristotelian theory about the heart being the "command center" of the body, and instead demonstrated that the nerves carried the impulses to the muscles from the brain.

Why do we breathe? The Greco-Roman philosophers had no concept of oxygen, and were left pondering the role of respiration and conjectured that there was a *pneuma*, a vital spirit, that must be drawn in to infuse the body. The psychic pneuma, Galen concluded, must well up from the net-like plexus of arteries at the base of the brain, what he termed the *rete mirabile*, and travel to the ventricles, the fluid-filled caverns in the middle of the brain. As Galen had established that impulses must originate in the brain, the empty space of the ventricles must be the domicile of the psychic pneuma. Galen's rete mirabile, the fount of the psychic pneuma, would become a major issue 1,300 years later, but for now, Galen had led a critical revolution in deciding that cognition originated in the brain.

Near the end of Galen's life, at the end of the 2nd century, "peace and stability collapsed, and for about a hundred years political conditions close to anarchy disrupted cultural and economic life."[16] Eventually, barbarian incursion on Roman soil destabilized the empire. In one of the most impactful developments in Western civilization, the emperor Constantine made Byzantium (which he renamed Constantinople, modern day Istanbul) his capital in 330 c.e. Rome and Constantinople were dual capitals for decades, but in 395 c.e., after the death of the emperor Theodosius, the empire was divided permanently between the Latin West and East. By the late 400s, Rome was in full collapse, and the Latin Middle Ages would last for one thousand years.

One cannot comprehend Western civilization without understanding the (temporary) survival of the Roman Empire in Constantinople for hundreds of years. "While the West was on its way to the Latin Middle Ages, the Greek East preserved its ancient heritage. Justinian even succeeded in reconquering Italy, Africa, and part of Spain, and reuniting the Roman Empire. But the reunification did not long outlast him. Culturally, Antiquity faded away slowly, but politically the East, even before the onset of the Arab conquests in 634, had become the Byzantine Empire."[17] During the final breakup of the Roman Empire, Greek culture (and medicine) continued to spread throughout the Middle East, first to Syria, then to Persia, and finally to the Mohammedan world. "Several of the Prophet's successors (Mohammed died in c.e. 632) . . . were great patrons of Greek learning, and especially of medicine. The Arabian scholars imbibed Aristotle and Galen with avidity."[18] As we shall see, Arab scholars kept the faith, so to speak, of Hippocrates and Galen, and their writings were translated from Greek and Latin into Arabic. These Arab language books, serving as repositories of ancient wisdom, awaited translation back into Latin at an appropriate time—a time of awakening—the Renaissance.

Homo sapiens have been on earth for some 250,000 years, but modern man has existed 8,000 years, which equates to 300 generations. On a single sheet of paper, you could write the word "great" 300 times; each "great" would represent a particular ancestor of yours, ending with our common biological "Adam," the first modern man.

So we are left contemplating 295 generations of vulnerability, completely at the mercy of nature; five generations blessed to exist under the auspices of "good medicine"; and two generations thriving in an era of modern medicine and what I will call the implant revolution. The Greco-Roman domination of medicine persisted into the 16th century, finally undermined by an elegantly simple innovation that revolutionized humanity's ability to communicate, allowing incremental advances in the understanding of the way our bodies work, and then, later, dynamic leaps forward with the invention of surgery.

ONE

Dilemma

"It has been the experience of all who treat tuberculous joints that bony ankyloses [joint fusion] is the most satisfactory result which can be obtained. No other result assures equal permanency of cure and freedom from late recurrences. The truth is that our means of treating tuberculosis are limited and only feebly effective. In the last analysis, it is the patient himself who masters the infection. We have no specific drug, serum, or therapeutic agent, the use of which will quickly kill the organisms. [Only two things] are of any material value: rest and sunshine."

—R. I. Harris, Toronto, Ontario, 1935[1]

"Dr. Neer became disenchanted with the end results of patients with fractures of the proximal humerus treated with resection of the humeral head. He mentioned this to Dr. Darrach who said, 'Smiley, why don't you do something about it?'"

—Charles Rockwood, MD

My heart sinks when I hear that Miranda has dislocated her shoulder again. I first met her a year ago, when she estimated that she had dislocated both shoulders dozens of times each. As a seizure patient, Miranda is susceptible to a particularly diabolical type of dislocation where the humeral head is forced backward (posterior) and out of joint, instead of the usual anterior dislocation, in which the humeral head is displaced forward and toward the chest wall. Most full-blown dislocations require manipulation (reduction) by a clinician, preferably under deep sedation in the Emergency Room, with the sobering realization that vast numbers of patients over the millennia simply lived with a chronically dislocated and crippled shoulder.

Miranda's latest dislocation was particularly discouraging for her, because as a twenty-five-year-old, she and her doctor had finally found a medication regimen that had eradicated her seizures. Finding the right anti-seizure medicine can be extremely tricky, balancing side effects against the burden, embarrassment, and inconvenience of a seizure. She had lived seizure-free for months, daring to hope that they were finally gone. But here she was in our ER, painfully frozen to her gurney, her arm protected against her abdomen, downcast and dispirited. She knew the drill: we'd start an IV, "knock her out" with powerful sedatives, and I'd maneuver the arm around while pulling powerfully on her forearm. It seemed plain that she was more disconsolate about her seizure than her dislocation, but then again, my primary job was to relocate the shoulder. Meeting people at their lowest and later helping them to be at their best is among the greatest honors of being a surgeon. A major part of facilitating that transition is providing hope, and I told her that we'd promptly get her shoulder reduced, but more importantly, I gently suggested that we should surgically address her shoulder in the future and make her dislocations a thing of the past. It was as though she hadn't realized that there was a cure for her problem; I saw a spark of hope, asking, "Is it really possible to keep my shoulder from dislocating?" "Yes," I assured her, "we are much better at solving people's shoulder instability problems today through a combination of techniques. Once we're done here today, let's set up an appointment in my clinic to fully talk about your shoulders."

Miranda eventually saw me in clinic, where we discussed surgery. After a detailed conversation she opted for surgery, and soon we addressed her stretched-out shoulder capsule, torn labrum (the gristly connective tissue around the shoulder socket that keeps the humeral head in place), and damaged bony surfaces, and were able to give new life to her left shoulder. In the months that followed, she progressed well, not dislocating either side, and more significantly, not experiencing another seizure.

Now, half a year after her left shoulder operation, Miranda has returned to my clinic, where I learn that she dislocated her shoulder again. "Her left or right?" I ask. "The right side—not the one she had fixed," my assistant Kristy replies.

Relieved that her operatively repaired left shoulder is still doing well, I knock on her exam room door and enter, finding Miranda sitting on an exam table, and I'm struck by how nervous she is. We know each other pretty well, but she's anxious, even fidgety.

"Miranda, how are you doing?"

"I had another seizure . . . I'm sorry," she blurts out.

I have seen this kind of apologetic reaction in patients who are subject to migraines, seizures, inflammatory bowel disease, and other episodic illnesses where the sufferer has little-to-no control over the disease. The self-reflection on causality, I think, makes them explore whether or not they are to blame for their infirmities.

"This last seizure was a really bad one. Normally I have a pretty strong sense it's coming on, but I had almost no warning this time. My boyfriend had never seen me have a seizure, and it was really tough for him to see my face so screwed up during the convulsions. You know, my friend videoed me once during a seizure and I couldn't believe how scary I looked. Now, he saw me look that way . . ." and she trails away with her eyes welling up with tears.

My hand on her shoulder, I console her, "You know it's not your fault, right, Miranda?"

"I just feel so bad about it. I also wet my pants and had to leave the restaurant with pee all over my jeans. I just don't know why I have to have all these damn seizures."

"Miranda, I cannot imagine how frustrating that must be. I feel so sorry that you suffer from these seizures. It simply isn't fair. I'm hoping that you and your neurologist can tweak your meds and get your seizures under control, and this I swear to you: I will do everything in my power to make both of your shoulders stable and pain free, so that even if you have another seizure your shoulders will be okay."

When treating a chronic dislocator who suffers from seizures, I often think about patients in antiquity, who were castigated for their epileptic fits, abused for their "demon possession," or suspected of witchcraft. The writhing and grimacing of a paroxysm, and the apoplexy that followed a seizure, lead the ancients to conclude that some supernatural power was governing the bodily temple of the victim. And just when the patient's existence couldn't be more precarious (with the inference of a hellish collusion), their earthly subsistence increasingly deteriorated with headaches, bodily injuries, tongue bites, confusion, and psychosis.

The rare early philosopher had insight that seizures were not underworldly, but instead were physical disease states. Only in the last century have seizure disorders become treatable, roughly in the same time frame that shoulder instability has become manageable. All medical pioneers shared a certain exasperation, an odium, for the way things were. Even today, when speaking with patients who are burdened with unjust conditions, I have a bitter sadness and vexation for their "dis-ease" that I know my medical forbears had in great measure, as well as a disgust for their poor understanding of what causes disease and how to treat it.

Dr. Charles Neer glanced at the X-rays of Mrs. Harrison's shoulder, recognizing in a moment that the elderly New Yorker's arm would be useless for the rest of her life. Frustration growing, Dr. Neer reckoned that this was the third severe shoulder fracture of the month, and he had nothing to offer the patient—at least nothing that would help—and his sense of impotence roiled beneath his tranquil exterior. He had been summoned to the emergency room to evaluate the seventy-year-old Manhattanite who had fallen in her apartment earlier in the day and had been conveyed to the Columbia Presbyterian Medical Center. Although his hospital was one of the first in the world to have a "fracture service," Dr. Neer knew that in 1951, he was powerless to help Mrs. Harrison, not with surgery, not with a plaster cast, not with a prayer.

With the discovery of X-rays in 1895, Wilhelm Röntgen had revolutionized fracture care—instead of doctors blindly treating crooked and shattered limbs, X-rays divulged detailed information about the location and "personality" of the broken bones. Soon, fracture taxonomy reports appeared in medical journals, and these would eventually guide treatment. Each bone in the body, in time, would have its own classification scheme, usually referred to by its primary author. In the first half of the 20th century, little appreciable progress in patient care had been achieved, but physicians had started to notice the predictable patterns by which bones break.

The "father of shoulder surgery" is Ernest Amory Codman, a firebrand who published *The Shoulder* in 1934, the first textbook solely dedicated to the treatment of shoulder injuries.[2] Codman instigated many crucial changes in medicine, including outcomes research, hospital accreditation, tumor registries, and the advancement of shoulder surgery. Despite his pioneering role in medicine, and particularly in shoulder surgery, Dr. Codman never published a journal article on shoulder fractures, arthritis, rotator cuff tears, or shoulder instability. After a tumultuous career, Dr. Codman died in 1940 at the age of seventy, and in the war-torn decade that followed, a few scattered reports on the treatment of comminuted, or "shattered," fracture dislocations of the shoulder were published. These articles (written in English, Italian, and German), published just half a century ago, are shockingly simplistic to the modern reader and would stand zero chance of publication today. In general, the authors concluded that surgery of smashed and fragmented shoulder bones was successful (enough) if the fragments were simply removed, leaving a blank shoulder socket that was intended to heal with a blob of scar tissue, with the hope that the resultant "flail arm" provided at least a modicum of function with the arm at the side. The journal publications in the 1940s included no measurements of angular motion of the shoulder joint, no pain scores, and minimal commentary about functional abilities.

A more scientific (less anecdotal) evaluation of the flail arm patients was needed, and the young Charlie Neer was the man to do it.

Born and raised in Vinita, Oklahoma, Charles Sumner Neer II was the namesake of a general physician and surgeon, who was born in New York, trained in St. Louis, and practiced frontier medicine in the Indian Territory that would become the forty-sixth state of the Union. The elder Dr. Neer was himself the son of a physician, and Charlie once wrote that his father "never once thought of me being anything other than a doctor."[3]

Vinita, Oklahoma, was the epitome of a frontier town when Charlie was born on November 10, 1917. Oklahoma achieved statehood in 1907, formed from the many independent Indian lands of the (western) Oklahoma Territory and (eastern) Indian Territory. Vinita, located in northeast Oklahoma (near the Kansas and Missouri borders), was in the center of Cherokee lands when the elder C. S. Neer moved there from Missouri to start his new practice.

C. S. Neer, senior, established his clinical practice on the major intersection of town (Wilson Street and Illinois Avenue), on what is now US Route 66. Utilizing literature search techniques, one can trace Dr. Neer's path from St. Louis to Springfield, Missouri, and then to Vinita; he had publications in the *Journal of the American Medical Association* in 1907 while a resident, then in 1908 while employed in Springfield, and then in 1909 after setting up shop in Vinita. Charlie was born in 1917 when his physician father was thirty-eight years old, and grew up an accomplished horseman and natural-born Oklahoma lad. Expecting his son to become a physician, Dr. Neer and his wife made the decision to place young Charlie on a train and enrolled him at the Shattuck Military Academy in Faribault,

Minnesota (today known as a major incubator of National Hockey League talent), where he would spend his prep school days as a tennis and football standout. The superior education at Shattuck prepared him for Dartmouth College, from which he matriculated in 1939, and then medical school at the University of Pennsylvania, graduating in 1942.

After an internship in 1943 in Philadelphia, Charlie's surgical training was interrupted by World War II. Like so many physicians during the second great war, life was placed on hold, and Dr. Neer served in both major theaters of war, in field hospitals in Europe (under General George S. Patton) and the Philippines (under General Douglas MacArthur), and at a general hospital in Japan.

Dr. Neer returned to the United States, and for the first time in his life moved to New York City in 1945. For the next half century, the country-born Dr. Neer lived in the busiest city in the world, establishing himself as one of the most influential surgeons who has ever lived. His arrival in New York coincided with waning European medical leadership, and he is one of the pioneers who planted the flag on American soil. His papers are the most quoted in all of orthopedic surgery, and his shoulder surgery trainees became the most influential thought leaders around the world. The manner in which shoulder arthritis, rotator cuff tears, shoulder instability, stiff shoulders, and painful shoulders are treated are all deeply influenced by his original works. And it all started with his truth-telling about our incompetence in dealing with severe shoulder fractures.

In the late 1940s, Dr. Neer completed his orthopedic residency at the New York Orthopedic Hospital (which would join Columbia Presbyterian on the Upper West Side of Manhattan in the early 1950s), and his mentors were the physicians who led the fracture service: William Darrach, Clay Ray Murray, and Harrison McLaughlin. In today's orthopedic departments, there are many divisions: foot and ankle, sports medicine, total joints, spine, tumor, hand, shoulder and elbow, and pediatric orthopedics; but in the 1940s, fracture care was just starting to be the first specialty of orthopedics, undergoing a major metamorphosis due to a combination of historic advances in metallurgy and antibiotics. As Dr. Neer entered internship in 1942, *penicillin was in its first year of use in the United States*, reversing a trend where any open fracture (bone poking through the skin) was potentially lethal.

The act of operating in the era before antibiotics made any elective operation risky. There was, therefore, almost no enthusiasm prior to the immediate postwar epoch for insertion of any type of foreign material into the human body. The track record of implanted ivory, bone, glass, metals, plastics, and rubber was abysmal: almost every occasion of implantation resulted in infection, necessitating removal. Today, we hear of fracture and trauma patients undergoing fixation of broken bones on a regular basis; this simply did not exist just a few generations ago. Like an invalid stroke patient, fracture patients were placed in bed, with weeks or months passing before getting in a wheelchair or standing bedside. With no possibility of surgically reassembling the bone fragments, pioneering surgeons were little better than ancient "bone-setters." Instead of "fixing fractures," doctors would treat their supine patients with heavy plaster bandages and a dizzying array of ropes, pulleys, splints, and overhead trapeze frames.

Dr. William Darrach was seventy years old, and newly retired from full-time academic surgical practice when Dr. Neer returned from the Pacific Theater. Dr. Darrach had been one of the world's first great fracture surgeons, and in the few years that their professional lives overlapped in New York, the elder surgeon left an indelible mark on Neer's career. Decades later, Charlie Neer would still refer to Dr. Darrach as "my Chief." When Dr. Neer was a resident, he prepared his first publication, "Intracapsular Fractures of the Neck of the Femur," which was published in the *American Journal of Surgery* in November 1948. This was copublished with Harrison McLaughlin, MD, then the chief of the fracture service at Columbia. Unusual for its time, the five-page article describes a retrospective chart and X-ray review of 130 fracture patients over a thirteen-year span (1932–44). All 130 patients suffered a hip fracture of the femoral neck, and all of them were treated with the Smith-Petersen nail, the metal plate and screws that were developed by the pioneering orthopedic surgeon from Harvard Marius Nygaard Smith-Petersen. Thoughtful data presentation of patient profiles, disease states, and rudimentary patient satisfaction was conveyed in fourteen tables. Noticeably absent are the outcome measures, hip range of motion numbers, and pain scores that modern orthopedic papers must exhibit. However, the brilliance of reasoning, arrangement, and conclusion reveal a prodigy in the making.

There are six conclusions in the hip fracture paper, the truths of which today are set in stone:

- The best time for reduction and fixation of a hip fracture is immediately. (There is no benefit in waiting for surgery.)
- Good treatment for impacted valgus fractures of the femoral neck is internal fixation and avoidance of bed stay. (Neer makes the assertion that patients do better when bones are stabilized and the patient is moved out of bed.)
- Open reduction, properly done, is surer, shorter, and no more dangerous than closed reduction and blind nailing.
- Open reduction does not increase the incidence of subsequent aseptic necrosis. (Surgery, by itself, does not cause bone death—it's the fracture that causes necrosis.)
- Results are known only after objective evaluation. (Neer echoes the great scientists and surgeons of the 17th and 18th centuries: *Take no man's word for it.*)
- Almost all of the bad results of hip nailing are still the results of bad hip nailing. (In this, Neer's final sentence of his first paper, he makes plain that *technique matters.*)

Charlie Neer completed his residency in 1949, shortly after the publication of his hip fracture paper. He immediately became an assistant professor in the Department of Orthopedic Surgery in Columbia University's College of Physicians and Surgeons on the Upper West Side of Manhattan, and served on the Fracture Service, treating fractures from the neck to the toes. Manhattanites could choose from several world-class hospitals that proudly boasted of new fracture services. The newly constructed Columbia-Presbyterian Medical Center in Morningside Heights (completed in 1928) served the upper portion of Manhattan, the Bronx, and even New Jersey, with the recently opened George Washington Bridge (1931) offering access to the bedroom communities across the Hudson.

Charlie Neer had arrived at Columbia at the perfect time. The mergers of a medical school and university, the building of a campus and a bridge, and the postwar boom provided an expanding patient population for his observations. Reflecting fifty years later, Dr. Neer said, "When I was a resident in orthopedic surgery [1946–9] at the New York Orthopedic, Columbia-Presbyterian Medical Center, the only procedures used to treat problems of the glenohumeral joint were fusions or resections to manage tuberculosis, infections, and old injuries. I became interested in severely displaced fracture-dislocations of the proximal humerus and made a study of lesions of this type that had been treated . . . with open reduction and internal fixation, closed reduction, and removal of the humeral head."[4]

The few resources available to guide orthopedic resident Charlie Neer in the treatment of shoulder fracture-dislocations provided no practically useful information. Ernest Codman's five-hundred-page-long magnum opus, *The Shoulder*, focuses on the supraspinatus tendon and bursa, while offering no effective treatment of shoulder arthritis and fractures. One can hardly blame the Boston surgeon for his anemic ministrations; he would die in 1940 without knowing about penicillin, screw fixation of broken bones, or joint replacement. Regarding surgical treatment of fractures, Codman only said, ". . . early operation is far more promising than if it is delayed for even a few weeks. Surgical skill in handling fractures of the head of the humerus will be displayed more in attaining rapid and comfortable recovery than in ultimately securing good results, for nature alone would produce them in most cases. Injudicious fixation is responsible for most delays and failures in the recovery of normal function."[5] And that's all—no technique recommendations, and certainly no comment on implants: there were none in 1934.

The other main textbook available to Charlie Neer during his residency was Arthur Steindler's *The Traumatic Deformities and Disabilities of the Upper Extremity*, published in 1946. Steindler, the chair of Orthopedic Surgery at the University of Iowa, had published a book that was the most comprehensive technique guide for shoulder, elbow, and hand surgery that had ever been written; by today's standards, it has almost nothing to say. For treatment of shoulder fractures in which the humeral head had broken and was dislocated, Steindler advised, "Incise along the axillary fold. Proceed bluntly through the subcutaneous tissues. Expose the head by blunt dissection and remove it."[6] Unimaginably terse, removal of the humeral head was the only option considered.

Shortly after Charlie Neer graduated, a breakthrough book by A. F. DePalma, professor and head of Orthopedics at Jefferson Medical College, was published in 1950. His book, *Surgery of the Shoulder*, was much more descriptive, richly illustrated, and practically useful than anything that had preceded it. Interestingly, there is no mention of penicillin or other antibiotics in the text, and no discussion of infections. A few pages in this lengthy tome dwell on fracture-dislocations of the shoulder, but as with other orthopedic textbooks of the day, the treatment of humeral head fractures is surprisingly crude. DePalma stated, ". . . removal of the head is unavoidable, despite the realization that the procedure causes great functional disability."[7] Later in the book, he softens, asserting, "with careful management, considerable control of the extremity and a surprisingly good range of painless motion may be obtained."[8]

The most authoritative works of Charlie Neer's early career all concluded the same thing: when faced with a severe fracture-dislocation of the shoulder, the only treatment available was extraction of the humeral head, and the only proper emotional response was a resigned, flimsy hope that a flail arm was better than an amputated arm.

Dr. Neer made his way up to the twelfth floor of the Columbia-Presbyterian Medical Center to visit Mrs. Harrison. She had been admitted to the Orthopedic Unit, awaiting surgery to have her humeral fracture fragments surgically removed in a day or two. Charlie Neer, bald from early adulthood but still athletically built, was accompanied by a few residents who were in their late twenties, boasting of no war experience. The small medical contingent shuffled into the elderly woman's hospital room, and Dr. Neer sat on the edge of her bed. Mrs. Harrison's X-rays divulged the severe nature of her injury: the upper portion of her humerus was in multiple pieces, and the humeral head, like part of a Granny Smith apple, was ripped in two. The fracture doctor needed to convey to the patient how serious her injury was and what the treatment plan was.

"Mrs. Harrison, you have a terrible fracture of your arm. The humerus bone is in many pieces."

With her arm swathed against her body with a linen sheet, and with broken eyeglasses and a fresh black eye from her fall, she was the very picture of a broken and vexed woman.

"There is no way we can save the ball of your shoulder, Mrs. Harrison. But I can't leave it where it is. We'll need to take you to the operating room, make an incision, and remove all the broken parts. The only way to treat your shoulder is to take out all the shattered pieces and sew together the tendons in your shoulder and close you up."

Dr. Neer was a man of few words and quiet contemplation. He paused, waiting for Mrs. Harrison to contemplate what he had proposed.

"Well," she haltingly started, "am I going to be okay? Will my arm be usable?"

"It's a little hard to say. This is a fairly rare injury, and we don't have much guidance in the medical literature, but I don't think you'll ever raise your arm above your head, and it will be difficult dressing and working around the house. I'm sorry to tell you this, but you'll mostly have to use your arm just by moving your elbow and wrist."

After a brief silence in which Mrs. Harrison pursed her lips, fighting tears, Dr. Neer resumed. "I've been interested in this very problem for several years. We don't do a very good job treating fractures like yours, and I have been spending a great deal of time trying to figure out how we can do better. My fracture colleagues around the world can't even agree how to describe these kinds of fractures, how common they are, and how to make a difference. But it all starts with seeing how our patients have done here in this hospital, and it's a project that I'm doggedly working on." With that, the retinue of residents accompanied Dr. Neer out the door, making their way to the orthopedic clinic.

The residents knew about Dr. Neer's new project, digging into old charts and X-ray jackets and reviewing the results of patients who had been treated for shoulder fractures at the New York Orthopedic-Columbia Presbyterian Medical Center since 1929, namely, since the hospital had opened its doors in the Morningside Heights neighborhood of Manhattan. This was no small task for the ambitious young attending surgeon, who knew well the dogged determination required in the bowels of the hospital's chart rooms, with their musty gray patient folders and hand-scribed surgical logs crammed into crenelated shelves, the new fluorescent bulbs purring overhead and the smell of mimeograph ink permeating the medical records department.

Similar to the hip fracture paper he had published a few years before, this chart review project required herculean effort, poring through logbooks from the Fracture Service and the operating room. Charlie Neer wanted to evaluate every shoulder fracture *and* dislocation (or both) that had darkened the doors of the New York Orthopedic Hospital over a twenty-three-year period (from 1929 to 1951), and determine how many of those injuries involved a fracture, and subsequent dislocation, of the humeral head. A young physician performing this project today would contact her medical records department, submit the ICD-10 code (the national standard diagnosis code, e.g., S42.241A for a severe fracture of the right proximal humerus), and the information technology department would, in a matter of minutes, churn out a list of every patient in that category, replete with their demographic information and hospital number. Armed with these particulars, a skilled data-miner could generate a treasure trove of information from any computer, opening the hospital's electronic medical records and imaging software. Dr. Neer, instead, needed to summon the skillset of an archeologist, scratching through opaque, coffee table–size logbooks with single-line handwritten entries of patients, with the briefest of descriptions of their names, dates of birth, and fractures.

Every available moment away from patient care was spent delving into the medical histories of shoulder fracture and dislocation patients over the life of the Columbia-Presbyterian Medical Center. Those patients who had suffered *simple* shoulder fractures were carefully tabulated but not investigated. He knew that most of those patients recovered reasonably well, without the need for surgery. Additionally, those who had experienced dislocations of the shoulder joint were chronicled, but again, not intensively evaluated. Dr. Neer was searching for the patients who had suffered the diabolical combination of a shattered proximal humerus and a dislocation of the humeral head. Slowly, the tedious work of searching through the medical records started to yield the occasional patient who had fallen victim to the terrible amalgamation. After months of chart review, Dr. Neer (and his helpers) had identified 1,796 total patients who had been afflicted with shoulder trauma at the hospital over twenty-three years. More than half of those patients (51.2 percent, or 921 patients) had suffered a fracture of the neck of the humerus. A total of 784 patients had dislocated their shoulders (44 percent of all patients seen at the hospital), and 71 patients had endured a fracture of the tuberosity (the large bump at the top of the humerus where the rotator cuff tendons attach). Of all the patients treated at Columbia-Presbyterian, only 20 had fractured *and* dislocated the proximal humerus, representing only 1.1 percent of all shoulder trauma patients. Less than one patient per year had fallen victim to the condition that would be the focus of Charlie Neer's first shoulder publication, but the most significant impact of his inaugural shoulder paper was to illuminate how poorly those twenty patients had responded to the treatment of the day.

"Fracture of the neck of the humerus with dislocation of the head fragment" was published in the March 1953 issue of the *American Journal of Surgery*, authored by Charles Neer, Thomas Brown, and Harrison McLaughlin.[9] After identifying the twenty patients of interest, an analysis was performed. The average age was fifty-six (what the authors described as being "midway between young and old"), and the typical mechanism was a fall from a standing height. Regarding treatment, in only two instances was closed treatment (no surgery) the final remedy. In three instances an attempt was made to surgically save the humeral head and reassemble the fragments together, and in a lone patient the surgeons had effected a fusion of the humerus to the shoulder socket.

Of the original twenty patients, sixteen underwent excision of the humeral head. In some of these patients, parts of the muscles and tendons were sewn to the broken top of the humeral shaft, not too different than someone duct-taping a car's side mirror back to its base after a vehicular accident. In the results section of the paper, the authors outlined average follow-up time and the level of patient satisfaction. In what is perhaps the most profound sentence of the publication, the surgeons concluded: "There was usually from 5 to 25 degrees glenohumeral motion following head removal regardless of whether or not a [reconstructive] procedure had accompanied the [humeral head] resection."

If a regular patient lifts his hand straight up in the air, evaluators would describe that as 160 degrees forward flexion; in the 1953 article, the typical patient barely had enough power to lift her hand imperceptibly away from the body. In other words, the wounded had essentially ended up with *ankylosed*, or fused shoulders.

Just a couple generations ago, chronic tuberculosis and trauma savaged many denizens of cities like New York, and a useless limb was all too common. Somehow, fourteen of the nineteen patients whose humeral heads were resected were "satisfied with their result and were carrying out their usual work without appreciable disability." Neer and his colleagues dared to disagree, saying, "Nevertheless, the limited motion and fatigue pain following resection has suggested the value of a replacement prosthesis to serve as a fulcrum for motion."

In the final column of the text, on page 257, is pictured a shiny, metallic object, which is described as, "a recently devised articular replacement currently being investigated." The concluding sentences of the publication assert, "Replacement prosthesis presents logical possibilities and may prove of value in dealing with major injuries of humeral head. Its true worth remains to be determined."

Like holding Edison's light bulb in your hands, unplugged, you couldn't be sure of what you were seeing. But, in time, surgeons would realize that the most important shoulder surgeon who ever lived, in his first article on the shoulder, had given a sneak preview of the future, not just for surgery of the shoulder, but for every joint. The ability to implant foreign materials in the body would awaken the imagination of engineers, biologists, and surgeons, and would usher in one of the most significant upheavals in human history—the implant revolution.

TWO

Paper, Prophet, and Printing Press

I struggle to remember what day it is. I came to the medical center Saturday morning, preparing to meet my fellow residents in the hospital cafeteria, and to "run the list," where we review our patient registry and assign the scut work for the day. As the junior-level resident, I knew most of the "dirty work" would be done by me over the weekend, and because the orthopedic hand team was on call for the trauma center, I knew I was at risk for a brutal sixty-hour slugfest. All it takes is a little bad luck for a call weekend to go to crap, and as soon as I sat down with my pancakes and my list, my trauma pager beckoned me to the trauma bay.

The first trauma patient had been in a motorcycle accident, and most of the bodily destruction was centered on his right arm. His right hand was a jumble of open fractures, exposed tendons, pumping macerated vessels, and dog-eared skin flaps. I knew at a glance his surgery would take hours, and the day was just starting. Additional hand-injured patients steadily arrived in the emergency room, and as day turned to night, there was no abatement of suffering. By Sunday morning, there was almost a chance I could take a quick nap, but someone's bagel knife mishap mandated that I descend down to the emergency room for a consult. The waves of traumatized patients, in a never-ending flow, pulsed through the ER, and my job was to stabilize, evaluate, and prepare the wounded for surgery and, if not otherwise detained, report to the operating room to surgically assist.

Late Sunday night, at my wits' end, not having slept all weekend, a twenty-four-year-old Central-Pennsylvania lumber mill worker was flown in on an air ambulance helicopter. Four fingers on his right hand were sawed off where they attach to the palm, and the giant saw blade had mangled all the digits into oblivion. All we could hope to do was surgically clean the wound edges, and over time, fashion a "mitten-hand" that he might use as a club. His family arrived later by car; it is always difficult to deliver the sobering news that nothing miraculous can be done to salvage a limb. But it is only 1996, after all.

Sunday night blurred into Monday morning, which meant Morning Conference, followed by assisting in the OR during total joint replacement operations. I still have managed no sleep, and my fatigue is oppressive. Very few people understand true exhaustion and the consequent short-circuiting of the brain and bone-deep achiness. I have always performed well during stretches of sleep deprivation, but at fifty hours, habituation and willpower start to mean nothing. Wakefulness requires extreme concentration; alacrity is an impossibility.

Like the urge to gag when food-poisoned or the impulse to squint with a flash of mirrored sunlight, the mind's insistence to power down at ultra-weariness means that your body will collapse, earthward, in a flash, underneath you. In what feels like a bus crash while you're napping, the paroxysm of awakening snaps your head around, your lungs gulping for air, your sea legs stumbling for firm footing in Wonderland, and your arms reaching for ballast. Often during residency, this seemingly drug-induced warfare with the primitive part of my brain (demanding rest, seeking solace, needing, above all else . . . nothing) would happen in the operating room while we were trying to operate. By dint of a miracle I have made it through a marathon of surgeries today, and now, as I piece together the events of the weekend, I remember the final thing I need to do today is to revisit that lumberman who lost the fingers of his right hand.

Dr. Pellegrini is my chairman, the person who presently controls my minute-to-minute existence. I've come to realize that every episode of the television show ER *and every movie about surgical residents, despite their best efforts to dramatize the draconian aspects of a chairman and his residents, vastly undersells the boss's power and the young medical doctors' helplessness and feelings of inadequacy. I meet "The Boss," as we call Dr. Pellegrini, and my chief resident Jeff Wood on the fifth floor, understanding that the patient's family has now gathered in his room. Walking down the dark hallway, I am the only one who has been awake for three days. One night, under similar circumstances of sleep denial, I was actually falling asleep while walking down an empty hospital corridor, crashing against the hand rails, stumbling like a frat boy on his way home from hazing. The extra jolt of adrenaline—accompanying my boss—keeps me upright, but I bitterly regret that I didn't bring extra socks or underwear with me Saturday morning. I am certain that I stink like a post-call intern, and the swampass from my three-day underwear and the sweaty feet entombed in cushionless, sodden socks make me even more desperate to trek home and collapse.*

Our three-man team meets with the patient and his family in his room, conveying the stark reality that his life is forever rocked by one accident. He knows, as a blue-collar worker, his future is permanently jeopardized. I would like to think I'm an empathetic person, particularly for a surgical resident, but in this moment I am cruelly reduced to a psychology class test-subject of sleeplessness; all I want is to lie down. I can't care about anything else, and (I'm ashamed to admit) I think to myself, this guy is part of the reason I was up all night. The patient's family, all working-class people who smell of cigarettes, fried foods, and musty dampness, grasp the situation, with heads bowed in reticent submission. We agree that another operation will take place tomorrow, to further clean up the stump of his hand remaining.

Exiting the room, with heavy sighs emanating from my chest, I determine to make a bee-line for home. Now in the dark hallway, I hear the patient's father calling out, asking for a minute. Jaw clenching, I know I could explode in desperate anger, "WHAT NOW? There is nothing more to be said."

The father, in worn-out flannel and dungarees, with muddy Red Wing boots and a mop of thick bristly hair, pauses. I'm thinking, I bet he's only fifty years old, when he hesitatingly starts, "Sorry to waste your time, but I got a question." Please—please for mercy's sake—be quick, I think to myself.

His leathery, sun-weathered skin and raspy voice belie years of chain-smoking and laboring outdoors, but his kind eyes reveal a humble decency. "I'm not a smart guy, and I don't know nothin' about doctor stuff, but . . ." he trails off. I wait, and my whole body aches with fatigue. "I've lived my life, I'm forty-three years old, and it kills me to see my boy with a wrecked hand and no future."

He extends his roughened, calloused hand, each finger thick with power from years of exertion, and softly asks, "Would it be possible to take the fingers from my hand and put them on my boy?"

Five thousand years ago, in South America, Africa, and Asia, primitive peoples simultaneously—and without communication—formulated the process of harvesting wild cotton, spinning it into cotton thread, and weaving it into material.[1] As beautifully detailed in Sven Beckert's *Empire of Cotton*, its ascension as the material that launched the Industrial Revolution is a study of global shipping, capitalism, slave trading, and the realization that cotton itself was an ideal multipurpose material. The saying "Success has many fathers" may suggest that multiple inventors vainly claim credit for another's innovation. Read a different way, the phrase highlights the fact that almost all discoveries and inventions occur to different people simultaneously.[2] Whether it's the airplane, the light bulb, scientific theories (evolution, relativity, calculus), toilet paper, or the hypodermic needle, "inexorable technological progress" means that great ideas come into full bloom, awaiting harvest, in multiple places at the same time.

The concurrent development of ideas can be explained by a certain *path dependence*, the concept whereby innovation occurs along a particular, predictable course. "There's not much point in mining uranium till you have invented steel, cement, electricity, and computing, and understand nuclear physics."[3] Inventions that are proposed far too early sound fanciful, like a "time machine," but innovation usually happens at just the right time, when all the necessary ingredients are available. The evolutionary biologist Stuart Kauffman coined the term "adjacent possible" to explain how biological systems are able to morph into more complex systems by making incremental and less energy-consuming changes in their makeup.[4] Steven Johnson, in *Where Good Ideas Come From*, applies the concept of the adjacent possible to science, culture, and technology. "The adjacent possible is a kind of shadow future, hovering on the edges of the present state of things, a map of all the ways in which the present can reinvent itself . . . each new combination ushers new combinations into the adjacent possible."[5]

This book, in essence, is about the adjacent possible. The rise of surgery, in retrospect, followed a simple pattern: enhanced connectivity among scientists and physicians fueled discovery and communication, small groups of investigators learned how the human body functions, doctors in the 19th century untangled the cellular basis of disease processes, and 20th century surgeons discovered remedies. Each advancement (with its own sub-advancements) rested upon an earlier breakthrough.

The first major foundation in the rise of medicine is, surprisingly, the invention of the printing press. The printing revolution (called an "integral part of the general history of civilization"[6]), was a classic coming-together of many technologies, but there awaited a major insight to make the printing press a reality—and it's not what you think.

Whatever environmental forces (ice ages?) necessitated the strengthening of social bonds among our primitive ancestors, there was a critical development of language and art that accelerated over the last thirty thousand years. But it is only in the last five thousand years that the written word has existed, which means that humans have spent 99.9 percent of our existence without writing. In the midst of the Renaissance, before science was invented, the greatest handicap humanity faced in conquering disease was the inability to share intellectual discoveries with a broad group of scholars. Hand-copied manuscripts, written on papyrus, were magnificently inefficient in conveying new information to investigators in far-flung municipalities. For medicine to flourish, and for surgery to become real, what was needed (to paraphrase Steve Jobs at the introduction of the iPhone) was a *breakthrough communications device*.

Coinciding with the invention of writing—around 3000 B.C.E.—the Egyptians made an ingenious utilitarian discovery for a ubiquitous plant: papyrus. Prior to domesticated crop production, the wetlands were replete with papyrus reeds—tufted three-sided emerald plants that held a peculiar interior that would change their society for millennia. Papyrus was used throughout the Mediterranean, but its production remained an Egyptian monopoly, and other than the Dead Sea Scrolls, its relics have only ever been discovered in Egypt.

The library at Alexandria was initiated by Ptolemy, the Macedonian Greek who became the ruler of Egypt in the 3rd century B.C.E. Besides being in a major cultural center and port, the library's great advantage was being close to the papyrus production centers. "Every ship that called in the port of Alexandria was searched for [writing materials], and any that were found were copied for the library. Ptolemy wanted works on any subject, poetry or prose, and three centuries later, the library was the repository of 700,000 scrolls."[7]

The ruler of Pergamum (in Asia Minor, the future home of Galen) aspired to build a magnificent library in the same era, but sensing a rival, Ptolemy refused to send papyrus to the Anatolian city. According to Pliny, the people of Pergamum were forced to innovate and create a new writing surface that was durable, thin, and in abundant supply. The invention was to be known as *pergamum*, and was made from animal hides that had been soaked in lime, scraped, and dried. The skins were then placed on a stretcher, further scraped and smoothed with stones. The final product is incredibly thin, and under the right conditions, is flexible and ages well.

Throughout Europe, pergamum retains its name in every language, but in English it is known as "parchment." The three main sources of parchment remain sheep, goat, and calf, but the finest material is known as "vellum," particularly when it is made from calfskin (most exceptionally, when it is fetal calfskin!). Parchment is still made worldwide, and uses include special manuscripts (like reproducing a diploma on a real "sheepskin"), collectible books, and bookbinding.

Soon after the life of Jesus, the Romans replaced the *codex*, a wooden tablet notebook, with parchment. Papyrus was not an appropriate substitute, since folding and sewing weakened it at the spine.[8] The rise of the parchment codex is linked to the rise of Christianity; all early Christian documents found in Egypt have been codices, whereas contemporaneous pagan documents were almost always scrolls. (The Latin word for scroll is *volumen*.) Unlike papyrus, "parchment could be made anywhere and preserved well in a wide range of climates. But like papyrus, it was labor intensive, and it was even more expensive to make—it could take as many as two hundred animals to make a single book. [The use of parchment] indicated that a document was important and meant to last."[9]

Johannes Gutenberg was born around the year 1400 C.E. in the city of Mainz, Germany. Founded as a Roman garrison shortly after the death of Julius Caesar, Mainz had grown into a small town of significance by the 15th century, and was one of the key Jewish centers of learning in Europe. The plague had swept through Mainz decades earlier, and as was typical for the era, the Jewish community was blamed (charged as "well poisoners") and hundreds of Jews were burned alive in the city square. The plague had reduced Mainz from twenty thousand to six thousand inhabitants,[10] leaving the Rhinelanders searching for scapegoats and vulnerable to excesses of the church, which was at its peak of corruption.

Gutenberg's family was involved with the striking of imperial coins in the local mint, and he grew up acquainted with the tools of the trade, including punches, molds, and dies. "The startling conclusion is that Johannes Gutenberg, from his childhood, was in the company of men who could carve a letter in steel that had at least six, and perhaps sixty, times the resolution of a modern laser printer, just at the time that King Sigismund gave Mainz the right to make imperial coins, with a consequent demand for new designs, and new punches."[11]

All of the constituent parts needed to craft a printing press were available to a tinkerer in the Rhineland in the early 15th century. Presses, with their massive wooden screws and crank arms, had been used since ancient times to make wine and extract oil, and more recently to squeeze paper dry. Punches were common among craftsmen for making medals, coins, armor, and decorations. Paper had arrived several centuries earlier from China and ink was well known to textile manufacturers. The time was ripe for an innovator who could connect the dots and start a revolution.

Gutenberg grew up in a family of goldsmiths, and would have witnessed the painstaking graving of individual letter punches. It is estimated that it would take a skilled punch-maker an entire day to make a single punch; it would require about three thousand punches for a standard printed page.[12] That would require a coterie of ten punch-makers working an entire year to make enough punches to print a single page. "A complete nightmare, economically a nonstarter, totally impractical, ten times worse than working with Chinese."[13] Johannes Gutenberg's big idea was not moveable type itself, and not even the punch: his breakthrough contribution was the ingenious idea to make a mold and to make the mold reusable.

The fabrication of a recyclable form, or mold, saved breaking the mold every time a letter was cast. Two blocks, in the three-dimensional shape of the letter *L* were nestled together around the *matrix*. An iron spring held the divisible form together, and this *mitered type* had the additional advantage of creating letters that had the same dimensions, thus creating a visually appealing print. The basis of Gutenberg's media revolution was therefore the process of: punch (*patrix*), matrix, hand mold, and type. He changed the world—not by "inventing the printing press," as is commonly concluded—but by inventing a dramatically improved way of rapidly crafting the reusable molds.

Gutenberg was a driven capitalist, but would never profit from his invention. In fact, it appears that he died with little money and scant celebrity. Experimentation with ink, press, and paper continued. Chemicals from gallnuts (bulbs formed on oak trees from wasp larva) were combined with soot, oil, and water to make an ink that was ideal for printing. Paper, and its "sizing" with animal fats, was also modified. Gutenberg and his partners tinkered with the degree of paper dampness, so that a perfectly moistened sheet of paper received well the ink from the punches. It was now time to produce his masterpiece, the printed Latin Vulgate Bible.

In every way imaginable, it is a masterpiece. "Gutenberg would need to match scribal Bibles in beauty and exceed them in accuracy, in two glorious, fat volumes totaling 1,275 pages. There might be a media revolution brewing, but it was essential not to look revolutionary, for otherwise no one would buy."[14] In essence, this was to be presented as a new form of writing, and since medieval scribes were so accurate, it was possible for this new printed book to pass as a magnificent and stately work of scribal art.

Having succeeded at last, with an astounding display of brilliance and perseverance, Gutenberg almost lost everything to his partners and colleagues, only by the skin of his teeth avoiding poverty and obscurity. And having produced one of the greatest publications, he ushered in a revolution—the Reformation—that blew Christian unity apart forever.[15]

When thinking of the European awakening of intellectual curiosity that started in the 14th century, it is too simplistic to exclusively think of Renaissance artists and their Medici patronage. On a scientific front, the rebirth was characterized by a nostalgia for classical thinking that was in large part fueled by a rediscovery of ancient texts from a most unlikely source.

During the reign of Byzantine Emperor Justinian (527–565 c.e.), no one on earth could have guessed that within a century, the preeminent power in the Mediterranean and near East would arise from within the Arabian Peninsula. Its founder would organize clans and tribes, introduce a new religion, unify the region between the Nile and Oxus under one language, and inspire the preservation of scientific and mathematical knowledge from ancient scholars. The orphaned prophet—who was most likely illiterate[16]—was born in Mecca, the village that already was a center of religious observation. Today, it is the focal point toward which Muslims face during daily prayers.

At the time of Muhammad's birth, in 570 c.e., Mecca was already a place of religious pilgrimage owing to the presence of the Black Stone, said to be a meteorite brought by Abraham. Prior to Muhammad's leadership, an annual truce was declared so that the warring tribes could gather together in Mecca to worship their pagan gods. Importantly, because Mecca was already a point of destination at the time of Muhammad's birth, the commercialization of the annual pilgrimage had already been developed.

Muhammad exhibited exceptional leadership skills and preternatural genius in unifying his region's tribes and clans, and in convincing them to abandon their pagan gods. He succeeded in creating the nucleus of the first Islamic society.[17] What seems like small-town intrigue centered around a dreamer in the desert would result in a religion and culture that would preserve Greek learning and foster new scientific discovery for centuries.

Islam scholar Seyyed Hossein Nasr has said, "In the same way that the aroma of the frankincense of this land reached the Roman Empire and medieval Europe, the spiritual fragrance of Arabia, holy to Islam, is sensed by Muslims near and far."[18] To further extend the analogy, Islam also infused the world with a curiosity about antiquity, preserved and translated the writings of the ancients, and helped erect the bridge to the Renaissance. The "millennium intervening between the fall of Rome and the scientific revolution was not an intellectual desert. The achievements of Greek science were preserved and in some cases improved in the institutions of Islam and then in the universities of Europe."[19] Another sobriquet for the Arab Peninsula was *Arabia Odorifera*, and in keeping with the historical use of fragrances to cover up the stench of rotting carrion, the intellectual perfume that emanated from Islamic writers provided some of the only "fresh air" during the Middle Ages.

At the time of Muhammed's death in 632 c.e., most of Arabia had been organized under his Islamic theocracy. After a few decades of tumult, in which the Sunni and Shia branches were established, the first dynasty—the Sunni Umayyad caliphate—was established in Damascus in 661 c.e. The Umayyads held power for almost a century, and during this time expanded across Northern Africa, Spain, and much of Central Asia. "From the formerly Byzantine lands that they now ruled they began to absorb Greek science. Some Greek learning also came from Persia, whose rulers had welcomed

Greek scholars before the rise of Islam, when the Neoplatonic Academy was closed by the Emperor Justinian. Christendom's loss became Islam's gain."[20]

The Golden Age of Islam began with the overthrow of the Umayyad dynasty by the Abbasid caliphate in 750 C.E. A new town, Baghdad, was built on the Tigris River by the Abbasid rulers and became the largest city in the world. The initial assimilation project by the Abbasids was the incorporation of Persian culture, and the Persians at the time revered Greek culture. As the Abbasid Muslims warmed to Greek philosophy, medicine, and science (if not poetry and drama), they eventually embraced ancient wisdom from other areas as well, including Egypt, China, and India. A flourishing and sophisticated society resulted, with educational and scientific advances across their empire that would serve as the intermediary from the Greek philosophers to the revolutionaries of the early Renaissance.

Al-Mamun (caliph from 813–833 C.E.) sent a delegation to Constantinople to acquire Greek manuscripts, and thus began one of the greatest intellectual transfers in world history; a tradition of translators, beginning with the physician Hunayn ibn Ishaq, and later his son and nephew, translated into Arabic the works of Plato and Aristotle, Galen, Hippocrates, and the mathematical works of Euclid, Ptolemy, and others. Historian Philip Hitti, comparing the staggering growth of wisdom among the Muslim savants to stagnant Europe, has said, "For while in the East al-Rashid and al-Mamun were delving into Greek and Persian philosophy, their contemporaries in the West, Charlemagne and his lords, were dabbling in the art of writing their names."[21]

The Golden Age of Arabic learning spanned the 8th to the 13th centuries C.E., and for the first time since Alexander the Great, the vast region was united politically and economically, and the "removal of political barriers that previously divided the region meant that scholars from different regions and ethnic backgrounds could travel and interact with each other."[22] The rise of Arabic science coincides with the spread of Islam from the Pyrenees to Pakistan, and the lingua franca of the day was Arabic, whether the writers were African, Spanish, Persian, or Arabic.

The House of Wisdom, founded by the caliph al-Mamun, became the world's center of learning. Whereas Alexandria had been the previous intellectual capital, with Greek and Roman manuscripts written on locally sourced papyrus, Baghdad become the new chaperone of philosophical and scientific inquiry, with conversion of all documents into Arabic, scribed on locally manufactured paper.[23] One of the early assimilations that occurred at the House of Wisdom was the adoption of Hindu numerals (1–9) as well as the base-ten system and the concept of "zero." An Arabic system of expressing abstract formulas (to the consternation of high school students everywhere) was introduced by al-Khwarizmi, which he termed *al jabr*, or algebra. The Abbasid Muslims incorporated the world's catalog of knowledge, including alchemy, mathematics, science, and law. As Islamic libraries flourished and dwarfed European libraries, the scientific and cultural stagnation of the Western Middle Ages ground on.

The earliest figure in the Arab tradition was Yaʿqūb ibn Isḥāq aṣ-Ṣabāḥ al-Kindī (known as al-Kindi, or in Latin, Alkindus), born in Basra (present day southern Iraq) of noble Arabic descent, and called the "philosopher of the Arabs." Al-Kindi was a polymath and was critical in translating Aristotle, the Neo-Platonists, and Greek scientists and mathematicians.

One of the most important medieval physicians was a Persian-born scholar named al-Razi (Latin: Rhazes), who was trained in Baghdad. Not confined to translating, Rhazes described smallpox and measles, and critically, was the first to seriously challenge the authority and infallibility of Galen. For instance, Rhazes postulated that fever was merely a defense mechanism and not an issue of humoral imbalance. His contribution was stunning; he was a "thinker explicitly questioning, and empirically testing, the widely accepted theories of an ancient giant, while making original contributions to a field."[24]

Another Persian-born Arabic speaker was ibn Sina (known as Avicenna, 980–1037 C.E.), widely considered the greatest physician since Hippocrates. Avicenna claimed to have memorized the Koran by age ten, and was a polymath, writing prodigiously on philosophical, scientific, and medical topics. He famously published *The Canon of Medicine*, a compilation of medical knowledge in a massive multivolume work that was later translated into Latin and would be a classic in the West for centuries. *The Canon of Medicine* was the main textbook throughout European medical schools (Montpellier, Bologna, Paris), even into the 17th century.[25] "Lecturing in 1913, Sir William Osler described Avicenna as 'the author of the most famous medical textbook ever written.' Osler added that Avicenna, as a practitioner, was the prototype of the successful physician who was at the same time statesman, teacher, philosopher, and literary man."[26] Avicenna, "the fountainhead of authority in the Middle Ages,"[27] was perhaps the greatest ambassador from the rich cultural enlightenment of the Islamic world.

Three thousand miles to the west of the House of Wisdom lay Andalucía, modern day Spain, which the Muslims termed al-Andalus. While eventually collapsing in 1492, Muslim rule in Spain had enveloped the Golden Age of Islam, and had precipitated vast cultural, scientific, linguistic, and architectural traditions that exist to this day.

Abu al-Qasim al-Zahrawi, also known by his Latin name Albucasis, (936–1013 C.E.) was born and raised near Córdoba (he descended from the Ansar tribe of Arabia), and is regarded as the greatest surgeon of the Middle Ages. "Because surgery was less burdened than other branches of medicine by ill-founded theory, [Albucasis] sought to keep medicine separate from philosophy and theology."[28] *Al-Tasrif* (completed about 1000 C.E.) was the result of almost fifty years of medical practice, and contained the earliest pictures of surgical instruments in history. For over five hundred years, his encyclopedia of surgery was the standard reference in the universities of Europe. Albucasis stated, "Whatever I know, I owe solely to my assiduous reading of books of the ancients, to my desire to understand them and to appropriate this science; then I have added the observation and experience of my whole life." If Albucasis scribed his eminent work in Arabic, how did it find its way into Latin?

Constantinus Africanus (Constantine the African) was born ca. 1020 in Kairouan, Tunisia, a city near the Mediterranean coast that had become one of the great centers of Islamic scholarship. Constantine studied medicine first in Tunisia, but traveled extensively (startling, for his time) to Baghdad, Syria, India, Ethiopia, Egypt, and Persia. While making his way back to Carthage (present day Tunis), Constantine passed through Salerno, Italy (near Naples), which, at the time, was considered the leading center of medical teaching in Europe. Unimpressed, Constantine returned to Tunisia, likely expecting never to return to Salerno. However, within a few years, he was suspected of sorcery and sent into exile. An avid book collector, Constantine the (Muslim) African brought with him his treasure trove of Arabic translations of classical Greek works, Islamic medical tomes, advanced international medical training, and his facility for many languages.

Constantine synthesized (at times, freely plagiarized) Arabic medical knowledge and finished a number of medical books in Latin, including treatises of surgery, prognostics, medical practice, the urinary tract, gastrointestinal disease, and medical instruments. His best known and most voluminous work was the *Liber pantegni*, the first fully comprehensive medical text in Latin. [29] By the time that Constantine was working on the *Pantegni*, he had converted to Christianity and was a Benedictine monk at the Monte Cassino monastery (between Naples and Rome), and he would spend the last decade of his life on his project of rendering medical textbooks in Latin.

Constantinus Africanus represents the change that was occurring in the world: a Mediterranean Muslim, convert to Christianity, who translated Arabic works into Latin presaged the return of Italian provinces to Christian control, the ascendancy of Latin scholasticism, and the domination of the West in medical education. Salerno would become known as the "first medical school in the world" (Greeks, Egyptians, and Arabs would contest this claim), and some would describe Constantine as the Muslim who ignited the Renaissance.

The second major figure in the translation movement was Gerard of Cremona (1114–1187 C.E.). While Constantine was an outsider who brought his external works and languages into Latin culture, Gerard was an insider (born in Cremona, Italy, the same city that gave us the Stradivarius) who left Italy for Toledo, still under control of the caliphate of Córdoba. Toledo was a city full of manuscripts and libraries, with ancient classics in Arabic and the newer works of the great Albucasis. [30] For the next forty years, Gerard translated treatises on mathematics, algebra, astronomy, philosophy, and medicine. It may be possible that a "second" Gerard of Cremona was active in medical translation; schools of translation were common, and when it comes to scholarly works of antiquity, many authors usually contributed. "Gerard's translation of the Great Arabic medical encyclopedias like Avicenna's *The Canon of Medicine* opened the eyes of medical scholars in the West to the fact that medicine was a rational science that could be studied logically and methodically, which had a sound foundation in philosophy and the natural order."[31]

The combined work of Constantine and the translators in Toledo sparked an interest in learning that been in hibernation throughout Europe for a millennium. The popularization of paper by the Muslims, combined with sourcebooks—now in Latin—fueled an intellectual awakening. For the thousand years that spanned the codex as the standard manuscript format until the 15th century, all Western books were copied by hand.[32] Scribes would often gather at a monastery's *scriptorium*, where copying of religious texts occurred under close supervision of a chief calligrapher. Until Gutenberg's innovation, book production remained exceptionally tedious and costly, but the mechanization of book production dramatically unlocked the ability to disseminate new ideas.

"The Renaissance was one of those few historical periods that discovered itself, rather than being defined by hindsight."[33] It would have been obvious to any curious soul that it was an era of unique access. The connectivity that pulsed through Europe exposed the great minds to classic works and the insights of other savants. In turn, mutual access to the cognoscenti inspired discussion, fueled dissent, and triggered widespread criticism of authorities.

The slow dissemination of information from master to apprentice was permanently altered in the 15th century. The foundations of medicine and surgery were built by Hippocrates and his followers 2,500 years ago; Galen expanded upon these writings in the early Christian era, but his authority was little challenged in Europe and only slightly meaningfully in the Islamic House of Wisdom. The work of Latin translators and the introduction of paper energized scholasticism and prompted the rise of universities in Salerno, Bologna, Padua, Paris, Montpellier, and Oxford. And in the mid-1400s, just as Constantinople fell to the Turks and churches were converted into mosques, the printing press unleashed a torrent of insights, observations, astrological tables, philosophical ponderings, religious arguments, political criticism, and thoughts about the human body—its form, function, and malfunction—and when surgery might work. The world's first great printed medical textbook would be lovingly produced in 1543 by a genius, age twenty-nine, and it would change medicine, and the world, forever.

THREE

Vesalius and
De Humani Corporis Fabrica

I stare at the bookshelf of my parents' library, trying to decide which volume of the World Book
Encyclopedia *I want to explore next. My dad is an academic veterinarian, and although
we don't have a lot of money, my family, like so many in 1976, has decided to "invest in the
future" and buy a home encyclopedia. It's the greatest invention in the world to me: twenty-
two inches of information, alphabetized and condensed, containing, as advertised, a world
of knowledge at my fingertips. In the days since the treasured series arrived, I've spent hours
leafing through the books, reading about the Appalachian Trail, the Congressional Medal of
Honor, and flags of the world.*

I reach for Volume 8, with the letter H *on the spine. I'm not looking for anything in particular, just waiting for something to grab my attention. As I plop down on our sunflower-
festooned canvas couch, I thumb the pages to survey the topics. Letting the pages fall open near
the back of the book I am arrested with wonder. Unlike all the other shiny paper pages, here
are four clear plastic acetate sheets, each representing a layer of the human body.*

*The first human form has no skin, and gazes left. The right side of his body, closest to
the book's spine, has the muscles intact over the chest, abdomen, right arm, and leg. His left
side bears only the rib cage painted on the first sheet, and I linger over his lungs and guts,
enthralled. Turning the page, I see inside the front of his body, its entire rib cage and muscles
visible. The front of the second sheet shows only the muscles of the left arm, but reveals the
lungs, liver, stomach, and intestines, all glistening and moist. Every organ is numbered, and
the corresponding legend makes a light blue column along the edges. Turning the second page,
to my great pleasure, I find the brain, visible within its bony home, the skull.*

*The third plastic page reveals lungs, heart, large blood vessels, pancreas, and kidneys. I flip
back and forth, memorizing the numbered organs and their habitations, not sure if they are
interconnected, but riveted by the thought that the guts and organs all have their own purpose.
The last sheet is mostly skeleton and nerves, and as I turn it, I see the back side of the body. To
my boyish glee, I find muscle number 159, gluteus maximus, which sounds like a forbidden term.*

*As I stare at the images I can't see how food gets from the mouth to the stomach, but the
accompanying text tells me the esophagus serves as the conduit for liquids and chewed-up*

bits, and the small intestine absorbs the food particles, which are further broken down by the
digestive enzymes from the pancreas. Whatever isn't broken down and absorbed moves on
to the large intestine, where water is drawn away, leaving "waste." My grade school mind
wonders, is "waste" the same thing as "poop?"

Nowhere else in this encyclopedia is there such a specialized set of illustrations, and the
message is clear to me: the human body is the most important subject in all these volumes.
Other topics still pique my interest, but these anatomical drawings will always be my greatest
fascination. In fact, I can't drift too far away from them; returning often to these acetate sheets.

In the 1400s, Gutenberg invented the printing press, Constantinople fell to the Turks,
Jan Hus and Joan of Arc were burned at the stake, the Medici rose to power, Columbus
sailed to the New World, the Jews were expelled from Spain during the Inquisition,
and the word "discovery" was coined.

When Columbus stumbled upon the New World in October 1492, he had no
word to describe the action of "encountering an unknown world." Columbus recorded
the accounts of his voyage in Spanish and Latin, but only Portuguese had the word,
discobrir. Prior to the late 1400s, authors struggled to convey the concept of *invention*
or *discovery*, and would rely upon paraphrases such as, "a new technique that never
existed previously."[1]

David Wootton, in *The Invention of Science*, posits:

> The discovery of America in 1492 created a new enterprise that intellec-
> tuals could engage in: the discovery of new knowledge. This enterprise
> required that certain social and technical preconditions be met: the exis-
> tence of reliable methods of communication, a common body of expert
> knowledge, and an acknowledged group of experts able to adjudicate
> disputes. First cartographers, then mathematicians, then anatomists, and
> then astronomers . . .[2]
>
> Therefore, the idea of discovery is inseparably linked with ideas of
> "exploration, progress, originality, authenticity, and novelty. It is a char-
> acteristic product of the late Renaissance."[3]

Copernicus (1473–1543), the Prussian astronomer who discovered heliocentrism
(placing the sun at the center of our solar system), was fortunate to live during
"the very decades when a great many changes, now barely visible to modern eyes,
were transforming the 'data available' to all book readers. A closer study of these
changes could help to explain why systems of charting the planets, mapping the
earth, synchronizing chronologies, codifying laws, and compiling bibliographies
were all revolutionized before the end of the 16th century."[4] The starwatchers
comprehended that the heavens could be described with charts and tables; the early

anatomists—cartographers of the body—would similarly map the intricate and predictable anatomy of humankind.

While it took a full century for print culture to assimilate scribal records of the ancient philosophers into coherently presented books, the simultaneous distribution of well-made figures and charts enhanced the works. It was one thing for a Florentine publisher to present, say, the philosophical works of Aristotle, it would be quite another for a 16th-century savant to revisit the works of Galen, the most authoritative physician-author in human history. The year 1543 witnessed the publication of two of the greatest books in human history: Copernicus's groundbreaking manuscript, published in Nuremberg in the year of his death, and an anatomy book by a twenty-nine-year-old, Andreas Vesalius, who would dare to challenge the great Galen. His book, *De Humani Corporis Fabrica (On the Fabric of the Human Body)*, a tour de force folio, would set the stage for a renaissance in medical education.

During the Sack of Constantinople, in 1204, crusaders (inspired by Pope Innocent and supplied by Venice) pillaged and purged the ancient city of its treasures, and carried them back to the Italian peninsula. These included works of art, sculptures, precious metals, and ancient manuscripts. The works of the philosophers of antiquity from Constantinople and other conquered lands "caused some to understand that there had once been an age that far outshone their own, one that emphasized the humanity of humankind, rather than its spirituality. As a result, there was now a new humanism in the air, which began to emphasize freedom of thought, rather than the selfless submission demanded by medieval philosopher-theologians. This humanism encouraged the exploration of human potential, and the expression of humanity, especially in literature, philosophy, and all forms of art."[5]

Constantinople surrendered to the Turks in 1453; the mass westward emigration of the eastern Christians was an important factor in the reawakening of Europe. The translation centers in Italy and Spain had their new Latin works (from Arabic), but the Byzantines brought their Greek manuscripts with them to the Italian city-states just as Gutenberg was perfecting movable type printing.

A small group of 15th-century craftsmen fled Constantinople and made their way to Venice; the Venetians had been major power brokers for decades in Byzantium, with their sophisticated ships, trading networks, accounting systems, and banking erudition. The artists and technicians who sailed for Venice included a group of glassmakers who found themselves in one of the greatest commercial trading hubs in the world. The making of colored glass had been around since Roman times, but the Byzantine workers had elevated their craft to previously unseen heights, and a new luxury good was introduced. In the Venetian marshlands, however, the glassmakers had an unappealing habit of "burning down the neighborhood,"[6] and were moved across the lagoon to the island of Murano. An "innovation hub" was created, and the "Isle of Glass" (which still produces exquisite glass today) became the paragon of the craft.

Steven Johnson describes the big breakthrough:

> After years of trial and error, experimenting with different chemical com-
> positions, the Murano glassmaker Angelo Barovier took seaweed, rich in
> potassium oxide and manganese, burned it to create ash, and then added
> these ingredients to molten glass. When the mixture cooled, it created an
> extraordinary clear type of glass. Struck by its resemblance to the clearest
> rock crystals of quartz, Barovier called it cristallo. This was the birth of
> modern glass.[7]

Surprisingly, modern, clear glass would enable several of the key innovations that
empowered the birth of modern science and would define the birth of the Renaissance.
The accidental making of clear glass would be followed by the (almost) accidental
manufacture of mirrored and curved glass, and the possession of mirrors and small
curved lenses would revolutionize medicine and science in unimaginable ways over
the course of the next century.

Once the Venetians had discovered the technique of making *cristallo*, the next major
challenge was making a larger flat piece of glass, not an easy task when you consider
that glassmaking always started with a glass-blown bubble that had to be rapidly flattened while cooling. Additional experimenting with basic ingredients from faraway
lands yielded ever superior glass, including herbs from Egypt and sand from Mediterranean trading partners. In an effort to make larger, flat panes of glass, they adapted
a method of blowing glass into cylinders, slicing the molten glass lengthwise and laying
it flat. Early mirrors had been made with a technique of adding silver flakes on the
back of cooling glass, but the differential coefficient of contraction between the glass
and metal made the glass fracture. The Muranese innovated an amalgam of mercury and tin, which resulted in less breakage of glass and yielded a shiny and highly
reflective surface.[8] While still a relative luxury good, mirrors became commonplace
enough that they became part of the fabric of everyday life in the early Renaissance
in Venice and Florence. "This was a revelation on the most intimate of levels:
before mirrors came along, the average person went through life without ever seeing
a truly accurate representation of his or her face, just fragmentary, distorted glances
in pools of water or polished metals."[9]

Therefore, a combination of developments occurred in the mid-1400s that set the
stage for a seismic sociological change. Within a few decades, Lucretius's poem "On
the Nature of Things" was discovered in a German monastery, crystal clear glass and
advanced mirrors were invented in Venice, Constantinople fell to the Turks—with
the original Greek manuscripts flowing into Italy—and the movable type printing
press was invented. Conventional wisdom holds that *individualism* was born in
1500 c.e.,[10] and it is no accident that the refinement of mirrors and the appearance

of the first self-portraits are coincident. "Self-consciousness, introspection, mirror-conversation developed with the new object itself," writes Lewis Mumford in *Technics and Civilization*.[11] Man could see himself for the first time, and as the personage came into focus, property rights and legal customs began to revolve around the individual, rather than the former collective units of family, tribe, city, or kingdom.[12] The new individualism and humanism of the mid–15th century would compel prodigies to turn their gaze inward, to explore the motives of the mind and the *corpus*, and, following Columbus's example, to *discobrir* the fabric of the human body. Our interior thoughts and our physical makeup became fertile ground for exploration at the fading of the Dark Ages, and the surveyors who sharpened their gaze on the human body had no idea about the new continents that lay before them.

The decline of human dissection in Alexandria at about 150 B.C.E., at the time of Herophilus and Erasistratus, presaged the extinction of the medical school in what had been the most advanced center of scientific study in the world. The incorporation of Alexandria into the Roman Empire in 30 B.C.E. further codified the opposition to human dissection, both by statute and general pagan religious sentiment.[13] As we have seen, it was Galen (129–199 C.E.) who became the unquestioned anatomical authority, even without ever dissecting a human cadaver or performing an autopsy. His investigations were animal-based, including barnyard animals and Barbary apes.

The prohibition of human dissection continued through the Muslim epoch of intellectual leadership from the 8th to the 13th centuries, with only scattered original anatomic investigations. "The anatomical knowledge of Islam was merely that of Galen in Moslem dress,"[14] and the great Arabic translators were merely recapitulating what Galen had claimed. Curiously, it may have been the practice of dismemberment, boiling, and cleansing of bones of crusaders who had died in the distant East for easier transport back home that laid the foundation for the revival of human dissection.

The renewed interest in medical learning in the Italian Peninsula, first in Salerno, and later in Bologna and Padua, inspired young researchers to ignore the prohibitory bull of Boniface VIII of 1299 and perform the first human dissections. The bull was "directed not against human dissection but against the practice of boiling dead bodies of those far from home [for burial in their own homeland] . . . the papacy never issued any statement specifically opposing dissection although there seem to have been instances in which overzealous local ecclesiastical authorities, by interpretation or misinterpretation, did oppose the practice."[15] It is simply not true to claim that the church forbade dissection; ironically it was the Roman pagans who enacted these laws that had lasting power until the 1300s, and it was their Italian descendants who most powerfully challenged and reversed the laws.

Mondino de Luzzi, a physician from Bologna, Italy, became the first important dissector of the Middle Ages, publishing the classic *Anatomia*, in 1316. This was the first modern book devoted solely to anatomy, and while it appears that Mondino relied heavily upon Galen's writings, it is clear that much of the book was based upon his own anatomic dissections. *Anatomia* is simple, concise, and systematic, and would be the guide to anatomists for two hundred years, helping spark medical curiosity across Europe. The University of Bologna was, therefore, the first home of the revived practice of dissection and study of the human body[16]; the revival would soon spread to Padua, Venice, and Florence throughout the 1300s, and later to Siena, Perugia, Genoa, and Pisa by 1501. Again, while the sins of the Catholic Church, particularly in the 14th and 15th centuries, were legion, the prohibition of human dissection was not one of them, as is commonly claimed.

It is no coincidence that the rise of anatomical understanding, humanistic self-awareness, and enriched artistic representation occurred simultaneously in the Italian Renaissance. In the early 16th century, Botticelli, Leonardo da Vinci, Michelangelo, Raphael, Dürer, and Titian coexisted, competed, and, occasionally, cooperated. In 1502, Giacomo Berengario was appointed chair of surgery and anatomy at the University of Bologna (situated halfway between Florence and Venice) and would become Mondino's successor, writing *Commentaria* (1521), an extensive work of almost a thousand pages (possible only because of the printing revolution). Berengario was the first physician "not constantly overwhelmed by earlier authorities, either Galenic or Moslem [sic]"[17] and put considerable trust in his own vision of the human body and its function. Importantly, he had a sincere interest in art, even owning Raphael's *John the Baptist*. *Commentaria* would be the first anatomy book to integrate (although crudely) text and illustration, and Berengario was the "first anatomist to have a fairly good idea of the true significance of anatomical illustration."[18]

Although surgery was still limited to the lancing of boils and rudimentary battlefield triage and temporization, the tide was swelling for an improved understanding of how the body works. With the advent of the printing press, the refinement of woodcut printmaking, and a new, scientific approach to investigation, the stage was set for a young anatomist/surgeon who would write one of the greatest books ever written.

Andreas Vesalius was born in Brussels, Belgium, in 1514, to a family well positioned in society, with a father (Andries) who was the imperial pharmacist and a grandfather who was a physician to the Archduke Maximilian. In a time when royalty was often on the move, the ambulations of the imperial train compelled Andreas's father rarely to be home. Andreas benefitted from an elite education, first in Brussels, and then as an adolescent in nearby Louvain. At the Castle School at the University of Louvain, the teenage Vesalius studied philosophy, including Aristotle, and the arts, and was thoroughly versed in Hebrew, Greek, and Latin. With a family tradition of medical studies, it is not surprising that he opted for medical school, and by 1533, Vesalius was on his way to Paris.

Vesalius enrolled in the medical school in Paris, with the expectation that he would earn his degree within four years. In retrospect, it seems amazing that a medical baccalaureate degree demanded four academic years' work. A modern surgeon asks, what were they studying and how did the program take so long? There was no such thing as a microscope, the concepts of physiology (the study of the dynamic functions of the body), pathology (the study of diseases of the organs and cells), and microbiology (the study of bacteria and viruses) were completely undiscovered, and surgery was as primitive as we may find in a stone age village in Borneo today. We must conjecture that they learned Galenic and Hellenistic medicine, replete with philosophy and misinformation. Vesalius was in Paris for three years, but was forced to leave before being granted a medical degree, as we shall see.

Prior to the arrival of Andreas Vesalius to the City of Light, barbers, surgeons, and physicians continued to wrangle for prestige and recognition. The longstanding prohibition against human dissection had contributed to physicians' disinterest in anatomical study of any kind. Because the study of anatomy was so strongly linked to surgery, there was little incentive for physicians to engage in serious scholarship of the body and certainly not to touch a corpse. Modern readers understand that today, all physicians and surgeons, whatever their specialty, started out as classmates in the same medical schools. But in medieval times, physicians and surgeons did not train together. The guild of surgeons trained independently from the faculty of medicine; the barbers were far below, with no schooling in Latin (and certainly not in Greek), and only occasionally benefitting from instruction by physicians and surgeons. Barbers first organized around monasteries, where they performed the tonsure haircut we associate with medieval monks; over the preceding millennium, barbers became expert with knives while providing haircuts, shaves, and Hippocratic bloodletting. In England, the barbers melded with surgeons from 1540 to 1745, eventually becoming irrelevant except for a shave and a haircut. The striped barber pole is the only reminder of their former job as bleeding patrons.

Like medieval priests exercising control over parishioners, "the employment of Latin seems to have been in the ancient tradition of power and control . . . through its possession of the keys to the esoteric mysteries."[19] After years of simmering tensions, an agreement was finally reached in Paris in 1516 that resolved the medical hierarchy, with physicians preserving their vaunted position and surgeons accepting a subservient station. Instead of the Parisians emulating the more advanced Bolognese and Paduans, who rolled up their sleeves, dissecting and investigating for themselves, the French physicians eschewed touching cadavers, instead lecturing high from their *cathedra* (high chair) while the surgeon performed the actual dissection.

Whereas surgery had achieved some measure of respect in Italian cities by the 1400s, in Northern European countries like France, Germany, and England, esteem for surgeons languished far behind that enjoyed by physicians. Guilds (like modern-day trade unions) were formed by both surgeons and barbers, and were critical in establishing membership rules and standards. The craft of barber-surgery would have

resembled the "surgery" of Greek and Roman times, limited to basic trauma stabilization of broken bones, sword and knife wounds, and the new injuries associated with
gunpowder's arrival from China.

The battlefields of 14th and 15th century Europe would bear witness to the new
power of gunpowder, and the "blast injuries" seen from guns and cannons represented
much greater trauma than had ever been seen. Ambroise Paré (1510–1590), himself the
son of a barber-surgeon, never attended formal medical school but would rise to become
surgeon to four French monarchs. Paré, the first great French surgeon, revolutionized
the treatment of war injuries, becoming influential through his writing—in French
and not Latin. Physicians in the early Renaissance found themselves helpless to treat
patients whose gunshot wounds were dramatically worse than any injuries humankind
had ever faced. Therefore, care was left to barber/surgeons, and in the pre-Newtonian
age, it was difficult to understand that it was the *energy imparted by the gunpowder-
propelled shrapnel*, and not some "poison" within the fragments, that imparted such
significant injury. Giovanni da Vigo (1450–1525), surgeon to Pope Julius II, theorized
in his publications in 1514 and 1517 that gunshot wounds were "poisoned by the
effects of gunpowder," and should be cauterized with hot oil to counteract the poison,
mimicking the ancient treatment of gladiatorial battle wounds. As one can imagine,
the searing effects of hot oil might staunch bleeding, falsely leading the traumatolo-
gist to conclude that care has been rendered, when in actuality the "zone of injury"
has perversely been enlarged and further trauma has been introduced. Unfortunately,
Vigo's writings had influence, leading battlefield surgeons to obediently pour hot oil
on blast injuries.

In his famous 1575 book, *Oeuvres*, Paré elegantly described his crisis during the
Siege of Turin of 1536. Late one night following a horrific battle, Paré's ration of oil
had been extinguished. He recorded:

> At last I ran out of oil and was constrained to apply a digestive made of
> egg yolk, oil of roses and turpentine. That night I could not sleep easily
> thinking that by the default in cautery I would find the wounded to whom
> I had failed to apply the said oil dead of poisoning; and this made me get
> up at first light to visit them. Beyond my hopes I found those on whom I
> had put the digestive dressing feeling little pain from their wounds which
> were not swollen or inflamed, and having spent quite a restful night. But
> the others, to whom the said oil had been applied, I found fevered, with
> great pain and swelling around their wounds. From then I resolved never
> again so cruelly to burn poor men wounded with arquebus [gun] shot.

Paré had serendipitously discovered a better way, accidentally performing a comparative study. More importantly, he published his results, contradicting the established

academic authority of his time. He would significantly influence early surgery, advocating ligatures (sutures) in tying of blood vessels, the use of prosthetic limbs following
amputations, and important changes in the management of childbirth. The advent
of book printing had arrived just in time for Paré to publish his works, and, as will be
seen repeatedly, war has been fertile ground for medical advancement.

Andreas Vesalius started medical school in 1533 (the same year that Paré arrived
at the Hôtel-Dieu de Paris, the oldest hospital in the world, located next to Notre-
Dame Cathedral) and his training was typical for the day. Galenic teaching was in
its ascendancy, and Vesalius's anatomical instruction was rudimentary at best. Here,
Vesalius first showed his supreme curiosity (if not his morbid oddity, on display
throughout his life), when he struck out on his own, admitting that he would not
have been successful under his professors' tutelage, "if when I was studying medicine
in Paris I had not put my hand to the matter but had accepted without question the
several casual and superficial demonstrations of a few organs presented to me and
to my fellow students . . . by unskilled barbers."[20] As will be seen again and again,
the invention of surgery was crafted by tinkerers, oddballs, lonely geniuses, inspiring
mentors, and stubborn misfits; Vesalius was all these things. He visited the Cemetery
of the Innocents in Paris on many occasions, picking through the decaying corpses and
maggot-cleaned bones, later recalling his long hours in the cemetery "gravely imperiled
by the many savage dogs."[21]

With the outbreak of war between Emperor Charles V of the Holy Roman Empire
and Francis I of France, Andreas Vesalius was forced to return to Brussels, given his
identity of a Flemish enemy-alien living in Paris. He quickly installed himself in the
medical school at Louvain (outside Brussels), and was soon scouting the locale for
bodies. Outside the walled city of Louvain, while searching for the bones of executed
criminals, Vesalius and a physician friend stumbled upon a cadaver, hanging upon a
gibbet (the upside-down L-shaped frame used to hang criminals). He inspected the
cadaver, concluding that the body had initially been burned and roasted over a fire of
straw, but had been "freed of flesh" by the birds. Observing that the bones were now
held together only by the dried ligaments, he recalled:

> Observing the body to be dry and nowhere moist or rotten, I took advan-
> tage of this unexpected but welcome opportunity and, with the help of [my
> friend], I climbed the stake and pulled the femur away from the hipbone.
> Upon my tugging, the scapulae with the arms and hands also came away,
> although the fingers of one hand and both patellae as well as one foot were
> missing. After I had surreptitiously brought the legs and arms home in
> successive trips—leaving the head and trunk behind—I allowed myself
> to be shut out of the city in the evening so that I might obtain the thorax,
> which was held securely by a chain. So great was my desire to possess those

bones that in the middle of the night, alone and in the midst of all those corpses, I climbed the stake with considerable effort and did not hesitate to snatch away that which I so desired.

Vesalius finished cutting away the ligaments and soft tissues after softening them with boiling water. He continued:

Finally and secretly I cooked all the bones to render them more suitable for my purpose. When they had been cleansed I constructed the skeleton that is preserved at Louvain.[22]

After a brief stint in Louvain, Vesalius was on his way to Padua, Italy, home to the most distinguished medical school in the world, and it was at the University of Padua where Vesalius would take his final examinations. The almost four-hundred-year-old academic documents at the University of Padua inform us that he "conducted himself very well in this, his rigorous examination . . . he was approved unanimously."[23] Amazingly, the following day, Vesalius was named the chair of surgery and anatomy. Despite his meanderings through three schools in four years, he had distinguished himself among his professors, and it was clear that something special was brewing in Padua.

A day after graduating, in December 1537, the new chair of anatomy and surgery started his first anatomical dissection, on an eighteen-year-old male, and the dissection would last for eighteen days. Vesalius would follow the traditional protocol established by Mondino: abdominal cavity first, then thorax, head and neck, brain, and then extremities. The greatest change witnessed by the audience was that Vesalius performed *every role:* lecturer, demonstrator, and dissector. The prestigious physician, acting as surgeon, had dismounted from his *cathedra* chair and positioned himself, knife in hand, over the body. He didn't have to read from Mondino or Galen, he knew their works *verbatim.* Just shy of his twenty-third birthday, Vesalius also introduced a new pedagogical device of posting illustrations, or charts, for his students. Here was a man who was obsessed with true knowledge transfer, and within a year he would publish his first book, the *Tabulae Anatomicae.* The drawings were nontraditional, reflecting what Vesalius observed and was trying to convey, with high mnemonic value. The *Tabulae Anatomicae* was printed in Venice, with six large woodcut illustrations of anatomy, measuring 19 by 13½ inches. In this 1538 book, there are the first hints at Galen's fallibility. Vesalius discovered inconsistent findings in Galen's descriptions, and the young anatomist was beginning his program of refusing to accept past authority until his own research proved it to be true.

Two years later, Vesalius published a revised edition of another anatomist's book, Johann Guinter's *Institutiones Anatomicae,* which would serve as a text to accompany the lectures and demonstrations in anatomy. In a sense, much of the book was plagiarized, with Vesalius revising the work of the primary author, yet adding content throughout. While it seems odd that Vesalius would publish a revised edition of another author's work, worse would happen to Vesalius when his words and illustrations would be entirely plagiarized verbatim by other publishers.

In the late 1530s, Vesalius began a program of comprehensive analysis of Galen, with Greek interpretation and scholarly evaluation of his anatomical descriptions. It was becoming obvious to Vesalius that Galen was not infallible, and emboldened by some of his professors in Paris and Louvain, he began preparations for a monumental work to challenge Galen's authority, while taking advantage of the new printing press technology and vastly improved artistry of early Renaissance northern Italy. During this investigation, Vesalius was housemates with the Englishman John Caius, who was also in his twenties, and had journeyed to Padua to study medicine. Caius had studied at the University of Cambridge, matriculating at Gonville College. It appears that Caius assisted Vesalius in his Greek translations, but maintained greater loyalty to Galen than Vesalius was willing to pledge. Historian C. D. O'Malley says, "Despite the fact that Caius belonged physically to the generation of Vesalius in which the dissection of human specimens was under way and the scientific treatment of anatomy had begun, he belonged spiritually to that previous generation in which the medical humanists had believed that Galen held the key to all medical problems and that accurate Latin translations from sound Greek texts were the greatest boon they could offer to the medical world."[24] Caius would return to London, succeeding enough to financially rescue Gonville College and lend his name in 1557 to the now renowned Gonville & Caius College at the University of Cambridge.

Vesalius was busy in both Padua and neighboring Bologna, entrancing students with his flair for teaching and dissection ability. "It is significant that wherever Vesalius traveled to give extramural lectures a wave of body snatching ensued."[25] Newly buried citizens and criminals were fodder for Vesalius and his "anatomies." A contemporaneous report states, "The mistress of a certain monk [in Padua] died suddenly . . . and was snatched from her tomb by the Paduan students and carried off for public dissection. By their remarkable industry they flayed off the whole skin from the cadaver lest it be recognized by the monk."[26]

Challenging the conventional wisdom with which he had been indoctrinated, Vesalius became the man who knew more about the human body than anyone who had ever lived. In the Age of Exploration and Discovery, voyagers had mapped the coastlines of South America, Africa, India, and East Asia; Vesalius had a similar program of exploration, and much to the betterment of mankind, an urge to convey that knowledge in a most excellent way.

The project of writing *De humani corporis fabrica*, hereafter referred to as *De fabrica*, began in earnest in early 1540, just after his twenty-fifth birthday. Vesalius intended for *De fabrica* to be a guide for dissection and to understand the human body. This was not simply a book *about* the body that could be read by a gentleman in isolation. This was an instruction manual for physicians (a later condensed book, *Epitome*, was intended for medical students), with step-by-step descriptions (and pictures) of the tools needed for dissection, the technique for boiling and cleaning bones, and the dissection process for each muscle, joint, organ, and nerve. He would sometimes spend weeks by himself in his Paduan home, writing and reflecting. Vesalius probably spent at least a year writing the text for *De fabrica*. His earliest publication had woodcut prints of drawings from his own hand, but in the end all the illustrations in *De fabrica* were drawn by professional artists. The printing revolution had given Vesalius and his team of illustrators the power of multiplication when all his predecessors were writing "one-off" books whose text had to be scribed by hand but whose drawings, redrawn one at a time, rapidly denigrated in quality with successive reproductions.

Vesalius, in his earlier works, had hinted at Galen's inconsistencies and the problem of his lack of human dissection. In *De fabrica*, there was no more hinting and only slight nuance regarding the master. In the introduction of *De fabrica*, Vesalius affirmed:

> At Padua, in that most famous university of the whole world . . . I gave lectures on surgical medicine, and because anatomy is related to this, I devoted myself to the investigation of man's structure. Thus I have already conducted anatomy very often here and in Bologna, and, discarding the ridiculous fashion of the schools, I demonstrated and taught in such a way that there was nothing in my procedure that varied from the tradition of the ancients.

He then referenced the gods of anatomy, including Galen, and then criticized their followers:

> To one earnestly concerned with dissection there is nothing in which they [earlier anatomists] seem to have had less interest than in the dissection of the human body. They are so firmly dependent upon I-know-not-what-quality in the writing of their leader [Galen] that, coupled with the failure of others to dissect, they have shamefully reduced Galen into brief compendia and never depart from him—if ever they understood his meaning—by the breadth of a nail.

Powerful words from the twenty-eight-year-old, but later in the introduction, he softened:

> At present I do not intend to criticize the false teachings of Galen, easily prince of professors of dissection; much less do I wish to be considered as

disloyal from the start to the author of all good things and as paying no
heed to his authority.[27]

In the words of Mark Antony, "I come to bury Caesar not to praise him." Vesalius
went on to reference over two hundred instances in which Galen was wrong regarding
"human structure and its use and function." The message was becoming clear: The
king is dead, long live the king.

Earlier in this chapter, mention was made of Ambroise Paré, who some consider
the first great surgeon. An argument can be made that Vesalius is the luminary that
elevated surgery from the lowly barber/surgeons by his emphasis on the skillful use
of the hands. In *De fabrica*, Vesalius mourned that man had neglected "that primary
instrument, the hand, so that [the manual aspects of medicine] were relegated to
ordinary persons wholly untrained in the disciplines subserving the art of medicine."[28]
Whereas in ancient times, the early physicians used the three methods of the regimen
of the diet, the use of drugs, and the use of hands, by degrees physicians of Vesalius's
time had "promptly degenerated from the earlier physicians, leaving the method of
cooking and all the preparation of the patients' diet to nurses, the composition of drugs
to apothecaries, and the use of the hands to the barbers."[29] Vesalius, wellborn into a
proper family, was making the case for intimate contact with patients, which would
require getting dirty and smelly. Physicians in ancient times "devoted themselves
especially to the treatment of luxations [joint dislocations], fractures, wounds,
and . . . freed warriors of javelins, darts, and other evils . . . of war."[30] Vesalius wanted
physicians to still use their hands to treat patients, "if it please the gods, [to be] like
the Greeks, to scorn the whisperings of those physicians,"[31] who had disdained the
art of using the hands to make dissections and treat patients. While Paré is rightly
credited with wiser treatment of war injuries, and more elegant treatment of severed
blood vessels, Vesalius's simultaneous reinvigoration of the study of the human body
and exaltation of the use of the hands makes him one of the most important figures
in the history of surgery.

With his anatomical text prepared, final planning was made on the illustrations.
The magnificent drawings commissioned by Vesalius, transmitted via woodcut
blocks onto massive pieces of paper, are recognizable to most of us. The master artist
Titian (1490–1576) was born, lived, and died in the Republic of Venice. In his many
productive years, Titian maintained a studio in Venice, and it is thought that the
superlative illustrations in *De fabrica* likely emanated from a young genius in that
studio. The illustrations fall into four general groups: the introductory illustrations,
the muscle men, the historiated initials, and the sublime anatomical figures. All
of the refined artwork would have first been drawn on paper, and then came the
arduous task of converting the drawings into reverse images on identical-size woodblocks. The woodblocks were made of pearwood, sawed with the grain, rubbed with

hot linseed oil, and then meticulously carved by the cutters with extreme delicacy. Following completion of the blocks, Vesalius wrote a letter to the printer in Basel, Switzerland. He had chosen Johannes Oporinus, a professor of Greek in Basel, who was well known to scholars for his attention to detail and extremely high production values. The letter and all the woodblocks arrived in Basel in September, 1542, following a transalpine cartage from Venice. Oporinus and his team (along with the help of Vesalius) then spent months organizing the manuscript and woodcuts, and by the summer of 1543, finished books were emerging. (While most woodblocks used for printing were recycled or thrown away, the *De fabrica* blocks survived for centuries, at times being lost for decades. There were whispers that the blocks were hidden at the University of Munich library, and after some investigation, they were discovered in pristine condition in a large trunk in the library attic in 1936.[32] Tragically, they were all destroyed during World War II bombings and none survive today.)

In *De fabrica*, there are two large, introductory images. The first is the title page and the second image is that of Vesalius himself. The title page image is one of the greatest woodcut images ever produced; its perspective, clarity, composition, and technical mastery would be wonders to behold if it were a mere drawing. To realize that a master cutter carved this image in relief is mindboggling. The scene is a public dissection, with a throng of people (at least eighty-five, by my count, not including the corpse, a dog, and a monkey) crowded around the dissection table with Vesalius demonstrating the innards of a cadaver. It recalls Raphael's *The School of Athens*, completed in 1511, and in fact, there is a figure on the right side of the image that resembles Raphael's *Plato* (by tradition, modeled after Leonardo da Vinci, who died in 1519), but here he is not pointing heavenward, as Plato did, but motioning toward the body. Is this Vesalius's ideal? If astronomers were mapping the heavens, was our anatomist mapping our bodies?

The *muscle men* encompass entire sheets of paper, and in fact, in the few remaining original copies of *De fabrica*, the sheets are larger than folio size, unfolding to become another one-third larger. These large drawings are macabre representations of the corpses in action, skinless, yet not lifeless, their faces contorted in agony. Successive illustrations show increasing degrees of dissection, with less and less muscle still attached. The muscle men are situated in a pastoral scene, perched high on a hill with the Venetian countryside in the background, with churches and village buildings dotting the horizon.

The historiated initials throughout the book offer a whimsical, if not ghastly, interlude between sections of the writing. The large capital letter that starts a new passage was formerly decorated by hand but after the advent of printing could be efficiently printed from woodblocks. Two-thirds of the letters of the alphabet are used, and all include putti, or angels, mischievously participating in bodysnatching, boiling bones, setting fractures, and more gruesomely, experimenting on a live pig.[33]

All are a reminder of the dreadful epoch that our ancestors endured but from which all future generations greatly benefitted.

The anatomical figures themselves are the main attraction. In his letter to Oporinus, Vesalius exhorted him to print everything as "handsomely and swiftly" as possible and to use the woodblocks as "correctly and elegantly" as he could. For the first time ever, the figures were inscribed with letters that correlated with the text. The reader could follow Vesalius in the text, encountering the letters that guided him to the specifics in the illustrations. Additionally, in the margins, there were numbers and letters that cross-referenced other drawings. Vesalius was breaking ground in many ways, presenting a visually stunning, didactic tour de force that, on occasion, challenged 1,500 years of authority. Through hundreds of pages and illustrations, the human body and its functions are presented in stunning, comprehensible figures; Vesalius on occasion opines about the physiology, the function of the organs, at times, challenging Galen. Noticeably absent, of course, is conjecture about the gods, evil spirits, and the seat of the soul; however, it would take several hundred years for advanced microscopy to reveal the secrets of the cells and their functions.

Vesalius stated: "I am aware that by reason of my age—I am at present twenty-eight years old—my efforts will have little authority, and that, because of my frequent indication of the falsity of Galen's teachings, they will find little shelter from the attacks of those who were not at my anatomical demonstrations."[34] To the contrary, his masterpiece was hailed immediately, and his renown as an anatomist and surgeon became unrivaled, perhaps for two hundred years.

I will admit it: I am an unrepentant book lover. The privilege of holding a rare priceless book in my own hands is extraordinary indeed. After months of communication to establish my credentials with archivist librarians, and after completing the requisite forms granting me access to the special archives room at the Wellcome Library in London, the day has finally arrived for me to investigate a 1543 copy of De humani corporis fabrica. *I stow my backpack and pens (no pens are ever allowed in rare book rooms) in a locker, and after scanning my scholar's ID through the security checkpoint, I make my way to the top floor of the building in one of the greatest medical libraries in the world. I do have some trepidation—despite all my preparations, I have concerns that I will have come all the way to London, only to be turned back at the last moment. My last email exchanges were with the archivist named Ross, and he approaches me immediately after I pass through the last security doors into the inner sanctum. "Dr. Schneider? Welcome to the Wellcome Archives. How 'bout having a look at* De fabrica?*"*

I take a seat, realizing he has done his own homework, investigating me online and ensuring that I am not some imposter who will come to the holy of holies and disrespect these inestimably valuable tomes. Ross reappears, carrying a massive 16-by-11-inch rich green book. I am gobsmacked by its size—it's really true—it's a beast of a book. The lustrous green leather exterior must certainly be less than a hundred years old, and I instantly wonder what the original binding looked like. The book is set down on the archive table, and before opening it, a religious rite of sorts takes place, of purification, solemnity, and homage. As Ross and I have not encountered a rare book together before, I sense he wants to ensure that I will appropriately care for this treasure. There are only about one hundred De fabrica *copies in the world, and this one is in pristine condition.*

De fabrica *lies next to me, unopened. Ross and I start to build a small mountain of foam blocks and white canvas bean bags. The goal with archival books is gentle handling of individual sheets of paper, assiduous care of the spine, minimal handling, and no accidents. Again and again, Ross places the book in the center of the angled black foam wedges and starts to let the book naturally fall open. Still sensing not enough support on the right side, the librarian adds another bean bag to the wedge. After several minutes of tinkering and adjusting, the book is ready for inspection.*

Staring at a page, I am struck by faint, colored lines that delineate the margins. They appear hand drawn, and are on almost every page. The particular book in my hands is from the original 1543 publication run, meaning that Vesalius himself may have handled this book. The four-hundred-year-old paper is in wonderful condition, with no foxing and no edge wear.

I want to see one of the large illustrations, and turning to the back half of the book, I see a folded sheet. Shocked, I see that this is an extra-large sheet of paper that is completely filled with a detailed illustration of veins and arteries of the body. I unfold it partway, and then again along the bottom. Considering that this diagram of veins and arteries to the entire body came from one wood carving is stunning. There are scores of letters and numbers labeled on the structures, and the remainder of the sheet is filled with the names of the vessels, all in Latin. I am spellbound.

Returning to the front of the book I encounter the famous muscle men, *and in sheet after sheet, I see the progression of dissection. With each sequence, less and less muscle is depicted. It's like the acetate overlays in my beloved* World Book Encyclopedia. *It all started here, with this book, the masterpiece of instruction and artistry that dared to criticize Galen, and Vesalius had thrown down the gauntlet and proclaimed,* I will see for myself and prove to myself through investigation what is true. *Over the next one hundred years, this resolve took root and ignited a scientific revolution.*

FOUR

The Rise of Science

"Let us hope . . . there may spring helps to man, and a line and race of inventions that may in some degree subdue and overcome the necessities and miseries of humanity."

—Francis Bacon, *The Great Instauration*, 1620

"Their first purpose was no more, than onely [sic] the satisfaction of breathing a freer air, and of conversing in quiet one with another, without being ingag'd [sic] in the passions, and madness of that dismal Age."[1]

—Thomas Sprat, *History of the Royal Society*, 1667

On a sunny spring morning in London 1715, a group of young scientists gathered at the world's greatest clubhouse for insatiably curious people. Weeks before, the astronomer Edmund Halley had issued a bold claim: he had carefully reviewed the star charts, calculated the courses of the sun and the moon, and was predicting, with startling bravado, that a total solar eclipse would engulf the city of London on the morning of April 22. Come to the Royal Society's home on Crane Court—he bade the *virtuosi*—and witness the spectacle with me.

There are apocryphal stories of ancient stargazers predicting eclipses; usually these prophecies were stated in months, if not years. Eclipses are not altogether rare; such predictions were not completely daring. Alternatively, Halley (the one member of the Royal Society who could interface with the brilliant and testy Isaac Newton) was inviting the geniuses of the firmament to gather in the old city on a particular day, hour, and even minute. There had not been a *total* solar eclipse in London since 1104, and those intervening six hundred years had barely witnessed any technological advancements. Most Englishmen believed in witches, werewolves, unicorns, and magic; although heliocentrism and the new mathematics had acceptance among the learned, there were no practical applications for the common man.

A total solar eclipse was a biblical event. For those lucky enough to witness a *total* eclipse on a cloudless day, the phenomenon of utter darkness for a few minutes is fantastical and, well, magical. We are, at once, Mayans staring at the sky, dutifully shoulder to shoulder with our fellow man, swept up in the vortex of planets, stars, and moons.

Halley calculated the path of totality and published a map of his predictions. By making known the imminence of the event, Halley hoped to limit terror and to maximize calculations from the learned on the British Isles and throughout continental Europe.

With quill in hand, Halley had spent weeks reviewing the data tables generated by "natural philosophers" in the preceding decades. Prior to 1662, there was essentially no sharing of scientific information among the scholarly, but the *Philosophical Transactions of the Royal Society* changed everything. Halley was able to gather data from years of astronomical records, and with industrious determination, was convinced there was an impending eclipse. An invitation was proffered just in time, and the well-to-do intelligentsia arrived at Crane Court, wearing wigs and waistcoats at the break of dawn.

Crane Court is a narrow alleyway off Fleet Street in the heart of the City of London (the City is the formerly walled enclave in the heart of modern-day greater London); the Court was on the outermost boundary of the Great Fire of 1666 that consumed almost the entire core of London. At the end of Crane Court was the building that served as the Royal Society's meeting house, and it was here that scientific history was forged with sunlight and shadows.

In the hours before an eclipse, you cannot see the moon nearing the sun—it is simply invisible to the naked eye as it is outstripped by our star. What is appreciable, particularly on a sunny day, is the diminution of radiant heat. An odd, windless, cloudless cooling occurs.

The eclipse started shortly after daybreak in London (just after 8:00 A.M.), and an hour later, totality occurred nine minutes after 9:00 A.M. For the next three minutes, darkness reigned over London, exactly as Halley had predicted, and precisely *when* he said it would happen. Because so many astronomers and scientists were poised with their equipment and recording devices, important observations were made about the surface of the moon and the exact duration of totality.

To the fortunate catechumens at Crane Court that day, what optimism must have filled their hearts and minds? Yes, there was certainly gratitude and more than a little awe directed toward Halley; those gathered must have reveled in their newfound powers of prediction, but the more significant contemplation must have been, is there nothing we cannot predict, experience, and conquer?

David Wootton has concluded, "A basic description of the Scientific Revolution is to say that it represented a successful rebellion by the mathematicians against the authority of the philosophers, and of both against the authority of the theologians."[2]

It was a fundamental principle of Aristotelian philosophy that there could be no change in [the heavens],[3] yet the last of the "naked eye" astronomers—Copernicus, Tycho Brahe, and Johannes Kepler—were able to observe exploding stars and to predict the motions of the planets. David Wootton argues that Tycho's nova (two stars fusing together in a thermonuclear reaction, eight thousand light years away) marks, quite precisely, the beginning of the Scientific Revolution. It was not the cause of the revolution, but it is the signpost that signals the start of the insurrection. Whereas Aristotle was preoccupied with *qualities* to explain the world (the four elements of earth, air, fire, and water), Tycho painstakingly gathered data and charted the skies. Instead of sophisticated dialogue and intellectual guile, the leaders of the revolution would use measuring instruments, numbers, data tables, and calculations. In short, they would *mathematize nature*, and the world.[4]

When Galileo Galilei's book *Sidereus Nuncius* (*Starry Messenger*) arrived in Prague a few weeks later, in April 1610, Kepler was desperate to digest what Galileo had seen. The book announced on the title page that Galileo was a *"Patritio Florentino,"* a gentleman from Florence, and that he was a *"Patauini Gymnasii Publico Mathematico,"* a mathematics professor in Padua. In the middle of the page, on its own line was the word "PERSPICILLI," Galileo's word for his telescope.

In the opening of the book, the forty-six-year-old Galileo describes how his telescope came to be. Ten months earlier, in May 1609, Galileo received word that a Dutchman had constructed a telescope with the aid of lenses and a tube. After some thought, Galileo "finally determined to give myself up first to inquire into the principle of the telescope, and then to consider the means by which I might compass the invention of a similar instrument, which a little while after I succeeded in doing, through deep study of the theory of Refraction; and I prepared a tube, at first of lead, in the ends of which I fitted two glass lenses, both plane on one side, but on the other side one spherically convex, and the other concave."[5]

As will be seen time and again in this work, the breakthrough in an area of science resulted from a tinkerer who relentlessly focused on the problem at hand and physically involved himself in the manufacture of tools, instruments, and measuring devices. The innovation that made telescopes possible was first the development of clear glass in Murano and the subsequent perfection of curved glass fabrication by the Dutch and Germans. Galileo himself prepared the lenses, perfecting the size and shape of each glass disk, turning a three-powered spyglass into a twenty-times magnification tool suitable for celestial navigation. In all, he fashioned over two hundred lenses, ending up with ten telescopes with magnification of at least twenty times.[6]

Realizing the functional military usefulness of his viewing instrument, Galileo approached the Senate and Doge (chief magistrate) of Venice. In the bell tower of Saint Mark's Basilica in Venice, Galileo demonstrated the superiority of using his telescope over the naked eye. On August 21, 1609, a faraway ship was invisible to onlookers, but with his 12x telescope, Galileo was able to show the Doge the potentially threatening vessel in the Venetian Lagoon. Pleased with the promise of naval superiority based on early detection of enemy warships, the Doge granted Galileo a handsome salary and a mathematical professorship at the University of Padua.

Galileo spent eighteen years in Padua (1592–1610), including an overlap of three years, between 1599 and 1602, in which he and Englishman William Harvey were both at the university. In an Italian university town at the turn of the 17th century, two men occupied the same sphere; two giants of science who would be the first physicist, observational astronomer, and experimental scientist; and the world's first physiologist, who explained how blood circulated through our blood vessels. Galileo and Harvey represent the best and worst of their time. Galileo is one of the last scientists to be tried by the Roman Inquisition, spending the last decade of his life under house arrest for his belief in heliocentrism. Harvey rose to fame as Physician Extraordinary to King James I and "Doctor of Physic" at Oxford.

"Thomas Hobbes, writing in 1665, thought that there was no astronomy worth the name before Copernicus, no physics before Galileo, no physiology before William Harvey."[7] Both Harvey and Galileo saw the world with mechanical sensibilities, perceiving orbits and revolutions; their insights into the ambulations of satellites, the motion of blood, and trajectory and velocity of moving objects were radical indeed, but were limited by simple, Euclidian geometry.

The Scientific Revolution had been launched, but what was desperately needed was an earth-shattering insight into how to numerically describe and predict the world. In the same year that Galileo died, in a small hamlet far north of London, Isaac Newton was born a fatherless, premature infant who would become perhaps the greatest genius who ever lived, and who would provide the rules and the calculus to contemplate our world.

The year 1610 found the city of London as the principal North Sea port, with an exploding population of a quarter-million people enjoying the English Renaissance of Elizabethan times. Literary giants like William Shakespeare, Ben Jonson, and John Donne held center stage; performances of *Macbeth* at the Globe Theatre delighted crowds (although there was no bathroom for the three thousand spectators). The Jamestown settlement was underway in the New World, and the Ulster Plantation had just been initiated by King James I, who authorized an English translation of ancient scripture for the Bible that would bear his name.

The fragile world of the 17th century left the European metropolitans "extremely liable to disease, physical suffering, and early death . . . with life expectancy not much over thirty years of age."[8] Plagues, epidemics, famine, overcrowding, rampant poverty, and poor sanitation meant that London was awash in both poetic brilliance and excrement; new institutions and outdated, filthy traditions.

When Galileo released his *Starry Messenger* in March 1610, it didn't take long to reach Gray's Inn, one of the four "inns of the court" in London. These professional associations of barristers (British litigators) and judges have been headquartered in stone and brick building complexes (resembling Ivy League college dorms) in the heart of London for hundreds of years. Even today, if a sightseer were to saunter about the neighborhood where the inns are parceled, you would likely encounter barristers in white wigs and red robes making their way to the nearby Courts of Justice. Like encountering young brokers in power suits near Wall Street and stumbling into sleepless, unshaven residents schlepping their way home post-call near Mass General Hospital, ensembles do "make the man," if not mark the man.

Into that powdered-wig world of barristers and jurists came the *Starry Messenger*; Galileo's central character was the supernova, but to a particular political philosopher living at Gray's Inn, the Italian genius was himself the envoy. That barrister was Francis Bacon, who never conducted a scientific experiment himself, but is considered by most to be the father of empirical science.

Francis Bacon was born into privilege in a stone mansion on the banks of the Thames in 1561. A child prodigy, he entered Cambridge University at age twelve; his precociousness prompted his portraitist to inscribe the motto, "If I could only paint his mind" directly onto the canvas itself. His brilliance was renowned his entire life. Bacon was meditative, scientifically curious, and a thinker of "soaring ambition and vast range."[9] Simultaneously a professional lawyer, politician, courtier, and royal adviser, he was also ruthlessly insecure and preternaturally striving for advancement.

Bacon's "active public life, under both Elizabeth and James I, was taken up with political business and legal reform. Bacon achieved high office as Lord Chancellor in 1618, until disgraced by corruption charges. His final years saw a furious spate of writing on natural philosophy, politics, and history."[10] Much of that writing occurred late in life when he was largely abandoned by his friends, his wife, and the king.

In three great works, the last of which was published posthumously, Bacon accomplished a project of rethinking how we think. Universities were stuck in a morass of Aristotelian futility, impotent in generating new knowledge. Worse, because their philosophies were grounded in his "first principles," it was impossible to challenge their conclusions. "Bacon rejected all existing knowledge as being unfit for making discoveries and useless for transforming the world."[11]

With cartographers creating new continents ex nihilo and our solar system made to reveal new planetary satellites, learned minds were open to new realities. The late Renaissance was characterized by openness, and "the idea of discovery [of truth] is inextricably tied up with ideas of exploration, progress, originality, authenticity and novelty."[12] For centuries, learning had been limited exclusively to gathering together what Aristotle had posited; there was almost no new knowledge, just new commentaries about his conclusions. Realizing the supreme limitations that this worldview had engendered, Bacon exposed the "inability of the logic of the day to make scientific discoveries or invent new sciences"[13] in *The Advancement of Learning* (1605) and *The Great Instauration, New Organon* (1620).

Arguing that our process of intellectual discovery was flawed, Bacon set about proposing a *methodology* to unearth new truths. This organized approach would lead to the *scientific method*—where future scientists would turn conjecture into a hypothesis, then perform systematic observations and measurements, thus drawing conclusions, and finally developing general theories *based upon experimental outcomes*, which would lead to new hypotheses and new experiments. No one more than Bacon emphasized methodology[14] (not even Descartes), and it was Bacon who introduced the concept of interpretation.[15] The ancient unscientific, mystical (even paranormal) arts of astrology and alchemy would be transformed into the technical sciences of modern astronomy, physics, and chemistry with the scientific method; however, the revolution in medicine would take a century more to shake off the bonds of Hippocrates and Galen for good in the late 19th century.

Criticizing the philosophers of old, "Bacon compared empiricists to ants, who 'only collect and use,' and rationalists to spiders, 'who make cobwebs out of their own substance.' With these two insects he contrasted the bee, who both gathers its material from garden and field and 'transforms and digests it by a power of its own.' The business of philosophy, he said, is to imitate the bee."[16] Profound insight from a man of letters, who had no scientific education, and was bereft of mathematical understanding or laboratory organization. Somehow, he perceived there were more scientific discoveries to be made, novel planets to detect, and ideas to formulate. He could not possibly have dreamed of the invention of calculus, electricity, or manned flight, but interestingly, he did imagine a new medicine.

In his final work, *New Atlantis*, an incomplete masterpiece published in 1627 (the year following his death), Bacon gave "remarkable composition to scientific utopianism."[17] He concluded that the great works for human use consisted of "the prolongation of life, the restitution of youth in some degree, the retardation of age, the curing of diseases counted incurable, and the mitigation of pain."[18] To achieve that, microscopy would have to be developed, germs understood, chemistry refined, and epidemiology conceived. Vaccination and prevention would have profound affects in the coming centuries, but to achieve "restitution of youth . . . and the mitigation of pain" mankind would need a series of breakthroughs to create the implant revolution.

Bacon predicted, remarkably, that in the future our society would be led by benevolent philosophers, with scientists playing a dominant role. He dreamed of a new college, which would be a great research institution, "a scientific establishment endowed with facilities for pursuing far-flung investigations into the secrets of nature."[19] In essence, the jurist was describing the modern research university. In Gorhambury, St. Alban's, an immense brick and stone mansion in the faraway reaches of outer London, warmed by log fire and illuminated by candlelight, Bacon dreamed of a new academy and a novel way of thinking. Today, Gorhambury lies in ruins, but its 450-year-old skeleton is accessible via a "permissible path" on private property. Standing among the ruins, gazing over the undulating, verdant hills, speckled with sheep and cows and interrupted here and there with patches of ancient trees, one can only wonder what Bacon would think of, say, Boston, with Harvard University and the Massachusetts Institute of Technology dominating the land north and south of the Charles River, with space-age engineering, DNA research, cancer mitigation, and computer programming and artificial intelligence proceeding at warp speed.

Francis Bacon proposed a moniker for his intellectual establishment: Solomon's House. While it is true that he had indulged in philosophical debates and legal disputations over the decades, he, like every scholar in the world, lacked an "establishment . . . pursuing . . . the secrets of nature." Remarkably, within two generations, the hypothetical Solomon's House would come to fruition in the assemblage of the Royal Society (not by coincidence in the City of London); but it would take a civil war, the beheading of the king, a fortuitous conglomeration of virtuosi, and the restoration of the sovereign to spark the world's first genius society.

After centuries of increasing power and broadening influence, the monarchs of England had authenticated their greatest dominion under the rule of King Henry VIII and later, Queen Elizabeth and her son, King James I. Church and state were united, land and assets had been seized, and the Divine Right of Kings was proclaimed to an unprecedented degree. King James's son King Charles I assumed the throne in 1625, ratcheting up the tension between king and parliament, with its eventual dissolution. Parliament and the king went to war, with Charles I beheaded in 1649. England was ruled by Oliver Cromwell and his cabinet for over a decade, but by 1660, Parliament restored the monarchy and installed Charles II as king. The Houses of Commons and Lords had never been so powerful, and to this day enjoy authority of rule throughout Britain.

In the half century before the ascension of Charles II, the Roman Catholic Church on the continent was still persecuting Christians for heretical scientific thinking (while Catholics in England faced similar life-threatening persecution). With the restoration of the monarchy and the English antipathy toward Catholicism, combined with the new king's interest in intellectual topics, those intrepid philosophers of the new *Scientia* (Latin: knowledge) had found their man.

For a few brief years, there was peace and quiet in London. A political and rational homeostasis had been secured in England following the interregnum, and as happens so regularly, the philosophical advancements that occurred *during* and *after* the revolution were massively consequential.

Francis Bacon had thrown down the gauntlet for science. "Bacon's main and permanent significance, therefore, is as a thinker about science: the conditions favorable to its growth; the changes and procedures required to ensure its progress; its contribution to the inauguration of a new regime of knowledge; and its technological and moral realization in works to improve the human condition."[20] Solomon's House, slowly at first, became a reality. On November 28, 1660, three dozen men met at Gresham College in London to hear twenty-eight-year-old Christopher Wren speak on astronomy. In attendance that night were some of the most important thinkers of the century, including Robert Boyle and Henry Oldenburg.

In due time, Charles II granted a Royal Charter to the bourgeoning scientists, and thus the name "Royal Society" is used. The complete, proper name is "The Royal Society of London for the Improving of Natural Knowledge." It did not take long for those early members to realize the gravitas of their "venerable learned society" (the "finest club" in the world[21]), and soon, regular meetings were occurring in London. Speeches and presentations were given, with King Charles II oftentimes in attendance, particularly for scientific demonstrations or dissections. "As to our work," wrote an anonymous member of the Royal Society in 1674, "we are all agreed, or should be so, that it is not to whiten the walls of an old house, but to build a new one." Tearing down the old and starting again from scratch is what revolutions are all about.[22]

There are other genius societies that originated in the same era, such as the Académie des Sciences (founded by Louis XIV in 1666), but it is undisputable that the *Philosophical Transactions* (still in publication) is the oldest scientific publication in the world, starting in 1665. "The Royal Society invented scientific publishing and peer review. It made English the primary language of scientific discourse, in place of Latin. It systematized experimentation. It promoted—indeed, insisted upon—clarity of expression in place of high-flown rhetoric. It brought together the best thinking from all over the world. It created modern science."[23]

Copies of the *Transactions* (produced as a newssheet) were sent, of course, to Oxford and Cambridge, eventually reaching a young man; solitary and obsessive, abstemious and born of exceptional concentration—Isaac Newton.

Isaac Newton's entry into the world, in retrospect, is messianic. He was born on Christmas morning in the bedchamber of a stone farmhouse. His illiterate father, a yeoman who had inherited the carefully constructed, local cream-colored limestone manor house and associated barns, died months before Isaac's birth.

A three-hour journey (on the A1 motorway), north of London to the farming lands and forests of Lincolnshire, lies the small village of Woolsthorpe-by-Colsterworth. Like thousands of hamlets and parishes throughout the English countryside, it is a clutch of homes surrounded by pastures and rolling hills. However, on the southeast corner of the settlement, a group of stone and brick buildings present a different appearance, perfectly preserved over the centuries. This is Woolsthorpe Manor, the birthplace and home of Sir Isaac Newton. The outbuildings and barn are to the east, but to the westward-facing ancient stone farmhouse lies an apple orchard, with one exceptionally old, craggy apple tree, still bearing fruit. The tree has survived pestilence, fire, lightning strike, and mayhem, and like its famous former lord of the manor, has stood the test of time.

The limestone manor house has two floors; Newton's bedroom is upstairs, with a large window facing the apple orchard. Low doorways, expansive fireplaces for cooking, and creaky, uneven floors give further reminder of the age of the structure. The artistic significance of the apple tree's appearance from his window is outweighed by the scientific, practical usefulness of a smaller second window in the room, facing south. This is the window that Newton used as his source of sunlight, perfect for capturing light during long English winters, particularly when he needed uninterrupted shafts of white light. The "chief architect of the modern world"[24] conceived the future in this simple country abode.

Isaac's challenging entry into the world turned cruel when his mother married an Anglican priest from a nearby parish whose precondition for marriage was to leave young Isaac behind. Raised by his maternal grandmother in Woolsthorpe, one wonders if his solitary inclinations and lifelong difficulty with relationships arose from childhood trauma. He would never marry and lacked deep friendships, begging the question of autism or other social-disconnection conditions. When Isaac was ten, his mother, again widowed, returned to the farmhouse, with three half-siblings in tow. Instead of adapting to life with his own family in Woolsthorpe, Isaac faced another challenge of being shipped off to boarding school eight miles up the road in Grantham, living with the local apothecary.

In Grantham, Newton became well acquainted with the works of René Descartes, the brilliant French philosopher who, along with Bacon, is one of the chief founders of the new Western philosophy. "I think, therefore I am" is Descartes's most famous quotation, but his lasting influence was in mathematics and physics, and most important, his insistence on a mechanistic view of the world and cosmos. He echoed Bacon with his insistence on empirical investigation and an emphasis on a scientific method. It would not be long until Newton was incorporating Descartes's techniques and building upon his theories.

As Newton was nearing manhood, it was becoming increasingly obvious that he would make a distracted and uninspiring farmer. (Similar to Abraham Lincoln and Albert Einstein, Isaac was often criticized for having "his head in the clouds.") Thankfully, arrangements were made for Isaac's admission to Cambridge University, three days' journey to the south. In June of 1661, he was installed in Trinity College, Cambridge University, completely dedicated to his scholastic mission. By his third year Newton had absorbed all that could be known. This sounds impossible to modern ears, but Newton was able to learn all contemporary math and physics in his first few years in Cambridge.

And then, in 1664, an astonishing and beautifully horrific upheaval occurred across England. Plague, the ancient scourge, began to claim lives and ignite fear from down south. Cities and universities began to drain across the country, and Cambridge, bowing to the Black Death, shuttered in 1665. Isaac returned home to Woolsthorpe, living with his mother and half-sisters for the first time. Medieval instincts called for isolation in the face of pandemics, and Newton may be seen as the greatest solo artist of all time.

In 1665, the Royal Society was gaining momentum and preparing for publication; Isaac Newton was secreted away in the upstairs bedroom of Woolsthorpe Manor. Unanswered questions plaguing Newton, he began an approximate eighteen-month project that would be the most productive and astounding by any theorist, ever. Taking fastidious notes in tiny script (on precious paper bequeathed to him by a former tutor), the solitary Newton unlocked the secrets of light, the meaning of gravity and the laws of thermodynamics, as well as the calculus concepts of integration and derivation that would make all modern mathematics and science possible.

What held the moon above us? Why did it not come crashing down to earth, or simply fly away? Why did all objects have a certain weight that pulled them to earth, and why always straight down? The ancient Greeks and early Renaissance philosophers had tried, in vain, to unlock the mysteries of orbits and objects; Newton employed all the known mathematics of the time and invented more.

Through an amazing coincidence, the sun is four hundred times the diameter the moon, but is also four hundred times farther away. This explains their similar "apparent size" in the sky and the mathematical prospect of complete solar eclipse. Both orbs occupy one-half degree of arc across our sky (it is 180 degrees from horizon to horizon), meaning that you could line up 360 moons across the sky. By sheer coincidence, or providence, lying in the Woolsthorpe apple orchard and gazing up at the ancient apple tree, an apple would subtend the same half-degree in the sky. Newton's thought experiment about falling objects and the pull of the moon by the earth started at Woolsthorpe, further bolstered by initial simplistic calculations about the size of the moon, the distance from earth, its velocity across the sky, and thoughts about the powers of attraction. While it would take many years to form a comprehensive lexicon about universal gravitation and the laws of thermal dynamics, the seeds were planted in the Lincolnshire countryside, with Newton gazing at the moon and sometimes staring straight at the sun.

The Royal Society lacked a formal meeting place of its own at first, but the ingenious members did not tarry in adopting a motto—*Nullius In Verba*, Latin for "take nobody's word for it." This contemporary Solomon's House would base all new knowledge on proof. Adopting Baconian empirical processes, the intelligentsia disputed old beliefs, tested new theories, and reported new findings. The center of the group from the beginning was Robert Boyle, a tall and thin aristocrat from Ireland who devoted his life to scientific experimentation. Boyle's law is familiar to every chemistry student as the explanation of the behavior of gases, and is remarkably insightful for the 17th century. Boyle's longtime lab assistant was Robert Hooke, an Englishman who was a brilliant technical scientist, manufacturer of equipment and devices, and grumpy sidekick.

Also present from the outset was Christopher Wren, a polymath anatomist, astronomer, physicist, and most notably, architect. Wren's monstrous energy and architectural cunning would be tried in 1666, when all the buildings at the heart of London were razed by the great fire that had been foretold by soothsayers who feared the implications of 666 in the year.

Geniuses all, none would claim to be the first scientists or philosophical wunder-kinder; what made them remarkable was the congregation of virtuosi, the world's original scientific organization. The story of "early science would have more to do with collaboration than with solitary contemplation,"[25] with the glaring exception of a young man in Woolsthorpe.

Whether in Woolsthorpe, Cambridge, or later, London, Newton was character-ized by one, most unusual trait: his power of concentration. When asked by a devotee how he had conjured the concept of gravitation, Newton replied, "By thinking on it continually." When faced with a theoretical dilemma, he did without sleep and almost completely without food. He took no pleasure in exercise, avocation, epicurean delights, or fraternity; only intellectual pursuits provided temptation. For a man who decoded the tides, it is remarkable that he never saw the sea, yet his ascetic lifestyle yielded profound insights into the physics of motion, light, and gravity. His radical curiosity about the mechanics of the world did not extend, physically, more than a one-hundred-mile radius from his origination.

Completely (intellectually) alone in Woolsthorpe, Newton began a project of math-ematical and philosophical exploration in 1665 that started with the proof of the binomial theorem, a pillar of mathematics to this day. How much farm work and tasks around the house did the twenty-three-year-old perform? We don't know, but it is clear that a huge amount of mental energy was expended by Newton.

For any reader who has struggled with high school and college calculus, it is startling to consider one man inventing the process of differentiation, and later, integration. (There was a protracted debate over primacy in the innovation of calculus between Newton and German Gottfried Wilhelm Leibniz (1646–1716), with Newton never surrendering the title.) Newton developed the mathematics of calculus to deal with the complex computations he was faced with when considering the motions of the planets and properties of gravity. Setting aside his new math weapon, he turned to the triangular glass prism he had purchased at a country fair outside Cambridge.

Rainbows in the sky have always had the same color pattern (ROYGBIV: red, orange, yellow, green, blue, indigo, violet); but nothing more than the order of the colors was understood. Why were they always the same? Newton retreated upstairs to his room where a small window faced the southern sky. Cutting a small hole into a board covering the window, Newton was able to isolate a shaft of light into the dark room. This afternoon light, particularly on a sunny winter day, was used to pass light through a prism, diffracting the sunlight into a rainbow of colors splashing onto a distant wall. In a stroke of genius, Newton arranged a *second* prism in the path of the isolated colors, curious to see if another dispersion of a second rainbow occurred. Dear Reader, what is your guess—does another rainbow ensue, or is it maintained as that selfsame color? Or another?

The answer: the same color emerges from the second prism. With this result, and others, Newton concluded that sunlight, or white light, is composed of the colors of the rainbow. The experiment to confirm this is to angle a series of prisms, or mirrors, into one focal point. Colors flow in, and white light emerges. "Newton's experiment of sunlight refracted by two prisms—so ingeniously conceived, carefully performed, and exquisitely narrated—came to be seen as a landmark in the history of science. It established a great truth of nature. It created a template for the art of reasoning from observation to theory. It shines as a beacon from the past so brightly as to cast the rest of the Society's contemporaneous activity to relative shadow." [26]

During the "miracle years" in Woolsthorpe, Newton also performed the foundational work on gravitational theory; exploring how the moon was held in balance above the world, spinning and rotating, without flying away or crashing down to earth. Using only his basic estimations and new mathematics, Newton was able to prove to himself that all objects have gravitational pull, and for the first time in world history, was able to grasp why objects fell, why water flowed, why cannon balls arced in the air as they were shot from a cannon, and why the celestial bodies traveled across the sky. These concepts would underpin his laws of thermodynamics in years to come, but for now, wandering around the manor, Newton could take pleasure in gazing up at the moon and understanding what forces held it in balance.

For that matter, Newton was allowed a full measure of satisfaction everywhere he looked. The moon, orbiting overhead, sunlight reflecting off rivulets in the Woolsthorpe pastureland, apples plopping onto the orchard sod, the arced trajectory of a stone thrown by a neighbor boy, and even the function of his own eyes all were, to him, the function of the mechanical laws he was discovering. A deeply religious man, Newton's insights convinced him that everything around him conformed to his conception of a "clockwork universe," an orderly reality that he could understand, if he concentrated with enough gusto. He later concluded, that during the "plague years of 1665–1666 . . . I was in the prime of my age for invention & minded Mathematicks [sic] & Philosophy more than at any time since." Among the Lincolnshire heath, Newton had tilted the earth toward a new philosophical understanding; and in intellectual investigation, there "is nothing remotely like it in the history of thought." [27] He had the added pleasure of knowing that *only he* comprehended the rules of the machine, and many years would transpire until he was essentially forced to divulge his discoveries to the members of the Royal Society.

Following Newton's annus mirabilis, in which he created modern mathematics, optics, and mechanics, he returned to Cambridge, quickly ascending to the title of Lucasian Professor of Mathematics, vacated by his mentor, Isaac Barrow. (The most recent, familiar Lucasian Professor has been Stephen Hawking.) Barrow departed to London, where he was an early member of the Royal Society—and the link between the Society and his protégé back in Cambridge. Over the course of many years, Barrow urged Newton to correspond with various members in London, and beyond, but the reclusive professor evaded contact with surprising dedication. After more than a decade, Isaac Newton finally made an appearance at a Royal Society meeting, in 1675.

As the years passed, Newton overcame isolationism and envy. At first, Newton's discoveries were steadfastly cloaked behind a murky veil, but a small circle of trusted friends, like Isaac Barrow, Robert Boyle, and Edmund Halley, were able to entice him to reveal his secrets. The first taste of his genius was an examination of the telescope he had fashioned by hand. Later, papers started to trickle in to Oldenburg, followed by months, or years, of silence. To a society dedicated to information flow, these tantalizing notes from the nearby mastermind were scintillating, if not frustrating.

Another new discovery had recently descended upon English shores: coffee. The first London coffeehouse opened in 1652, and by 1663, there were eighty-two coffeehouses within the old Roman walls of the City.[28] Just as the Royal Society was emerging, "coffee came to be portrayed as an antidote to drunkenness, violence and lust; providing a catalyst for pure thought, sophistication and wit."[29] The Enlightenment was coming into full bloom, and characters like Newton were the focus of debate in the coffeehouses and homes of the great city.

Isaac Newton published his masterpiece, *Philosophiae Naturalis Principia Mathematica*, referred to simply as *Principia*, in 1687. Fortunately for mankind, Halley, the skillful diplomat, was able to cajole the reluctant Newton to share his ideas about mechanics and mathematics. *Principia* is Newton's magnum opus, one of the most important works in the history of science. In it, Newton had painstakingly laid out, in overwhelming intelligence and insight, the laws of physics, explaining gravity, celestial motions, and why things work the way they do. Like Solzhenitsyn's *The Gulag Archipelago*, it was hailed by everyone, read by few, and understood by only a tiny handful. It set the stage for the Scientific Revolution.

Newton would rise to become president of the Royal Society and Master of the Mint of England. He died at age eighty-four having never married and with no heirs. Deeply religious and dedicated to alchemy, he was described by economist John Maynard Keynes as the "last of the magicians, the last of the Babylonians and Sumerians, the last great mind which looked out on the visible and intellectual world with the same eyes as those who began to build our intellectual inheritance rather less than ten thousand years ago."[30] However magical were his inclinations, he was able to see a new "system of the world," presenting the mathematical framework necessary for all future advanced calculations, including the mathematics required to launch spaceships to mingle with comets and planets.

This is not a book about comets and spaceships, but it is about the birth of science that gave support and foundation to the rise of medicine. And it was the 17th-century scientists, most notably Bacon and Descartes and Newton, who theorized that our world could be scientifically investigated and codified. "Bacon may have been among the earliest, if not the first, of Western philosophers to give to the concept of a law of nature the meaning it came to acquire in the natural sciences. When he refers to law in defining forms, it seems to be detached from any association with a divine lawgiver, providential design or oversight, or teleological purpose."[31] Descartes and Newton would rightly claim that their discoveries were universal truths and laws; this mentality would open the door to the modernization of medicine, and within a few decades of Newton's death, another unlikely character would arrive in London to become the world's first scientific surgeon.

Although the Royal Society had only existed for a few short years, the members immediately understood that a tidal wave of inspiration and change was sweeping through their lives. In what can only be considered a bold move, if not ostentatious, the members decided to write a history of their founding. In 1667, Thomas Sprat wrote *The History of the Royal Society of London for the Improving of Natural Knowledge*, just five years after it received its royal charter. On the frontispiece, an engraving of the bust of King Charles II receiving laurels, and flanked by Francis Bacon and the first president of the society, William Brouncker.

In the preface of the book, Sprat dedicated the book to their royal patron, King Charles II, declaring that glory was due to the King for freeing man from the "bondage of errors." Later, Sprat considered what the ancients had decided about the philosophers of old:

> What Reverence all Antiquity had for the Authors of Natural Discoveries, is evident by the Diviner sort of Honor they conferr'd on them. . . . *That a higher degree of Reputations is due to Discoverers, then to Teachers of Speculative Doctrines, nay even to Conquerors themselves.* [italics added]

Finally, Sprat concluded, proposing what glory and remembrance awaited the king and not just the inventors:

> Nor has the True God himself omitted to shew his value of Vulgar Arts. In the whole History of the first Monarchs of the World, from Adam to Noah, there is no mention of their Wars, or their Victories: All that is

recorded is this, They liv'd many years, and taught their Posterity to keep Sheep, to till the Ground, to plant Vineyards, dwell in Tents, to build Cities, to play on the Harp and Organs, and to work Brass and Iron. And if they deserve'd a Sacred Remembrance, for one Natural or Mechanical Invention, *Your Majesty will certainly obtain Immortal Fame, for having establish'd a perpetual Succession of Inventors.* [italics added]

Prophecy fulfilled. The Royal Society's fellows have split the atom, discovered hydrogen, the double helix, and the electron. They've invented the World Wide Web and established peer review. The world's first scientific organization, truly a house of Solomon, paved the way for modern science and in so doing, established the proper foundation for the transformation of medicine in the 1800s into a scientific discipline. But first, a wild, unschooled Scotsman would arrive in London, unbelievably becoming the world's first scientific surgeon.

FIVE

Harvey and Hunter

"When I was a little boy . . . I wanted to know about the clouds and the grasses, why the leaves change colour in the autumn. I watched the ants, bees, birds, tadpoles, and the caddis worms. I pestered people with questions about what nobody knew or cared anything about."[1]

—John Hunter

"When I heard this Man, I said to myself, 'This is all day-light.' I felt that what I had previously been taught was comparatively nothing . . . and thought I might, like Mr. Hunter, venture to Think for myself."[2]

—Henry Cline

The Surgical Intensive Care Unit, or SICU, is a labyrinthine block of three large nursing units, designed for the postoperative care of the sickest or most unstable patients. For the past four weeks I've been on thirty-six-hour shifts on alternating days, and even though I'm a surgical intern (in my first year of my surgical residency), I've never ventured back to the operating room. Instead, in my role as an SICU intern, I attend to the post-trauma and post-surgical patients who require a higher level of care than can be delivered on a typical "floor" nursing unit. On those units, no one is intubated and their intravenous (IV) medications are relatively simple and not life-sustaining on a minute-to-minute basis, like they are in the Intensive Care Unit.

In my first few days, I was completely overwhelmed and terrified. It was instantly obvious to my fellow interns and myself that we were unqualified and inexpert at keeping the SICU patients alive; in fact, we were all petrified of killing a patient who was clinging to life with one small misstep. The SICU nurses were right in doubting us: despite our newly conferred medical doctor degrees, we held no practical knowledge. Query us about reciprocal translocation of genetic material in chronic myelogenous leukemia and we'll wax lyrical about switched-on tyrosine kinase; but ask us about a simple ventilator setting and you will be greeted with awkward blank stares. In time, however, we all improved. As the month wore on, the mechanical ventilator (the "vent"), IV drips, lines and tubes, lotions and potions, and bed settings became intelligible and powerfully manageable.

Now in my last week in the SICU rotation, I have spent many hours at the bedside of Travis, a sixteen-year-old Pennsylvania boy whose car full of high school teammates was blindsided by a cement truck after football practice two weeks ago. Three young men were instantly killed, and Travis arrived in the trauma bay in cardiac arrest. My fellow residents resuscitated him right there in the emergency room, making the "trauma zipper" incision from his throat to the pubic bone, clamping his aorta to prevent it from completely rupturing and to barely avert death. In the two weeks since that accident, Travis has undergone multiple operations, coding three different times and facing multisystem organ failure.

There is a stunning level of control over a patient's body in the SICU. Before residency, I had not fully comprehended how powerful our machines and medicines are in controlling the concentration of gases in the body, the blood pressure, the heart rate, and the degree of wakefulness. The physiology I learned in medical school, the revelations about the organ and cellular functions, is now becoming muscularly effective. But not entirely. Travis was dead on arrival, brought back to life with valiant swiftness by the trauma surgeons, and has endured several operations since then, all while battling failing kidneys, Clostridium difficile (C. diff) bowel infections, alarming edema and bloating of his arms and legs, and little evidence of brain wave activity. He has started to smell like death. I initially doubted the claim of the ICU nurses that they could smell death, but now I get it, the pungent smell of C. diff. mixed with the mildewy essence of pseudomonas pneumonia so common in ventilated patients. During a central line change a couple days ago, one of Travis's nurses predicted to me that he wasn't going to make it, and I'm a bit mortified today to admit that he resembles one of those drowning victims that washed up days after going missing.

Yesterday, Travis's parents asked for permission to bring his loyal Irish setter, Honey, to the SICU. They claim it will be good for Travis. Normally a dog lover, I have serious reservations about bringing a family pet into this hallowed place. Where infections are such a preoccupation, why risk it? Initially rebuffed, the family petitions for the right to bring Honey, and winning their case, have arranged to bring Travis's best friend to his bedside. I haven't slept in a couple days, and I've come to realize that midday following an all-nighter is the greatest challenge, burdened with a supreme heaviness of fatigue and anhedonia. To be honest: I don't give a rip if Honey comes here or not; I don't have the energy to care.

Travis lies in his ICU bed, breathing tube in his mouth, secured with pink electrical tape around the tube and splayed onto his cheeks, IVs connected to electronic pumps that hum with activity while providing him with fluids and medicines, his body supported on an inflatable cushion to help prevent bed sores, and his surgical wounds on his chest, abdomen, arms, and legs covered with gauze dressings. He is bloated and lifeless, and although his chest expands and relaxes, it is the work of the machine that provides this isolated motion. Since the moment he was life-flighted to our trauma hospital he has not moved a muscle, and I don't think he ever will. I am getting more impatient by the moment.

With some pomp and circumstance, Honey makes his way into the SICU #2, accompanied by Travis's parents and sister. I wonder if the dog will even recognize Travis. I stand at the sliding glass door entrance to the room, readying myself to keep the dog from accidentally stepping on medical tubing or electrical cords. Honey paces into the room, and with an alert snap of his head, fixes his gaze upon Travis. He deliberately takes another few steps closer to the bed, his head approaching the bed rail. Something very intentional is happening, and with my curiosity piqued, Honey sits down, observing. I glance at his family, and his mother's hands are to her mouth, his father a study of concentration. Honey rises, and manages to probe his snout through an opening in the bed rail, and touches his nose to Travis's hand, cajoling a response. There are now about ten residents and nurses jockeying for position, trying to witness this surreal moment between boy and dog, and Honey spontaneously yelps a plaintive bark, and assumes his posture on his haunches. And now, did I detect a wiggle of a finger? Is Travis moving his hand? We are poised, nobody moving, everyone watching that left hand. Except now I see Travis slowly moving his head to the left, eyes still shut and swollen, and although it's the tiniest of movements it's obvious that it's intentional. Honey barks again, and like starting an old car after years out of service, Travis is slowly reigniting. His head lifts almost unperceptively, and murmurs of "yes" and "no" and "oh my god" are whispered out loud. Travis continues to respond, moving every limb as the minutes pass by; this further encourages Honey to bark, nuzzle, paw, and whine as Travis is resurrected by some mysterious force. Eventually, Travis manages to lift an eyelid, and we are so flummoxed there is nothing left to say. Sensing our earlier doubts, Travis's father locks eyes with mine and says, "I told you there was something about our son and that dog."

Six months later, I am walking along a public corridor in the clinical towers of the hospital. Internship continues to be a grind, and my preoccupation is to make it through the remaining months until I transition to orthopedic surgery residency. I guess I'm in daydream mode, eyes glazed over until I realize that two people are making a beeline my way. I don't think I've met the younger one, but I recognize the older gentleman, but I can't recall why. Turning to his son he says, "Travis, this is one of the surgery residents who saved your life." I am dumbfounded—I don't recognize Travis at all—and the only mark I see on his body is the tracheotomy scar at the base of his neck. His dad adds, "Dr. Schneider was also there when Honey brought you back to life." Struggling for words, all I can say is, "It's nice to meet you Travis."

Five hundred years ago, Western Europe was still mired in primitive conditions no better than Romans and Greek citizens had enjoyed two millennia before. Muddy roads, animal excrement, plague diseases, and swelling cities subjected mankind to more misery than even our foraging ancestors had endured. While the printing press had revolutionized communication flow, there was precious little new knowledge prior to the groundbreaking advancements ushered in by the astronomers in the mid–16th century. A suffering patient in the late 1500s would have been attended to by a physician with almost no appreciation of how the body functioned and certainly no understanding of individual organ function. At the dawn of the 17th century, there was still not a single human being who grasped what breathing accomplished, how nutrients were taken up from the food we ate, and why our hearts pounded in our chests.

In the year 1600 c.e., a young Englishman abandoned Cambridge medical school in the midst of his studies, determining to venture to Europe's home of scholasticism and to the Venetian city of Padua. William Harvey (1578–1657) had earned his bachelor's degree several years earlier at Cambridge's Gonville & Caius College, but sensing greater prospects in Italy, made his way to Dover to cross the English Channel. While his companions embarked on a ship bound for Calais without incident, William Harvey was singled out by the governor at the port.

"You must not go, but must be kept prisoner," the governor told Harvey. The young Cantabrigian was furious, but was forced to watch as his friends sailed away on a packet boat into the evening. During the voyage the boat was caught up in sudden storm, capsizing and leading to the death of all those on board. News reached Dover of the cataclysm, and as the only passenger not allowed to board, Harvey sought out the governor who had detained him. Why had Harvey been the sole isolate, alone on the English shore while his friends drowned?

The governor informed Harvey that "Two nights previously I saw a perfect vision in a dream of Doctor Harvey, who came to pass over to Calais; and I was given a warning to stop you." Although Harvey was completely unknown to the governor, this premonition had saved his life, and he often told this story as evidence of a special providence and mission for his life. [3] The world's first physiologist had perhaps been identified by the gods, and although it would take years to publish his manifesto about the function of the cardiovascular system, his journey of exploration was officially underway at the onset of the Age of Experimentation.

Like Isaac Newton after him, William Harvey was from the English yeoman class, freeholders who cultivated small estates and the educations of their progeny. The 16th and 17th centuries were auspicious times for upward mobility; William Harvey's patrilineal sheep-farming predecessors had made provision for his future success, and while his father had little education, the benefits of his financial success surely paid dividends in William's and his brothers' futures.

The man who would rise to become *physician extraordinary* to the King of England originated as a Kentish boy who was endlessly curious about the spiders, horses, dogs, pigs, and hens on the family farmstead. Like William Shakespeare (just fourteen years older), Harvey was educated at grammar school, well versed in Greek, Latin, and Hebrew. When Harvey arrived in Padua, he easily absorbed the information in Latin, the lingua franca of sophisticated scholars. Harvey's timing as a new medical student in Padua was propitious, with 16th century forefathers Vesalius, Falloppio, and Eustachi having established Padua as the greatest medical center of learning in the world.

Padua also held the greatest advantage for international students, with the establish-ment of "Nations," in which the French, English, German, and English expatriates associated with each other within codified structures and leadership. Within months of his arrival, Harvey won election as the "councilor" of the English Nation, which afforded him special privileges, including a front row seat at anatomical dissections in the newly constructed anatomy theater at the Palazzo del Bo, Europe's first (and still, oldest) anatomy theater.

Standing on the dark granite cobblestones in front of the Palazzo del Bo, the oldest building at the University of Padua, one cannot immediately decipher what distinguishes this edifice among the other medieval and Renaissance structures that surround it. The five-hundred-year-old Palazzo, a rosé-hued, three-story stone building with a center courtyard, has colonnaded arches and stone-framed windows, behind which lie some of the most famous classrooms in the world. In Padua, Copernicus argued for heliocentrism, Galileo lectured about the orbits of the planets, and Vesalius reinvented the scholarship of anatomy. The lecture halls, examination rooms, and anatomic theater are centuries old, and were the staging grounds for some of medicine's greatest innovators.

Entering the courtyard and following the Italian signs that guide visitors to the entrance of the steward's office, I am disappointed to find it closed for construction. I check my email on my phone, confirming my appointment for a private tour of this historic venue. The police officer turns me away in broken English, telling me my presence is not permitted. I try to tell him that I have a prearranged engagement, but to no avail—I must go. Frustrated, I walk across the street to the Caffè Pedrocchi, one of the oldest coffee shops in Europe, to plot my next move.

After some deliberation over a scrumptious espresso, I cross the street again, plead my case, and sense my insistence is becoming quite annoying to the Polize. Again, with the wave of his hand, he insists the Palazzo is closed and no one may enter.

Leaving the constabulary obstruction behind, I meander away, examining the locked metal gates at the bottom of a stairway leading up to the Palazzo classrooms above me. Tantalizingly close, I imagine the ancient exam rooms up there somewhere, and heaven only knows where the glorious anatomy dissection theater is.

"Sir—are you the American surgeon who wanted to see our classrooms?" I turn and see a young curator who speaks with a strong Italian accent.

"Yes!" I exclaim, realizing the barriers have come crumbling down at once. And with that, Francesca swings her large ring of skeleton keys up to my face, and with a smile, turns to the venerable gates and skillfully unlocks the aged mechanism with a heavy key, and we march up the steps to the first level of the Palazzo del Bo.

Half a millennium old, the largest classroom is adorned with the family and nation coats of arms of scholars who have been here for centuries. Francesca points out the names I have studied for years, legends of anatomy, medicine, and science. Alone in the grand hall where Galileo instructed, I respectfully approach the lectern and look out across it. Hundreds and hundreds of years later, there is still magic in the air in this place where some of the greatest minds in the history of thought presented their ideas.

In the next room, a timber-ceilinged classroom reserved for final oral examination for the medical degree (to this day), a U-shaped configuration of tables surrounds a small wooden chair. All of the green leathered chairs behind the tables face the simple, lone wooden chair where the candidate faces a barrage of questions testing her worthiness of a degree from Padua. Hanging on the cream-colored plaster walls are the aged paintings of the cognoscenti, no doubt adding to the anxiety of the applicant. I inspect Eustachi's expression, detecting an aura of supreme shrewdness; I wish he could know what he and his fratelli started.

Francesca solemnly turns my way, and with a warm smile, asks if I'm ready to visit the anatomy theater: I eagerly say yes. Leaving the timeworn room behind (which faces the street below), we walk through an undersized door and enter a low-ceilinged small room with almost no lighting.

Craning my neck and bending forward, dodging heavy, hand-hewn timbers set at an angle, I follow my guide's voice in the inky blackness. Francesca flips a switch over in the corner, and I become aware of a tiny constellation of lights around me. I still don't understand the framing around me, but as my eyes become accustomed to the low-light conditions, I realize I am standing in the opening of a great funnel, forty feet high with multiple levels of rising, concentric elliptical circles. I am in the world's oldest anatomy theater, and I am standing where the body would have laid on a table.

The theater structure was built into the large, empty room, and was made of logs and hand-fitted planks and boards. Rising at a steep angle, the succeeding levels supported a circular ring of boards just deep enough to support the human foot. On the inner side of the ring on each level was a wooden rail at about the level of a man's knees. This theater was built for standing, and the banisters and balustrades (carved of walnut) kept even a fainting man from falling forward.

The Padua anatomy theater was built in 1594 for Hieronymus Fabricius, the custodian of the anatomical heritage of Vesalius and the man who would train Harvey. As the councilor of the English Nation, Harvey was positioned in the front row of anatomy demonstrations, always held during the colder months of the year where a rotting corpse was less pungent. Much of what Harvey was taught was the old Aristotelian medicine, which he would continue to practice once he returned to England. Steeped in Hippocratic humourism, Harvey was of an age of unscientific analysis and primitive understanding of organ function; in fact, he would remain a Galenic physician till his death.

To understand how revolutionary Harvey's 1628 book—*De Motu Cordis, On the Motion of the Heart*—was, one has to contemplate what his professors taught him about the heart and the blood vessels. Galen's conclusions about the generation of blood, the function of the heart, and blood flow were sacrosanct—and entirely and (by today's standards) ridiculously wrong. From Greek and Roman times all the way to the early Renaissance, no researcher could reliably and precisely dissect any mammal and ascertain what the purpose of the vessels and the heart was. There was no way of properly sedating an animal while slicing open its chest, so any animal experimentation gruesomely involved strapping down the poor subject for a few seconds and hurriedly carving open the thoracic cavity as it bled to death. This left fleetingly few seconds to ponder about the flow of blood and dynamic function of the heart. It is little wonder that no one understood how it all came together.

Galen and every anatomist after him were confronted with the labyrinth of the large vessels, some as large in diameter as a garden hose, coursing hither and yon about the oddly shaped muscular organ. Anatomy students today are given textbooks with color-coded drawings delineating arteries and veins and the circuitous route of the flow of blood, but in ancient times an open chest with a pounding heart and indiscernible vessels presented an enigma.

For 1,500 years, medical students were taught that there were two distinct and parallel vascular systems based upon the thickness of the vessel walls. Dear Reader, if you and I were in an anatomy lab today I could show you the chief differentiating feature of blood vessels everywhere in the body: on the one hand, a flimsy vessel with thin walls—a vein—on the other, a blood vessel with thick, stout walls—an artery. No matter the location of the vessels (whether they are in the abdomen or in a limb), veins and arteries always fall into these two main categories.

Galen (incorrectly) concluded that arteries had the innate ability to pulsate based upon the thickness of the vessels. Worse, classic Galenic teaching held that the liver was the source of blood. While correct in maintaining that blood was supplied with nutritive properties from the digestion of food, Galen was wrong to teach that blood ebbed and flowed in both directions in every vein in response to the attractive powers of individual organs and muscles. To make it all work, Galen theorized that all organs "attracted" blood to themselves, and "consumed" the blood and its vital spirit.

Unable to understand the function of the lungs, Galen and his apprentices established that the vital spirit, or "pneuma," entered the lungs with each breath; how else to explain the impulse to breathe? The divine breath, they concluded, is what vivified the pulsating, arterial blood, changing its color from dark purple to scarlet. Additionally, blood was said to have naturally flowed from one side of the heart to the other through large pores in the walls separating the large chambers.

The final, flawed observation about the function of the cardiovascular system was that each organ consumed the blood, imbibing the vital spirit. Any surplus blood merely evaporated. In conclusion, all parts of the body pulled blood toward themselves, drawing in the pneuma and vaporizing away all the superfluous blood. If blood continuously pulsed into dead-end organs via the arteries, how else to explain the function of veins other than claiming that it flowed in both directions in the veins?

Thus was the function of the heart and vessels presented. And it was all breathtakingly wrong.

William Harvey graduated from the medical school in Padua in 1602, and after a brief return to Kent, moved to London, where he would live for the next fifty-five years (with an interrupted stay in Oxford during the English Civil War). He promptly married a well-placed woman who was the daughter of the physician to King James. While not a London native, he steadily climbed the ranks of society, qualifying in 1607 as a fellow of the College of Physicians, the small coterie of the most highly regarded doctors. Even within the college, Harvey rose in stature, becoming the treasurer within a few years of admission.

The "lowly and intensely intellectual country boy"[4] adopted the manners of sophisticated life in London, climbing the social ladder and navigating the court of King James. Like his patient Francis Bacon, Harvey attempted to curry favor with the king, eventually being named "physician extraordinary" in 1618, the same year Bacon was named Lord Chancellor. With the death of King James, Charles I became king in 1625, soon naming William Harvey "physician ordinary," an even more powerful and prosperous title. His newfound rank and affluence garnered something even more important: the time and resources to perform investigations. There can be little doubt that his exposure to Francis Bacon, combined with his best-of-class medical training, prepared Harvey to more scientifically evaluate the cardiovascular system.

As Harvey was advancing in rank and title, his practice thrived. He and his wife were childless, leaving him with greater liberty to perform investigations. By 1615 Harvey was the Lumleian lecturer in anatomy, enjoying the reputation of a skilled dissector and gifted lecturer, stylishly brandishing a sixteen-inch silver-tipped whalebone wand that he used as a pointer during presentations. His elegant attire of "black cloak, full doublet, ribbed stocking of black silk, and long high-heeled boots fringed at the top" set Harvey apart from less important Londoners.

Harvey's professional success fueled his avocational diversions, and by his mid-thirties, he settled into a routine of daily medical practice and evening home-based private research. His childhood fascination with plants and animals endured, and his London collection of aquatic life, terrestrial specimens, and barnyard animals became the provenance of future discoveries. An important transformation was occurring; Harvey was, in essence, bringing Padua to London, and his home laboratory was becoming a sort of Palazzo del Bo.[5] A revolution was brewing, and within a few years of research, Harvey would make one of mankind's greatest discoveries.

Completely self-motivated and endlessly curious, Harvey dissected almost nightly, even completing a program of the comparative anatomy of the anus of various species of birds. An ornithologist in a 21st-century NIH-funded lab could stomach such a topic if it led to a fount of published articles and tenure, but as a private program of exploration?

Another Paduan anatomist, Realdo Colombo (1515–1559), a pupil of Vesalius, had been a pioneer in describing the flow of blood to and from the lungs, stating, "Blood is carried to the lung by the pulmonary vein, and in the lung it is refined, and then together with the air it is brought through the pulmonary vein to the left ventricle of the heart."[6] Although not completely correct, Colombo's breakthrough insight was to deny that only air returned in the great vessels from the lungs; it was "refined" blood that returned from the lung in the large pulmonary veins. Colombo was also refuting the Aristotelian and Galenic hope that blood bypasses the lung and travels across pores in the middle of the heart.

With Colombo as his exemplar, Harvey set about to investigate the heart and its machinations. He had witnessed the pulsations of blood in fish and small animals, but a critical breakthrough occurred when he secured the corpse of a hanged criminal through his connections with the College of Physicians. Hauling the body to his private research area in his Ludgate home, Harvey placed the body on his dissection table, and with candlelight, sliced open the chest and cracked open the ribs.

After draining the fresh blood in the great vessels of the chest, Harvey found himself face-to-face with the dead man's heart and lungs. When an anatomist encounters a heart, the most prominent artery that immediately arises out of the topmost part of the heart is the pulmonary artery, the large diameter vessel that transmits oxygen-poor blood from the right side of the heart to the lungs [see picture in image section]. Harvey took a piece of string, and wrapping it around the pulmonary artery, ligated it tightly, preventing the flow of fluid. Carefully cutting open the right ventricle, Harvey next inserted a metal tube into the heart chamber and tried injecting water into it. With the pulmonary artery tied shut, there was no way for the water to pass through it, and even more important, water did not pass across the thick-walled septum that separates the right and left ventricles. Opening the left ventricle, Harvey observed not a single drop of fluid, and stated, "By my troth, there are no pores."[7] Galen was wrong.

Releasing the ligature from around the pulmonary artery, Harvey again injected water into the right ventricle, and within seconds, water poured into the left ventricle. This blood-tinged fluid had obviously passed from the right side of the heart, through the lung tissue, and then returned to the heart in the pulmonary veins and into the left side of the heart. In an instant, he knew that Colombo had been correct in asserting that the heart pumped blood to the lungs, and that the blood returned back to the heart vivified and scarlet.

William Harvey had convinced himself of the interrelationship of the heart and lungs, but he remained confused about the action of the heart. Should he continue to doubt Galen (true sacrilege among physicians), or believe that the heart, like all organs, swells and draws blood to itself? After dissecting the hanged man, Harvey returned again to his marine fauna, some of which had translucent skin that allowed him to peer at their tiny beating hearts. The germinating skepticism about a heart that flexes open to siphon blood to itself was growing into an ironclad incredulity the more he dissected small living things.

Harvey sliced open the chest cavity of a fish, and observed the minuscule heart rapidly pulsating, even witnessing the flow across the transparent aorta. Placing his finger on the heart, he could feel it contracting, further convincing himself that the heart was more like a muscle and unlike a bellows that opens to suck air (or blood) into itself. Experimenting on an eel, Harvey carved open the fish's chest and cut out its beating heart. Placing it on his dissection table, the heart continued to throb, and even when he cut it into smaller segments, each piece contracted. Completely convinced that the heart was indeed a muscle whose active phase was contraction, Harvey was well on his way to solving one of life's great mysteries.

Over the course of many years, William Harvey conducted countless experiments on animals, often turning to dogs. By modern standards, it is cruel and defenseless how hundreds of animals died without any consideration for their sensations or consciousness. In fact, it is daunting and troubling to review Harvey's vivisection experiments, no matter how important were the discoveries. In the 17th century, bearbaiting, cockfighting, and public animal cruelty were commonplace, and it wasn't until 1835 that the English Cruelty to Animals Act was passed. Until that point, most Englishmen viewed animals as incapable of feeling pain, and they felt as emotionally conflicted about a dog dying a tortured death as most of us feel about swatting a mosquito today.

Harvey's many experiments were convincing him about the flow of blood in and around the heart, but he was still confounded over the genesis of blood and where it vanished. In a truly great moment for science, Harvey realized that some sort of calculation was necessary to investigate the function of the heart. With astronomy, physics, mathematics, and biology in their infancy, Harvey pioneered an entirely new branch of science.

Concluding that the active phase of the heart was *systole* (Greek, "contraction"), Harvey realized he could estimate the amount of blood that was pumping through the heart. In retrospect, his calculations were very conservative, but his results led him to the proper conclusion. Harvey guessed that the heart was not able to fully squeeze all the blood out of its chambers with each contraction, so he calculated that only a fraction of the full diastolic (or relaxed) volume was ejected with each contraction; he estimated that amount to be no less than a dram weight of fluid (one-eighth of an ounce), a vast, but safe, underestimate.

Estimating that the heart beats at least thirty times a minute, and multiplying that times sixty minutes an hour and twenty-four hours a day, a (very conservative) dram in weight of blood ejected with each contraction would mean that a massive 50,000 drams of blood was pulsated into the arteries each day. If that massive amount of blood was continuously manufactured by the liver, as Galen and all ancient physicians had always postulated, it was simply inconceivable to Harvey that the flimsy world of

Hippocratic medicine was genuine. The world's first physiological computation led to the obvious conclusion that Galen was preposterously wrong, even before Lavoisier's conservation of mass would become law two hundred years later. There simply couldn't be a fire hydrant of blood gushing from the liver each day. Today, we know that an average 150-pound human has a stroke volume of 70 milliliters, which equates to a daily cardiac output of over 7,000 liters, or almost 2,000 gallons coursing through one's heart. Harvey would surely pound his fist on a table when realizing that over forty barrels of blood flows through our hearts every day.

If the liver was not continuously producing blood, then where was it coming from? And, perhaps more critically, what happened to the blood once it reached its end destination? Did it indeed evaporate?

Harvey's final experiment was likely triggered by his mentor in Padua, Fabricius. Shortly after Harvey had left Padua, Fabricius published *De venarum ostiolis* (*The little doors of the veins*), in which he investigated the function of the valves. Not comprehending that the flow of blood in the veins is from the extremities back to the heart, Fabricius had trouble determining the function of the "little doors," or valves of the veins. He wrapped a cloth around a subject's arm, causing the veins to swell; Fabricius then applied fingertip pressure over a vein near the tourniquet, but found that he was not able to force blood down toward the hand. He concluded that the doors "held up and delayed" the flow of blood, a clear misinterpretation of the venous valves. His blindness to the truth was grounded in his false Aristotelian foundation, but his publication did inspire his pupil to perceive the truth through a repetition of his experiment.

Harvey repeated the experiment of his mentor, affixing a cloth tourniquet around his servant's arm, and confirming Fabricius's observation that blood could not be forced back into the hand because the "little doors" prohibited backflow. Unencumbered by a blind belief in Galenism, Harvey was free to interpret the findings of his little experiment. (If you have visible veins in your arm, you can easily reproduce the experiment by tightly wrapping a scarf or rubber band around your limb above the elbow, and attempting to force the blood toward your hand. The blood simply can't go "backward" in the limb because of the valves inside the vein. Releasing the tourniquet allows the blood to continue its path to the heart.)

Harvey was on the verge of his great synthesis. Contemplating the vast quantity of blood pulsating across the heart, considering the "vivification" of the blood in the lungs and the transmission of the scarlet blood to the body's tissues via the arteries, combined with this new insight of the one-way passage of blood from the limbs back to heart, Harvey had his eureka moment. He later wrote, ". . . and when I had a long time considered with my self . . . I perceived . . . the blood did pass through the arteries to the veins, and so return into the right ventricle of the heart."

Harvey finally had it. The blood was forcefully expelled out of the powerful left side of the heart, pulsated through the aorta and arteries into the entire body, and through some mysterious exchange, the same blood itself was conveyed into the venous system to return to the heart, where it would be pushed into the lungs for aeration. It was, therefore, a closed system. His investigations were performed before the development of the compound microscope, and his naked eye could not see the microscopic branching of the ever-smaller arteries that exist in every organ of the body in what is called the "capillary bed."

Harvey, at the summit, recalled, "And so, I began to bethink my self if the blood might not have a *circular* motion [emphasis mine]. . ." Hippocrates, Aristotle, Galen, and all our ancestors were blinded to the truth of the function of our heart. Harvey, the man who knew Galileo in Padua, had declared that our blood orbited in our bodies through a double-circuit, one to the body and the other to the lungs. Instead of blood simply pouring into end-organs, it stayed within vessels as it passed through the organs and muscles, and the new instruments of microscopy would unveil this secret shortly after Harvey's death. In a word, *circulation* embodies Harvey's great revelation.

Marcello Malpighi (1628–1694) was an Italian scientist who first described the microscopic vessels that serve as the connection between tiny arteries and minuscule veins. His first breakthrough came when he was examining still pulsating blood in the lung of a frog, and with the aid of a simple magnifying glass, saw the tangle of tissue that he perceived was the intermediary tissue between arteries and veins. The circulatory pathway had finally been visualized, and Malpighi called them capillaries.

William Harvey endlessly presented his ideas throughout England and Europe on the cardiovascular system, eventually publishing his classic tome, *De Motu Cordis, On the Motion of the Heart* in 1628. One of the most important books ever published in the scientific domain, Harvey's magnum opus established him as a key pioneer in understanding the mechanical fabric of our bodies and a fellow coconspirator in René Descartes's clockwork universe.

Descartes and Harvey altered the intellectual outlook of the 17th century, and together with Francis Bacon's empirical inductive enquiry, natural philosophy was converted into a scientific program of investigation. Robert Boyle, one of the key figures at the founding of the Royal Society, described the human body as a "hydraulic engine . . . fram'd and contriv'd by nature." Harvey's "mind was incredibly sensitive to the intellectual and cultural spirit of his age, and his ideas were expressive of that spirit."[8] His investigations were the first mathematical and physiological interpretations of the human body, and were foundational works in the burgeoning age of experimentation. His natural heir was not the scientists and natural philosophers who formed the Royal Society within years of his death, but another Brit who was endlessly enamored of nature and who came to London almost as a feral outcast and is today interred at Westminster Abbey.

In England in 1540, the guilds of two unions were united by King Henry VIII to establish the Company of Barber-Surgeons. After years of acrimony, surgeons broke away in 1745 to form their own Company of Surgeons, which would later become the Royal College of Surgeons. Barbers and surgeons struggled for recognition, and in the mid–18th century, both groups could offer little more than a haircut, simple abscess drainage, or a hopeful (or hopeless) letting of the blood. Nonetheless, the emancipation of the surgeons in 1745 did present at least one significant challenge: they did not have their own anatomy theater, and this left them scrambling for a venue and proper proctors.

William Hunter (1718–1783), a Scotsman and renowned physician, obstetrician, and anatomist, grew up near Glasgow before receiving a stellar education in Edinburgh, Leiden, Paris, and London. Shortly after finishing his medical training at St. George's Hospital in 1744, William Hunter began a course in private anatomy classes, advertising, "Gentlemen may have the opportunity of learning the Art of Dissecting during the whole winter session in the same manner as at Paris."[9]

William Hunter's anatomy school was the first of its kind in London (amazingly, the first chartered medical school in London was founded in 1785, although there had been informal schooling at St. Bartholomew's Hospital for centuries). Hunter's anatomy school opened in a rented Covent Garden apartment in 1746 (in 1749, moving nearby to 1 Great Piazza, Covent Garden, now an Apple store!) and was an immediate success. The triumph of the school and Hunter's growing practice mandated assistance in the procurement and preparation of corpses, and in what seems like desperation, William asked his brother John Hunter, ten years his junior, to move to London and assist him at the school.

William Hunter, "attired in brocade and lace, and sporting a full powdered wig, dined with fellow Scottish intellectuals [and bridged settings such as] coffeehouses and theaters, dissecting rooms and salons . . . and the contrasting worlds of science and the arts."[10] Younger brother John could have hardly been more different, having dropped out of school as a thirteen-year-old (perhaps suffering from dyslexia), gaining notoriety as an "awkward, uncultured, and largely uneducated country lad . . . with a shock of red hair."[11] Ten years separated the two brothers, but there had been an abundance of tragedies since they had last seen each other when John was just twelve years of age. John Hunter was the last of ten children, born to a sixty-five-year-old father. Of John's nine siblings, six had died by the time he traveled to London, leaving William as his only stable guide in a world fraught with instability and disease.

John Hunter had rejected the traditional English grammar school upbringing, choosing instead to roam the countryside, investigating the flora and fauna of South Lanarkshire. If his later life is any indication, it seems that John Hunter was incapable of recoiling in response to putrefaction, essentially immune to all things unsavory or repugnant. Instead of a preparatory education at Eton and an Oxbridge bachelor's degree, John Hunter came to London with insatiable curiosity, congenital skepticism, and a battle-born durability.

John Hunter arrived in London in 1748 as a twenty-year-old, and despite his lack of formal education, was poised to serve as his brother's assistant at the anatomy school, now two years old and thriving. They hardly knew each other, and a safe conjecture is that William was searching for a lackey to deal with all the dead bodies. Ideally, John would also perform preparation of the corpses for anatomical lectures, and as the new term was set to begin in September 1748, William and John met at their Covent Garden venue to dissect a dead man's arm. The autodidact's practiced hand with the knife had been acquired through years of self-directed morbid curiosity. To William's great surprise, John was an instinctive dissector, and on the occasion of their first anatomic investigation, William told his unschooled and callow brother that he had the makings of an excellent anatomist and should never want for employment.[12]

Today, every medical school in the world has fully embalmed cadavers that have been rigorously treated by experts with fixative agents that prevent decay and putrefaction. The corpses' blood vessels are flushed with embalming chemicals that perfuse all the tissues in a body. A cadaver can be stored at a cooled temperature for years without rotting, and a semester's anatomy lab can be passed without ever smelling decomposing tissue, and the first-day shock of seeing a dead person fades away in succeeding days as you realize you will never encounter maggots or pools of pus.

The anatomy school at Covent Garden faced one major dilemma: the corpses themselves. By the mid–18th century, many continental countries had relaxed the Roman-era prohibitions regarding human dissection, but England still had rigid proscriptions regarding the procurement of bodies. A minuscule allotment of the bodies of condemned men was granted to the Company of Barber-Surgeons in London in the 1500s, and little had changed in the years that the Hunters' school had started. The summer months were off-limits for dissection due to the rotting stench of warm bodies, but autumnal changes signaled the start of another dissection season. As many as fifty people were hanged each year in London for even trivial offenses, such as filching a watch.

For centuries, prisoners were kept at Newgate Prison (near St. Paul's Cathedral and present-day London Stock Exchange) and hanged at Smithfield (near St. Bartholomew's Hospital, one of the oldest hospitals in the world and the site of the hanging, drawing, and quartering of Scottish independence leader William Wallace), or more commonly in Hunter's era at the Tyburn Tree. In the northeast corner of Hyde Park, near the Marble Arch, on Bayswater Road, is a small plaque in the middle of a small traffic triangle that commemorates the location of the Tyburn Tree, the three-legged gallows that hosted the capital punishment of thousands of criminals from the 12th century till 1783. Hangings at Tyburn were a spectacle, attended by the masses as a celebration of justice, and the wriggling, hooded victim struggling to breathe (and wishing to die) served as a macabre form of entertainment.

John Hunter's primary task in the twelve years he partnered with William was the procurement of bodies. John stood only five feet two inches, but his broad shoulders and strong hands from years of hard work made him an able combatant in securing the precious bodies that were freshly dead. The free-for-all and carnival atmosphere called for an indecorous ruffian, and John Hunter—precise dissector and future legendary British surgeon and paragon of scientific intellect—was the ideal down-and-dirty steward of the corpses.

The Murder Act of 1752 was passed by Parliament, codifying the expedient hanging of criminals and disallowing the burial of convicts after hanging. This would serve to increase the numbers of cadavers, but there were still not sufficient bodies available. Where else to get fresh pablum?

Shockingly, John Hunter became one of history's great grave robbers.

So, hungry for fresh bodies, John Hunter turned himself into a body snatcher par excellence. "So in October 1748, with William clamoring for more bodies for his pupils, John Hunter almost certainly set out himself under cover of darkness from the Covent Garden school, armed with a shovel and crowbar, to scour local burial grounds for freshly dug graves. Most likely, he commandeered parties of students, probably bolstered by several rounds of ale in a tavern beforehand to help in his grisly undercover work."[13]

In London and Edinburgh, and later in the Colonies, professional grave robbers ransacked cemeteries and traumatized families. Also known as "resurrectionists," the midnight burglars would shovel their way down through the freshly turned soil, dig-ging a narrow shaft at the head of the grave until they encountered the wooden coffin. Crowbarring the cheap wooden lid until it snapped, and exposing the upper portion of the body, the body snatchers then used ropes to drag the body from its supposed final resting place.[14] The team of miscreants would load the body into a carriage and deliver it overnight at the anatomist's back-door basement entrance.

John and William Hunter worked together for twelve years, the elder brother gaining notoriety among the well-heeled and royalty, eventually becoming obstetrician to Queen Charlotte (who gave birth to fifteen children, including the future Kings George IV and William IV). John, on the other hand, flourished first as a body man and later as an expert anatomist. A fresh supply of corpses was John's vocation, and his "youthful liking for a drink, lack of social airs, and colorful language evidently endeared him to his sinister suppliers."[15]

Even in their first winter together, a surprising transformation occurred: John's preternatural skills with the knife were laid bare, and by the spring of 1749, "William declared his protégé sufficiently accomplished to take over all the dissecting work at the school."[16] Thief by night, the resurrectionist was blossoming into an anatomist with authority, and after a year at the school, contemplated the impossible—brandishing his knife on the living. William had followed the typical route of an esteemed physician in London, matriculating from prestigious institutions and proving himself to the intellectual community. The path to becoming a surgeon would be vastly different for John, as there existed no formalized schooling, testing, nor accreditation for aspiring surgeons.

More than 150 years would pass before any semblance of a surgery residency existed either in Europe or America; John Hunter scrambled in 1749 to form a relationship with the most esteemed surgeon in the London area in hopes of transforming himself into a surgeon. William Cheselden (1688–1752), head of the newly formed Company of Surgeons, had himself become a surgeon through the study of anatomy and apprenticeship under practicing surgeons in Leicester and London. (In 1800, a royal charter was granted to the College of Surgeons, and for the first time, a college degree and acceptance into the Royal College of Physicians was required before training to become a surgeon.) For a few short years, before Cheselden's death, John Hunter was apprenticed to him at his home office and at the Royal Hospital in Chelsea. Instead of viewing Hunter as an unqualified and unsophisticated country boy, Cheselden identified Hunter as an anatomy enthusiast and heir to his surgical governance.

Years later, William Hunter recalled, "Were I to place a man of proper talents, in the most direct road for becoming truly great in his profession, I would chuse [sic] a good practical Anatomist, and put him through into a large hospital to attend the Sick, and dissect the dead."[17] William was surely referencing his own brother as he wrote this, favoring a candidate who came from a proper anatomic training as opposed to an adolescent who had tagged along with an unlearned surgeon in a perpetual line of ignorant and barbaric imposters.

Although John Hunter trained under the finest surgeons, first Cheselden and later Percivall Pott at St. Bartholomew's, he never lost his bizarre, outlandish, and even savage tendencies. To appreciate how dedicated Hunter was to anatomy and the dissection of the dead, consider that he regarded the *sense of taste* as a critical component in the physician's armamentarium. "The gastric juice is a fluid somewhat transparent, and little brackish to the taste."[18] Gloves for dissection would not be available for another century, but one wonders what would compel Mr. Hunter to draw his fetid fingers to his mouth and sample the residue of any exudate or detritus. He was at the precipice of comprehending the organ basis of disease (as pioneered by Giovanni Morgagni), and apparently nothing was off-limits in his quest to understand the body. "The semen would appear, both from the smell and taste, to be a mawkish kind of substance, but when held some time in the mouth, it produces a warmth similar to spices, which lasts some time."[19]

Many who attend medical school exhibit a normal aversion for putrid smells, necessarily prohibiting the conquest of surgical and gynecological specialties. Ask any general surgeon or gynecologist who has a standard practice when was the last time they faced a repulsive smell so pungent they were gagging and retching, and they will readily regale you with a tale of woe. A selection process that weeds out applicants with an incapacitating disgust for malodor happens in every medical school. Those who can "stomach" putrescence may continue on; those with "weak constitutions" (read: normal) go into specialties like dermatology and radiology. Even grizzled and gruff veterans of the operating room cannot overcome the deeply programmed "disgust as a disease-avoidance mechanism"[20] in severe cases, that in the end serves to protect us from potentially life-threatening infections. Posterity is grateful that John Hunter seems to have had no capacity for revulsion, which, combined with his spirit of inquiry and dogged determination, propelled him to be the preeminent surgical figure of his time.

John and William Hunter worked together for over a decade, and during that time, made gargantuan discoveries regarding the human frame. Their description of the lymphatic system would perhaps only be surpassed by the revelation that the blood vessels between the placenta and the uterine wall *did not share blood*, as had been assumed, after careful dissection by John Hunter on pregnant women deceased late in term. After years of dissections on mid- and full-term London women, William published in 1774 *The Anatomy of the Human Gravid Uterus Exhibited in Figures*, displayed in massive "elephant"-folio in one of the greatest works of anatomy ever conceived. The dissections were all the work of John Hunter, the words were William's, and the magnificent, if not sobering, artwork was by Jan van Rymsdyk.

After several years of training with Cheselden and Pott, and after serving as a surgical pupil at St. George's Hospital during the summer of 1754, Hunter began to treat his first patients. The coarse country boy was transforming into an objective scientist, and from the beginning, he maintained fastidious notes with observations and outcomes.

Historian Wendy Moore describes an early patient, a young man (a chimney sweep) who had contracted gonorrhea that had resulted in a urethral stricture—a blockage in the urethra—that resulted in painful urination:

> Establishing the patient's medical history, Hunter brought to bear all his natural scientific curiosity—embarking on the experimental approach to surgery that would characterize his whole life—on the sweep lying in pain and frustration . . . Initially, Hunter attempted the classic approach to unblocking a stricture, presumably learned from Pott, which entailed attempting to push a "bougie"—a cylindrical bung made of wax or sometimes lead into the urethra to force a way through. When this failed, he characteristically decided to experiment and, importantly, record his results. Hunter conjectured that he might shift the blockage by burning a way through, using a caustic salve on the end of a bougie . . . [Hunter first used mercuric oxide, which caused much inflammation and pain,

then attempted] with remarkable forbearance on the part of the sweep, Hunter fastened a piece of "lunar caustic"—silver nitrate—onto the end of [a hollow sliver rod cannula] and probed the urethra . . . [as Hunter recorded] "three times at two days interval, he came to me and told me that he had made water [urine] much better; and in applying the caustic a fourth time, my cannula went through the stricture; a bougie was afterward passed from some little time till he was perfectly well," Hunter jubilantly recorded. It was a victory for experimental medicine. His approach—trying a traditional method, analyzing the outcome, forming a hypothesis aimed at improvement, and implementing his results—would become a standard practice throughout his career. Ultimately, it would form the foundation for his scientific revolution of surgery.[21]

Like William Harvey, born 150 years before, John Hunter was an avid naturalist and keeper of animals. Hunter had a clear advantage over Harvey, as the British Empire held a stranglehold over the seas during Hunter's lifetime, permitting an exotic collection of animals from around the world, including Asian buffalo, a lion, a jackal, a dingo, and two leopards, which he kept at his country home west of London (in Earl's Court, Kensington). His outlandish assemblage of animals in close proximity to the tony confines of London has led some scholars to believe that Hunter was the inspiration for the children's book character Dr. Dolittle.[22]

Endlessly curious, John Hunter gathered up newly deceased human specimens and live animal oddities. Investigating for his own benefit (he would not publish in the Royal Society's Proceedings journal until 1767), he made breakthrough discoveries about the cranial nerves in the head, tear ducts, and the descent of the testes in young developing male humans. Writing of the time, Benjamin Franklin said, "This is an age of experiments,"[23] and no one characterized the epoch better than Hunter, who performed some of the first embryological research, in characteristic meticulous analysis and note-taking.

Most anatomists of the 18th century believed that every living being began life as a tiny miniature version of itself, steadily increasing in size in the womb or in an egg. John Hunter recognized this as preposterous, and began a project of exploration, wisely using chickens as his source material. Stealthily plucking eggs from underneath hens at frequent intervals, Hunter then cautiously cracked open each egg, using forceps to peel away the outer membrane and examine the minuscule embryo. While still fleetingly alive, he placed the little life in a warm bowl of water and examined it with the naked eye and with his microscope. Eventually placing all his specimens in spirits, he was able to establish a timeline of development over its three weeks of development within the egg. Others would eclipse Hunter in publishing their embryological findings, but his researches on this topic would exemplify his fastidious scientific technique.

Within a few years of Hunter caring for his own patients at St. George's Hospital, his reputation was sterling enough to attract his own pupils in anatomy and surgery. An early student who was tutored by Hunter was William Shippen Jr. (1736–1808). The elder Shippen, who was a self-educated physician in Philadelphia (and founder of the University of Pennsylvania and Princeton University), recognized the value of a European medical education for his son, and sent him on a seven-week ocean voyage to England in 1758.

Young William Shippen had graduated from the newly founded Princeton University in 1754,[24] and then spent four years in apprenticeship with his father. With only a clutch of medical textbooks available to the Shippens, the decision was made to dispatch the progeny to the ". . . finest Anatomist for Dissections, Injections, etc. in England."[25] "Billie" arrived in 1758, just ten years after John Hunter had himself appeared in London as an unsophisticated and ignorant country bumpkin, and now he was in charge of tutoring one of the Colonies' most promising offspring.

William Shippen underwent full immersion in anatomical studies, living at the Covent Garden home of John Hunter and spending the winter session of 1758–59 under the spell of the Scotsman. A daily journal entry describes William's day, "Rose at 6, operated till 8, breakfasted till 9, dissected till 2, dined till 3, dissected till 5. Lecture till 7, operated till 9, sup'd till 10 then bed." Weeks passed as William was fully indoctrinated in the methodical dissection of the human body, and on many nights, William and John Hunter chatted away, no doubt enjoying a bit of grog. "Shippen spent more and more time at Hunter's shoulder in the dissecting room, in awe of this charismatic teacher, who seemingly had a knack of inspiring young pupils with a fascination for anatomy."[26]

The first medical school in the Western Hemisphere was in Philadelphia, at what was initially called the College of Philadelphia (later "University of Pennsylvania"), and was cofounded by William Shippen and another John Hunter disciple, John Morgan. Philadelphia boasts the oldest hospital in America, the first medical school, and the first "genius society," the American Philosophical Society. While it is the cradle of modern American surgery, it is John Hunter who may claim paternity over the teaching of anatomy and surgery in the new world.

The remarkable transformation from hinterland simpleton to anatomy sage was largely based upon Hunter's obsessive and energetic dedication to body snatching, dissection, and inculcation. Generations of medical students trod their way to London to be swept up in the vortex of Hunter's "companionable" nature in which he "drank his bottle, told his story, and laughed with others."[27] But more than an amiable nature explains his success. His compulsive search for the truth about the structure and function of the body led him to systematically ". . . question every established practice, develop hypotheses to advance better methods, and test by means of rigorous observation, investigation, and experiment whether these methods worked."[28] To

truly achieve advancements in a procedure-based science like surgery, John Hunter
needed a volume of cases, and while metropolitan living had provided a concentration
of diseases among the growing populace, his future prospects as a reputable surgeon
were limited without some other distinguishing experience.

In what historians consider the first world war, Great Britain and France engaged
in battles on land and sea in the Seven Years' War, resulting in Britain's supremacy
and France's decline after 1763. Five years into the conflict, in 1761, Britain's war
secretary William Pitt decided to focus his attention on the west coast of France and
the small French island of Belle Île. Months before, in the fall of 1760, John Hunter
was commissioned as an army surgeon, following Hippocrates's ancient wisdom, that
"He who wishes to be a surgeon should go to war." Boarding the hospital ship *Betty*,
John Hunter surely knew that tragedy awaited, and after the weeklong voyage to the
Atlantic coast of France, the first assault commenced disastrously. The flotilla of 130
British ships faced a nearly impregnable island, with almost insurmountable cliffs set
along a jagged coastline.

Hundreds of British forces were killed or injured in the opening days, and those
clinging to life were ferried to the *Betty*, where the sailors and soldiers who had suffered bullet and shrapnel wounds were tended to by Hunter and his fellow surgeons.
"While the storm tossed and dispersed the wooden ships, Hunter battled to save the
wounded, bleeding, and dying men in his primitive surgery."[29] An age-old question
has always been, why does the meaning of "casualty" include both those *killed* in battle
and those who are *injured*? The answer is that prior to World War II, most seriously
injured combatants were likely to die in a rather short period of time, so a "casualty of
war" was essentially done for. "With no anesthetics to numb the pain, many patients
died of shock as the amputation knife cut through their flesh, or bled to death while
the surgeons probed around in their wounds to extract debris. Their bodies were given
hurried burials at sea. Others would join them days later as sepsis took hold in their
wounds while they lay in the hammocks of the foul-smelling sick bay."[30]

After a second assault on the island two weeks later, the British were successful in
establishing a hospital of sorts among its cottages. Here, John Hunter began a program
of investigation on the proper treatment of war wounds. Whereas most of the other
surgeons adhered to the ancient practice of exploring traumatic war wounds, Hunter
began to realize that less was more. His fellow surgeons were actually inflicting greater
pain and hastening death by inserting grimy fingers and filthy tools into wounds.
Exercising great restraint, particularly when considering his years of dissection and
his renowned dexterity, Hunter opted not to probe and enlarge bullet wounds, instead
"being very quiet" in the treatment of wounds.

A chance development occurred in the first week on the island that cemented
Hunter's belief that conservative care was better and the humbling realization that
surgeons were violating the central tenet of Hippocratic care, *primum non nocere*,

"above all else, do no harm." In the midst of battle, five French soldiers had been shot during an exchange of fire, but had avoided capture. Although some of their injuries were serious (including chest- and thigh-penetrating bullet wounds), the men ensconced themselves in a barn and evaded capture until being discovered four days later. Hunter tended to the men, realizing that the French troops who had received no care at all were faring better than the British forces who were unlucky enough to have been "treated." In the end, in what amounts to a controlled experiment, Hunter concluded that all the secluded soldiers healed without incident, despite not receiving what was considered standard surgical exploration. The folly of aggressive care was revealed by a side-to-side comparison of two treatment arms, made possible by the high volumes of casualties among the belligerents.

There can be little doubt that Hunter's experience with battlefield medicine was exhausting and exhilarating and, at the same time, very educational. The sheer volume of trauma cases honed his skills and provided further insights in treating wounds and how to think about infection. He stayed on the island for over one year, and then spent time in Portugal before returning home. He arrived back in London in 1763, just as Giovanni Morgagni's book *De Sedibus* was gaining notoriety, eventually being translated into English in 1769. An awe-inspiring book, *De Sedibus,* for the first time, connected symptoms with diseased organs, and was the foundation for modern pathology and medicine. Hunter's inquisitive nature, scientific programming, and the newly enlightened way of thinking about the human body was a serendipitous amalgamation that made basic surgical treatment, for the first time, a possibility.

Or was it? Hadn't primitive surgery been attempted in the ancient empires of India, Persia, Egypt, and Assyria? Although scholars in those early empires had misinformed ideas about the function of the human body, how was it that they had the ability to perform painful surgery on primeval peoples? The answer is opium.

Opium is an ancient substance from the poppy, a flowering plant native to the eastern Mediterranean that was cultivated throughout the Middle East and India. Ancient Greek and Egyptian medical texts mention the use of opium, and wherever early surgery was performed, its cultivation is known. The poppy flower's migration across empires, and its latex product (containing the opiate alkaloids morphine and codeine), has impacted trading and shipping, the modernization of medicine, and the spread of religion. Whereas the incense, spice, and silk trade routes had largely been based on land or sea routes close to the shore, the Age of Discovery led to worldwide ocean voyages and the international trading of tea, tobacco, sugar, cotton, and drugs.

The Dutch East India Company and the British East India Company pioneered intercontinental trade in the 17th century, specializing in the spice and tea trade in their first century of business. The Chinese luxury goods of porcelain, silk, and tea were in high demand from European countries, but there was little reciprocal demand for European goods in China. The solution was the British commercialization of opium production in India and the sale of the highly addictive opium to Chinese markets. With the flourishing opium market in the East, there was an inevitable influx of the elixir back to Mother England and, not surprisingly, a demand for morphine and its cousins—heroine and codeine.

Thomas Sydenham, considered the father of British medicine, prepared a proprietary opium tincture that he termed "laudanum," a blend of opium, sherry alcohol, cinnamon, saffron, and cloves. Laudanum became a staple of almost every English physician and surgeon, and allowed Hunter to venture into previous prohibited lands in the century before ether and chloroform. In fact, while it may be said that others may claim priority in being the first important surgeon (men like Paré and Pott), Hunter alone is the last great surgeon to operate without comprehensive anesthesia, and whatever success he enjoyed was, in part, owed to opium.

The two years abroad achieved for Hunter what he might have hoped—an acceptance among the well-placed and philosophically interested. He never forfeited his disdain for etiquette nor his love of drink and coarse language, but his "energy, breadth of interest, and ebullient manner"[31] won him favor among the members of the Royal Society, who elected him as a fellow in 1767 "as a person well skilled in Natural History and Anatomy."[32] Even within the prestigious society, Hunter ascended into a secret splinter group that engaged in sometimes bellicose confabs in London coffeehouses, with luminaries such as James Watt, James Cook, and Nevil Maskelyne joining in the brouhahas.

John Hunter passed the oral examination at Surgeon's Hall in 1768, earning the diploma of the Company of Surgeons and finally the right to call himself a surgeon. His pupils William Shippen and John Morgan had already started the medical school in Philadelphia, and were professors of surgery and medicine when Hunter had ultimately achieved the bona fide title of surgeon. Within five months of earning his diploma, a corresponding change was triggered for John Hunter, the appointment as full-time surgeon as St. George's Hospital (now The Lanesborough Hotel). At age forty, the arc of accomplishment and increased notoriety continued upward, now abruptly.

Hunter continued to build up his menagerie of animals and surgical specimens. His appetite (and forbearance) for the diseased and desperate further propelled his practice, particularly as a St. George's Hospital surgeon. The hospital's location in close proximity to Buckingham House and St. James's Palace strengthened William's entitlement as Queen Charlotte's physician and obstetrician, but it seemed impossible that John Hunter could ascend to such a lofty title.

John Hunter's enthusiasm and masterful teaching skills translated well from the simple anatomy school at Covent Garden to the prestigious confines at St. George's. The uncouth and uneducated Scotsman had been transformed into the most advanced scientific thinker among surgeons in the English capital, and an amazing five hundred pupils would fall under his spell of skepticism and investigation in the coming decades.

His dedication in animal experimentation and the assiduous collecting of surgical specimens slowly convinced him that the ancient notions of disease and humoral imbalances were ill-conceived.

The ancients were gravely mistaken when they intervened in the midst of disease without understanding the disease process. Hunter's predecessors stumbled in the dark, but he conceived simple experiments to comprehend healing, inflammation, and disease processes. Warmed by a log fire, armed with the most primitive instruments, and scrutinizing by candlelight, Hunter carried out investigations on an assortment of animals and conditions. Nothing was off-limits—not even his own body.

Georgian London was overwhelmed by sexually transmitted diseases (including syphilis and gonorrhea), the result of casual sex in the chief port city of the empire. Still a century from a consensus on germ theory, venereal diseases were the most convincing examples that microscopic terrors were to blame for the transmission of disease from one paramour to the next, but there was still no identification of the culprit. Hunter was determined to investigate one aspect of disease transmission: could the crusty discharge from a syphilitic patient alone be the carrier of disease from lover to lover? Additionally, Hunter was trying to determine if gonorrhea and syphilis were actually two separate diseases, or were they simply different manifestations of the same disease? To investigate this, Hunter conceived a plan to inoculate a symptom-free patient with no history of venereal disease with a sample of the milky-white discharge from a gonorrhea patient. But where to find a volunteer who would knowingly submit to the application of the discharge from the festering wound of a venereal victim . . . onto the innocent's penis? How about a body snatching surgeon?

Hunter recorded, in the third person in exacting detail, on the results of his stunning experiment of self-contamination. He never identified the patient, but most scholars agree that Hunter was himself the guinea pig. In May 1767, Hunter recorded, "Two punctures were made on the penis with a lancet dipped in venereal matter from a gonorrhea [milky discharge]; one puncture was on the glans, the other on the prepuce. This was on Friday; on the Sunday following there was a teasing itching in those parts, which lasted till the Tuesday following."[33] Several colleagues and students recorded contemporaneous accounts of Hunter admitting that he had given himself a chancre, sobering evidence that John Hunter would stop at nothing to delve into the mysteries of the human body.

As the days progressed, there was a "teasing itching" in the penis that blossomed into infected scabs that soon generated a discharge and "a little pouting of the lips of the urethra, also a sensation in it making water, so that a discharge was expected from it."[34] Still not sure if he was experiencing gonorrhea or confronting the early stages of syphilis, he realized within ten days that syphilis was now embryonically emerging within his body, with ulcerated chancres growing at the end of his penis and the development of glandular swelling in his groin. Was Hunter mad, or so insatiably curious that his utter lack of self-preservation led him to needlessly condemn himself to a lifelong pestilence? Dear Reader, ponder the verity that gonorrhea is often self-limited and does not recur; syphilis, on the other hand, is a lifelong infestation of a bacteria that visits its victim in waves over the course of years in three primary manifestations, including a brain and spinal cord infection that leads to madness in the final, or tertiary, stages. This is the bacteria that Mr. Hunter was intentionally smearing on self-inflicted lacerations on his male member.

Seven months after infecting himself, John Hunter developed "copper coloured blotches" on his skin and was experiencing ulcerations on his tonsils. For the next three years he was daubing mercury on his sores until a quiescence finally occurred. It is not known if he ever suffered from tertiary syphilis, but his final demise as a sixty-five-year-old was from cardiac disease, a possible connection to syphilis in up to one-third of untreated patients. Hunter's horrifying tolerance of auto-experimentation demands repudiation and simultaneous recognition of his dedication in the face of ignorance.

While wrestling with Hunter's recklessness, students of medical history must acknowledge that surgeons have often been immune to revulsion, even to the point of self-sacrifice. A certain resistance to disgust coincides with an odium of disease; these oddly balanced impulses explain many breakthroughs in healthcare. Like firefighters who abhor fire, yet derisively run headlong into a conflagration, healers despise—even dread—microbes while thrusting themselves in the middle of pandemics. John Hunter serves as the bridge between medieval surgical bleeders and the first surgeon-scientists. Explorers and pioneers are prepared to live in the wilderness in primitive habitations; Hunter was willing to dwell in the morass of unenlightened surgery, establishing a foundation of surgical science. "Hunter believed all surgery should be governed by scientific principles, which were based on reasoning, observation, and experimentation."[35]

Hunter's lectures were often iconoclastic, and he was one of the first surgeons to suggest that "bloodletting was not just largely ineffectual but potentially dangerous . . ."[36] The next generation of famous British surgeons, legendary in their own right, are the ones who best capture the impact of Hunter on modern surgery. Henry Cline, the president of the Royal College of Surgeons, first attended Hunter's lectures out of simple curiosity, and after first hearing him, said, "When I heard this Man, I said to myself, This is all day-light. I felt that what I had previously been taught was comparatively nothing . . . and thought I might, like Mr. Hunter, venture to Think for myself."[37] John Abernethy, another renowned London surgeon from St. Bartholomew's, concluded, "I believe him to the author of a great and important revolution in medical science . . . of this I am certain, that his works produced a complete revolution in my mind."[38]

The complex John Hunter would be named surgeon to King George III in 1776, the year of another revolution. The man who some say is the basis of *Dr. Jekyll & Mr. Hyde*—from his front-of-the-house prestigious medical practice on Leicester

Square and nefarious backdoor entrance for middle-of-the-night deliveries of stolen bodies—is undoubtedly the founder of surgical science. His massive collection of surgical specimens and medical curiosities were donated to the Royal College of Surgery from his penniless estate, exhausted from a lifetime of exploration, spending, and consumption, not dissimilar from his American philosophical contemporary in Monticello.

John Hunter died in 1793 while attending a hospital staff meeting at St. George's Hospital in London. Involved in a debate over house staff, Hunter's last moments were characterized in fury; he "died, as he had lived, in rebellion, speaking his mind."[39] The primitive surgery he practiced, without anesthesia and with no understanding of sterility, was mostly associated with removal of stones, cysts, and rudimentary treatment of traumatic wounds. A half century after his death would herald the arrival of ether and usher in a new way of placating man before the surgeon's knife.

The prototype for Langenbeck in Berlin, Billroth in Vienna, and Lister in the United Kingdom, Hunter finished his career without washing his hands, or dreaming of germs. In one of medicine's great ironies, the giant of surgery with almost no formal education had the greatest impact on the scientific training of surgeons. He theorized that "A hospital . . . should be not only a charitable institution offering aid to the poor—indeed, even a place where surgeons gained experience before trying their luck on wealthier clients—but also a center for educating the surgeons of the future."[40] "My motive," Hunter concluded, "was in the first place to serve the Hospital and in the second to diffuse the knowledge of the art that all might be partakers of it."[41]

When another young Philadelphian arrived in London shortly before Hunter's death, the pupil's father asked Hunter what books his son would need to read, "Hunter simply led the way to his dissecting room, where several open cadavers lay, and declared, 'These are the books your son will learn under my direction; the others are fit for very little.'"[42] Hunter, of course, was correct. Almost nothing in any medical textbook in the 18th century regarding the causes and treatment of disease is accurate or beneficial.

With surgical science in its infancy, Galenic medicine was in its death throes. A century before, at the death of William Harvey, a sobering recollection of Hippocratic faith was exercised. Suffering a stroke in 1657, William Harvey found himself unable to speak, with "dead palsy in his tongue."[43] Summoning his apothecary, Harvey motioned to have his tongue sliced open, completely in line with Galenic humoral conjecture that "letting of the blood" would hasten the return of speech. For the man who unlocked the secrets of circulation, the mysteries of cerebral hemorrhage and stroke were still profound, and the barbaric gashing of his tongue "did little or no good . . . so he ended his dayes [sic]."[44]

John Hunter ostensibly died of a heart attack in a hospital, which today would surely result in a "rapid response" by a team of lifesaving physicians, nurses, and technicians in almost any advanced medical center in the world. While no treatment was available to Hunter as he rapidly died, no pointless Galenic treatment was foisted upon his hapless soul. The "Founder of Scientific Surgery," as is inscribed over his grave at Westminster Abbey, had revolutionized anatomy, education, and surgical science, thus raising surgery from a shameful diversion to a profession with a future.

SIX

Pathology

Though we cut into the inside, we see but the outside of things and make but new superficies to stare at . . . Nature performs all her operations in the body by parts so minute and insensible that I think nobody will ever hope or pretend even by the assistance of glasses or other inventions to come to a sight of them."

—John Locke

The frigid, metallic reality of the morgue at the University of Kansas is still disquieting to me, even as a veteran of over twenty autopsies. Instead of charging ahead into my third year of medical school, I have accepted a research fellowship in the Bone Research Laboratory with the world-class pathologist and researcher H. Clarke Anderson, MD. In addition to investigating cancer cell lines and bone morphogenetic proteins (BMPs, the signaling chemicals that initiate and control bone growth), I am obligated to take autopsy calls with Dr. Anderson. Here, in the dead of winter, I have been summoned to the morgue to investigate the cause of death in a local milk deliveryman.

The morgue is altogether different from the anatomy labs. As first year medical students, we became habituated to the cadavers in the anatomy room—their rigid, embalmed structures slowly revealing themselves to us in successive weeks. Every day, my two dissection partners and I would follow the instructions in our dissection manual and explore a particular anatomical detail in our cadaver, a seventy-four-year-old female. In time, the novelty of being surrounded by seventy corpses vanished, and questions of their backstories and humanity faded.

Arriving at the morgue, the thirty-eight-year-old, moderately obese and powerfully built deliveryman lies on the stainless-steel autopsy table, completely naked. The autopsy suite has three sturdy tables, purpose-built for autopsies, each with a central plateau pockmarked with irrigation holes and a circumferential trench with flowing water to wash away blood, body fluids, and the vestiges of infection and contamination. The subject is positioned on the central table, which is equipped with a foot-controlled recording device and microphone to capture the comments of the pathologist as the autopsy progresses.

Like every hospital morgue, this workstation is sequestered in the basement, away from foot traffic and patient care areas. The paucity of inhabitants and caregivers imbues a sense of loneliness and fear, even during the day. This is accentuated with the knowledge that dead bodies are stored in the lockers, refrigerated and ready for inspection. In the bowels of the hospital, with an ominous stillness and a monotonous ebbing of water over the tables, there is no lifesaving; there is only death-explaining.

Dr. Anderson explains that this worker was found outside his house in the early morning dawn, next to his running truck, facedown in the snow. An ambulance raced to his home, his grief-stricken and panicked wife overcome with anxiety and helplessness. After a failed resuscitation in the emergency room, he was declared dead by the ER physician and trans-ported to the basement morgue. Every state has its own laws that guide the local coroners in deciding whether to order an autopsy; in this case, an unwitnessed death at home under unusual circumstances has prompted this man's final physical examination.

We are dressed in blue scrubs and disposable paper gowns identical to those worn in the operating room. Waterproof gowns offer a layer of protection and a little bit of warmth, and protective eyewear is donned to keep body fluids from splashing into our eyes. Our simple tray of instruments ready, it is time to "see for ourselves," the literal meaning of the word autopsy.

I place a gloved hand on the man's torso, and he is unnaturally cold, the combined effect of him dying in the elements and his refrigeration in a cadaver locker for the last few hours. Dr. Anderson grasps the autopsy knife, outfitted with an impressively large blade made spe-cifically for making the elongated cut in the front of the chest. I glance again at the decedent's face, studying his contorted, bluish visage flattened on the right side from lying prone in the snow as he was dying. He was alive a few hours ago, but his rigid and motionless corpse looks counterfeit; only cutting into his barrel chest will convince me he's a man.

The knife blade is placed at the sternal notch, the divot at the top of the chest bone where a thin layer of skin covers the windpipe. Cutting into the flesh, Dr. Anderson draws the knife in a straight line, down the front of the chest and along the abdomen, toward the belly button—and then taking a curved detour around it—and ending at the pubic bone.

Instruments are used to spread the skin edges apart, and I ready the bone saw to open up the sternum. I have learned how to use the saw, and Dr. Anderson allows me to operate the tool to cut the chest bone down the middle. This is the same device that heart surgeons use, and as a medical student, I'm thrilled to get to use real surgical tools years before I normally would. A rib spreader is placed between the bone edges, and winding a simple crank on the spreader forces the rib cage to gape wide open.

The heart and lungs announce themselves as the sternum is spread. The crimson-colored muscular heart is partially shrouded by the billowy lungs; gray, diaphanous, and boggy. The heart and lungs are bound together, yin-and-yang, different in color, structure, function, and heaviness. The thoracic cavity, demarcated as a cage of bony ribs and floored by the muscular diaphragm, encases the heart and fragile lungs. The diaphragm is a thickened, powerful membrane that partitions the thoracic and abdominal cavities. The diaphragm has three large, inch-wide holes that allow passage of the esophagus, aorta, and vena cava. By carefully cutting through the fleshy diaphragm, we can preserve the integrity of the vessels and esophagus, not causing leakage of blood or food contents.

Once in the abdominal cavity, we encounter the liver, the size of a small football, the kidneys, the intestines (stomach, small intestine, and large intestine), the spleen, bladder, and pancreas. Because the walls of the intestines have not been carelessly cut, there is no spillage of partially digested food, or further along the tract, in the large intestine, feces.

As medical investigators, we have several choices in how we examine these organs. The most old-fashioned method would be to simply poke around, handle the organs with our hands, and use our powers of observation to guess what hastened death.

A more advanced technique would be to cut out the organs as they lay in the thoracic and abdominal cavities. With the Rokitansky method, groups of organs are removed at once, and inspected on a side table. Virchow's technique is to remove the organs one by one in situ, cutting into them and preparing tissue samples for microscopic analysis, a dramatic leap forward compared to manual manipulation alone.

Today, we will use a technique that, indelicately stated, reminds me of how I learned to "gut a deer" while hunting with my father and brothers on our family's ranchland in Wyoming. The Letulle technique involves exposing all the organs and intestines from throat to anus, and dissecting them cleanly away en masse. While cutting away the soft tissue connections from the cavity walls, the pathologist lifts out the heart and lungs, all the intestines, and all the abdominal organs while still maintaining all the connections to each other; what remains are hollowed out, empty cavities. On the dissection table, all the organs are much more easily evaluated, with easy inspection of every aspect of every organ, since they are no longer tethered down to the body walls. This method is favored by my mentors, and no doubt arose from someone who was from a hunting family.

After removing the mass of organs and tissues, Dr. Anderson and I struggle to transport the slippery jumble of tissues onto the dissection table. Here, we slowly start to cut into the organs, looking for gross abnormalities. Small tissue samples are collected, and we plop the masses into small plastic containers with labels, filled with formalin, and screw on the orange lids.

The head is approached last. The hair on top of the corpse is divided, and after making a long incision along the top of the skull, the skin is easily peeled down on either side, exposing the skull. A specialized saw is used to cut the bone all the way around the top of the skull, and after chiseling away the final soft tissue connections, the membrane around the skull is encountered—smooth, thick, and opalescent. This is the dura mater (Latin: tough mother), and after cutting through it, the brain is encountered. Severing the nerve connections and the spinal cord at the bottom of the brain, the gelatinous mass is lifted out by Dr. Anderson, and he (unbelievably) hands me an entire human brain.

The brain of this gentlemen is firm and pink, robustly characterized by circuitous folds and wrinkles that shrink with age, but in this young man are so plump it seems that his brain was *too big for his skull, stuffed in by the promethean life-giver. Inspecting the brain matter with duteous fingers, pushing the gyri this way and that, peering into the sulci, and investigating for ruptured vessels or evidence of tumors, we satisfy ourselves that a gross examination of the organ reveals no immediate abnormalities.*

Standing opposite each other over a narrow table, Dr. Anderson and I stabilize the brain with our hands. He grasps a twelve-inch-long knife that looks suitable to carve a turkey, and begins slicing the brain like a loaf of bread. Each slice of brain is about a third of an inch thick, and we lay out the sections on a large pan. This results in a tray full of brain slices, almost like a pan of large cookies, permitting thorough inspection of the entire brain. This technique was developed decades before there were CT or MRI scans, and was the only way of delving deeply into the brain structure of deceased individuals.

Having harvested all the organs and intestines, we turn again to his heart. Everything else normal, our suspicions center on acute myocardial infarction, or a heart attack. With a massive heart attack that results in immediate death, there are no grossly visible (to the human eye) changes to the heart muscle. In someone who has battled cardiac ischemia for days prior to death, the myocardium begins to turn pale—even yellow. In our deliveryman, the heart looks normal, but now we focus on the coronary vessels.

The heart is our body's pump station, and even though all of our blood comes pulsating through the cardiac chambers, it doesn't perfuse the pump's muscle. The cardiac muscle demands its own vascular supply, coming in the form of the coronary arteries that branch off the aorta as it leaves the heart. These two main arteries, the right and the left, branch out and send little arterioles deeply into the muscle to provide oxygen and fatty acids (used as fuel). The coronary arteries are visible on the exterior of the heart, and using surgical instruments, we dissect the arteries for microscopic analysis.

In our hearts, the left coronary artery bifurcates into two main branches, just an inch from the aorta, and these two branches feed the most important part of the heart, the left side, which must powerfully contract to propel blood to the entire body. Dissecting further, we isolate the left anterior descending artery (the LAD, or "widow-maker" artery), and cutting it free, take the two-inch-long, spaghetti-thick vessel and dunk it in formalin. It felt rigid and crunchy, and Dr. Anderson's thoughtful and pensive gaze at me remedies an investigator's vehement skepticism about the cause of death for this man. I think he knows we have found the killer—a dislodged cholesterol and fatty clot that blocked this most important, if not proportionally sized vessel—and his sensitivity for this man and his family overcomes scientific dogma and pathology.

After "fixing" the small bits of tissues overnight in their formalin cups, the samples are further dissected and we place the critical samples into small blue plastic cassettes for microscopic preparation. These little cassettes are about the size of Tic Tac containers, and are loaded into a machine that exposes the contents to increasing concentrations of alcohol, half an hour at a time, and then xylene, with the aim of halting all cellular degeneration or bacterial growth while removing all the water and fluid in the original tissue samples. The apparatus then dunks the plastic cassettes into paraffin, creating a microscopic time machine where the tissue samples have been frozen in time in a chunk of white wax.

I hand the stack of waxy blocks to our lab tech, who loads them onto the Leica microtome, a machine that creates extremely thin wafers of wax that come reeling off the blocks in a little chain, like paper Christmas ornaments. She mounts the wax sections on glass slides, which are then taken to the staining station. Until the dark purple and red stains of hematoxylin and eosin (H&E) are applied, the tissue slice on the glass slide is nearly invisible, but in time, the tissue section will pop with clarity and differentiation. The tech picks up the slide and begins dunking it into small metal cups of chemicals and dyes. Sometimes for thirty seconds and sometimes for less, she works her way down the row of cups, and after drying, the slides are ready for viewing.

I can't resist the urge to look at the slide of the widow-maker coronary artery first. I position the slide on the microscope stage—the little platform on my Zeiss microscope with a hole in the middle—that allows for the light to pass from underneath. Lowering my head over the eyepieces, I rotate the focus ring until the cells come into clear view. The artery, cut in cross section, is completely occluded with thrombus and atherosclerosis, and I am staring at my patient's killer, the very real representation of a heart attack, frozen in time. This clot prevented precious blood flow to the most critical part of his heart, and deprived of fuel and oxygen, the cardiac muscle ceased to pump, leading to his collapse in his driveway.

Death and disease had been shrouded from humanity's comprehension from our earliest reasonings; in the span of a few generations my forefathers used discipline, skepticism, the microscope, and chemical dyes to lift the veil and unlock the secrets of illness. Here in the morgue, armed with the tools of the pathologist, I can explain, scientifically, why this man died, even if I can't comprehend how precarious our existence is.

By the mid–17th century, Padua, Italy, could lay claim as the crucial birthplace of knowledge and learning of the Italian Renaissance, even outshining the University of Bologna, the oldest university in the world. Sons of Padua included Vesalius, Falloppio, Harvey, and Galileo, and as the 18th century approached, a recent graduate from the University of Bologna school of medicine arrived in Padua, consumed by a lifetime devotion to a project that would seismically change medicine forever.

Giovanni Battista Morgagni (1682–1771) graduated from Bologna at the age of nineteen, soon forming an intellectual society for students and new graduates called the Academia Inquietorum—Academy of the Restless. As a new graduate, Morgagni had encountered a new book by Theophilus Bonetus the *Sepulchretum*, a compendium of thousands of cases of clinical histories and correlative autopsies. These were compiled by Bonetus from the burgeoning medical literature by a vast array of authors that, unfortunately, was disorganized and haphazard, rendering it almost unreadable. The young Morgagni "nevertheless, pored over the Sepulchretum . . . and it became clear to him that because the concept upon which it was based epitomized a fundamental truth,"[1] set about (at first) revising the book, but later constructing an entirely new work based upon his own cases.

Morgagni began his project as a newly minted physician, probably twenty or twenty-one years of age, and began collecting information on the patients he was treating and the results of their autopsies. One by one, he built a compilation of cases, with careful observations, astute clinical interpretations, and the occasional physiological experimentation to bolster his clinical conclusions. "To this enormous undertaking he brought his considerable talents as a practicing physician, his towering preeminence as an anatomist, his resourcefulness as an experimental physiologist, and his infinite patience with detail."[2]

How patient was Morgagni in preparing his book? While treating his patients in Bologna and Padua, he spent the next *six decades* collecting information, organizing the material, and writing the book that would change the way that physicians looked at patients and thought about the essential nature of disease. Morgagni published his book *De Sedibus et causis morborum per anatomen indagatis* (*Of the seats and causes of diseases investigated through anatomy*) in 1761 at the advanced age of eighty while still treating patients.

De Sedibus is written in conversational style, as if to a friend, and is organized into seventy letters to a young physician (perhaps fictional). The sum total of these seventy letters encompasses seven hundred cases, and is organized into five books: Diseases of the Head, Diseases of the Thorax, Diseases of the Belly, Surgical and Universal Disorders, and Supplement (including index). For each case, "historical background is given, the evolution of contemporary thinking is reviewed, authorities are quoted, their opinions discussed, and the logical development of the professor's conclusions, step by step, become clear."[3] Compiled over decades, the cases are meticulously organized and indexed, so that a young student could search the book by symptom, such as "chest pain," and investigate a similar case, searching for truths and possible effective treatments.

One hundred years earlier, Galileo had challenged the unscientific and superstitious view of the heavens. Morgagni occupied a world where physicians throughout the world were still ensnared in the ancient Hippocratic traditions, perseverating over humors, the seasons, miasmas, bad air, and celestial judgements. Morgagni's *De Sedibus* dealt the final death blow to humoral medicine, and turned the mind of the physician to "specific derangements of particular structures within the body."[4] The anatomist and physician Morgagni concluded, in perhaps the most iconoclastic statement in medical history, that symptoms were quite simply the "cry of suffering organs." Succinctly and beautifully captured, this new appreciation of disease focused the attention of the physician on a particular organ or body part. Instead of gazing at the stars or considering imbalances of mysterious fluids, Morgagni realized that disease was the manifestation of dysfunctional (and often painful) organs.

Hundreds of cases over the course of sixty years had convinced Morgagni that disease followed observable patterns. In the 1700s, pharmacology was in its nascence; physicians (and especially surgeons) were largely impotent. But Morgagni was gaining insight about organ-based illness, and with experience, he could predict what he would find at autopsy. He probably saved not one life, but as the decades passed, he grew confident that he could predict what he would encounter at the autopsy table. *De Sedibus* was quickly translated into French, English, and German in the same era of the American Revolution; the United States was born at the same time that physicians around the world were concluding that a constellation of symptoms were pointing to a specific ailing organ.

A truism exists regarding advances in medicine: to best comprehend how an organ (and its constituent cells) actually function, evaluate the organ following an injury or during disease. Descartes had proposed that the human body was merely a machine; Morgagni's influence was to view the coordinated physical-mechanical structures (that normally worked in faultless harmony) as a watchmaker or machinist whose job was to diagnose faulty parts. Physicians would now be in the business of carefully listening to the cries of sick organs, fastidiously observing and examining patients, and then empirically theorizing about what was killing them.

Morgagni is not simply the father of anatomic pathology, but also the figure who is credited with initiating modern medical diagnosis. "The full consequences of what he worked out were harvested in London and Paris, in Vienna and in Berlin. And thus, we can say that, beginning with Morgagni and resulting from his work, the dogmatism of the old schools was completely shattered, and that with him the new medicine begins."[5] It seems odd that we don't know his name better, but Morgagni is among the most important figures of the Enlightenment. His disciples include Jean-Nicolas Corvisart (1755–1821) and Pierre-Charles-Alexandre Louis (1787–1872), physicians who helped establish Paris as the mecca of medicine in the 19th century. The pendulum would eventually swing back toward the east, when a critical physician from Vienna fully adopted the concepts of the pathologic basis of disease, performing over 30,000 autopsies in his career, all without the most powerful tool in the history of medicine.

A wave of political upheavals pulsed across Europe in 1848, affecting almost every European country and many of their colonies worldwide. This revolutionary activity challenged feudal lords and royalty, establishing greater democratic rights for the lower classes, even witnessing the introduction of *The Communist Manifesto* that same year. In particular, the revolutions in the German states and the Austrian Empire had significant repercussions in medicine and academia, much like American college campuses in the 1960s. The stodgy, outdated lords of medicine held off revolutionary-minded fledglings in Vienna; the battle of the "new versus the old, the intellectually liberal versus the conservative, the true scientific understanding of disease versus the fuzzy theoretics of the old medicine" played out at the University of Vienna's school of medicine. The chief agitator was the Bohemian Carl von Rokitansky (1804–1878).

Rokitansky grasped the significance of Morgagni and his French devotees, digging deeper and searching indefatigably for the root causes of disease and death. The more he studied disease, the more profound was his understanding of function. If Morgagni is the father of anatomic pathology and medical diagnosis, then Rokitansky is the man who literally built the house of pathology.

The Allgemeines Krankenhaus (or General Hospital) now exists as the University of Vienna undergraduate campus, but the expansive courtyards and stately buildings remain intact, even if they are filled with Bohemian Austrian students and not suffering patients. In the northwest portion of the campus stands the Center for Brain Research, an imposing three-story stone building on Spitalgasse, guarded by a brick and steel barrier and topped by a grouping of Greco-Roman figures and the Austrian double eagle shield. Below the figures is a gold-embossed inscription in Latin that is the only indication that this building used to serve a completely different function. Fifty feet overhead, one can read INDAGANDIS SEDIBUS ET CAUSIS MORBORUM, an obvious nod to Morgagni's revolutionary book, meaning "Investigation of the seats and causes of disease." This was once Rokitansky's Pathological Institute, a concert hall of medicine, where he demonstrated over 30,000 autopsies in his long career.

The structure and organization of the institute, where every single patient who died at the general hospital was examined postmortem, helped relocate European medical leadership to Vienna. A collection of physicians came together in Vienna to birth the specialties of pathology, dermatology, psychiatry, ophthalmology, and surgery, and Rokitansky's influence on the young physicians' method of clinical reasoning and scientific observation was significant. Behind the Institute is a purpose-built lecture hall where Rokitansky could lecture and demonstrate, and where the legendary Viennese surgeon, Theodor Billroth operated (as depicted in the painting by A. F. Seligmann, displayed in the Galerie Belvedere in Vienna).

It is astonishing to consider that Rokitansky performed those thousands of autopsies without a microscope. Like realizing that Copernicus had no telescope, Rokitansky performed all of those autopsies grossly, with simple manual examination of the organs and tissues. This severely limited his ability to take the conceptual leap about what caused disease, but his significant world leadership is without question. Physicians, like their astronomy brethren, needed to magnify their objects of interest to see further and illuminate their minds. It would take a few serendipitous findings to turn microscopy into a formidable tool, powering the greatest biologic insight of the Enlightenment.

At the genesis of the Royal Society, the microscope took center stage. The problem solver, tool builder, and skeptic Robert Hooke published his groundbreaking *Micrographia* in 1665, just a few years after the Society had been founded. The book was the first major publication that included depictions of microscopic views, shattering

assumptions about the unseen world. Hooke was an expert draftsman, and his drawings fascinated readers and agitated the imaginations of his fellow geniuses around
Europe. Perhaps most famously, his drawing of a flea, produced in his book on a massive foldout sheet, showed every minuscule hair and shingled plate that revealed the
flea *not* to be a tiny, gnat-like, defenseless creepy-crawly, but instead a body-armored
miniature beast with a carapace. Size matters, but microscopists were poised to reveal
that structure reveals function, and although it would take another two hundred years
to firmly prove the point, it is the tiniest living beings that pose the greatest threat to
mankind. Hooke's flea was the ideal object to focus man's attention to the microscopic
world, just big enough to be visible to our naked eye, but small enough that all detail
was out of reach, safely in its hirsute haven. Through a strange coincidence, just as
Hooke was illuminating the character of the flea, the last great plague was terrorizing
London in 1665; only later would it come to light that the flea was the carrier of the
plague bacteria, and innocently, Hooke might have been playing with fire in his dissections and depictions.

Hooke spent considerable time investigating the structure of plants as well. His
microscope had just the right amount of magnification for him to detect the minute
building blocks that comprised the organization of the plant—coining the term *cell*
for the little "rooms" he saw in the microstructure of cork—and the term would be
adopted for all plant and animal microscopy going forward. The cellular basis of
life would not be uncovered until the mid–19th century, and it would take modern
chemistry to make it real.

Most of us have an image that immediately pops into our mind at the mention of
the word "microscope." It is a tilted black metal tube mounted on a U-shaped stand,
holding a glass slide on a platform. Today, of course, there is an electrical cord that
supplies the energy for the light bulb at the bottom of the microscope, illuminating
the slide from underneath. There are also focus rings and adjustment knobs to move
the platform and glass slides. This has been the form of the microscope for centuries,
but the world's first microscopist Antoni van Leeuwenhoek had a "bead microscope,"
a seemingly bizarre and limited tool that actually allowed him to see living cells
in a way that no one ever had.

Leeuwenhoek was a Dutch surveyor and cloth merchant, and was used to using a
telescope to see distant landmarks and a magnifying lens to count threads in material
(this brings to mind the "thread-count" in sheets). Utilizing a single, tiny glass bead
with a highly convex edge and placing it on a small metal paddle, Leeuwenhoek was
able to view tiny objects held in place with wax on a needlepoint very near the glass
bead. Small screws permitted movement of the object up and down, and forward and
backward. For decades, he corresponded with the Royal Society, submitting drawings
of the microscopic world, describing the unseen that had been revealed to him with
his primitive yet practical tool.

Leeuwenhoek initially published articles in the Royal Society's *Proceedings* about bee stingers, lice, and the hidden world in a drop of pond water, but within a few years, published a work on the startling appearance of sperm. In 1677, he wrote to the Royal Society, offering his willingness to submit a paper, "If your Lordship should consider that these observations may disgust or scandalise the learned, I earnestly beg your Lordship to regard them as private and to publish or destroy them as your Lordship sees fit." By 1678, he published his article on the nature of the "seed from the genitals of animals," including drawings from the sperm of rabbits and dogs, taking special care to write that when he examined his own semen, "That what I am observing is just what nature, not by sinfully defiling myself, but as a natural consequence of conjugal coitus . . . "[6]

The truths of conception had been debated for thousands of years, and the early microscopists eagerly sought to investigate what the constituent parts of semen looked like. It was too difficult to determine what was happening in the interior of the womb for these primordial scientists, but the specter of the wriggling sperm, so similar to a tadpole or the microscopic protozoa, all equipped with flagella for propulsion, was a verity that had been guessed at since humans had been able to ponder, "Where does life come from?" Just as important, the sperm did *not* look like miniature animals ready to travel into a uterus. (At the advent of microscopy, many wondered if they would find little puppies or kits in the cells of dogs or rabbits.) The travelers, instead, looked like purpose-built little machines, ready for a voyage into the womb, even if their mechanism was hidden from the virtuosi's insight.

As microscopy improved, an amazing transition occurred. As Bacon had predicted decades before, the sequence of innovation would be to catalogue, then sift, and then, with "suitable application of intellectual machinery," to arrive at a knowledge of invisible structures.[7] As Catherine Wilson argues, "science destroys the image of the familiar world and substitutes for it the image of a strange one, wonderful to the imagination and at the same time resistant to the projection of human values."[8] As new realities came into view, scientists were forced to change ancient conclusions and adopt new theories, but strangely, after a feverish half-century of microscopic discovery, a lull settled over the minds of the investigators of the infinitely small.

Regarding imagination, the Enlightenment author Bernard de Fontenelle's philosopher-hero observes that ". . . our minds are curious, and our eyes bad . . . we wish to know more than we can see . . . Thus do true philosophers pass their lives, in not believing that which they see, and in endeavoring to divine that which they see not."[9] The development of clear glass, the innovation of lens manufacture, the assemblage of compound microscopes, and the wide-ranging publications of the microscopic illustrations eventually led to complacency. How many doodles of fleas, sperm, and insect eyeballs can you stare at? By the late 18th century, microscopy stalled, even as the Industrial Revolution was exploding across the world. To comprehend just how passé microscopy had become, consider that the revolutionary Carl Rokitansky conducted 30,000 autopsies without ever attempting to examine the tissues with the one instrument that could have utterly transformed his practice. Some science writers, like David Wootton, have pondered why 17th- and 18th-century physicians and scientists were unable to advance tissue microscopy, but there is a rather obvious explanation for pause of advancement: the lack of dependable dyes that had the power to bring the tissues to life.

If we were together right now in a pathology laboratory, we could dissect and prepare tissue for microscopic evaluation, settling upon a step where the extremely thin piece of tissue was mounted on a glass slide, ready for viewing. If we mounted that slide on a microscope platform and switched on the light, we would peer down the compound lens tube and see a faint outline of cells and supporting tissues, but with almost no ability to discriminate or comment upon the structure or function of cells. If you had never seen a painting of Van Gogh's *Sunflowers*, and I presented you a low-resolution black-and-white rendering of his poignant and melancholy artwork, there would be little impact. Conversely, an intimate experience, face-to-face, with Vincent's canvas, confronted with the turquoise and Tiffany-blue background and stunning canary and butterscotch yellows of the petals, painted with heavy brushstrokes of dolloped pigments, you would agree that Van Gogh had achieved a "symphony in blue and yellow."[10]

Those who would criticize the microscopists of the 17th and 18th century would do well to remember the paucity of color and the lack of electric lighting during that epoch; while there were simple plant dyes, chemistry before the mid–19th century was so limited that chemical reagents for trial-and-error experimentation did not exist. A happy accident in the east end of London would bring color to an otherwise drab, scientific world.

Antoine Lavoisier (1743–1794) was a scientific genius, committed to methodically analyzing chemical reactions and determining why fires burned, why we breathe, and why substances react. After fastidious experimentation and thoughtful analysis, he reinforced the notion of the *conservation of mass*, saying, "Nothing is lost, nothing is created, everything is transformed." If he doesn't hold the sole title as the Father of Chemistry, he is the Father of Stoichiometry, the concept that chemical compounds are composed of molecules in exact ratios, and that new compounds can be formed via chemical reactions, either into larger novel compounds or smaller constituent molecules.

Lavoisier, a nobleman who profited dramatically from the inequities of the old French aristocracy, was the first person to organize a list of the elements and to develop a language of scientific nomenclature to describe the building blocks of the physical world. Like a trained chef who understands the uses of baking powder, baking soda, sugar, and eggs, Lavoisier was beginning to grasp how the elements interact with each other and why metals rust, and how plants take in minerals from the soil and chemicals from the air. His genius insight was to view the world as amalgamated from its ingredients, its atoms, and he influenced his French and European followers to conclude that the world could be described by its building blocks. (Sadly, Lavoisier did not survive the French Revolution, beheaded at age fifty. One of his pupils did escape to America before suffering a similar fate: Éleuthère Irénée du Pont, patriarch of the chemical dynasty.)

Before the periodic table could be formulated (by Russian chemist Dmitri Mendeleev in 1869), a chance discovery before a "prepared mind" helped transform grammar-school chemistry into a specialty on par with mathematics and physics. William Henry Perkin entered the Royal College of Chemistry in London as a fifteen-year-old in 1853, and although Lavoisier is the pioneering giant of chemistry, young William made a discovery in his flat in East London that set in motion modern chemistry and revolutionized biology, medicine, and the pharmaceutical and fashion industries.

Tasked by his professor at the college to synthesize quinine (the only effective anti-malarial at the time), Perkin returned to his home on Cable Street in London's Shadwell area with reagents, flasks, and instruments in hopes of creating the prized drug originally sourced from a South American tree. On Easter break in 1856, by himself in a home laboratory, the eighteen-year-old Perkin started with the basic ingredient, *coal tar*, a black liquid byproduct of heating coal in the absence of air. Coal tars were a common waste product in the new Industrial Revolution, and Perkin began oxidation experiments with the mucky stuff in his upstairs flat. Finding no success, he added potassium dichromate, creating a dark watery precipitate. While cleaning out his flask with ethylene alcohol, the precipitate turned a dark purple, which he initially named "Tyrian Purple," later changing the name to "mauve." [11]

Purple has been the color of royalty for millennia, and in Roman times, twelve thousand mollusks were required to produce enough Tyrian (Phoenician) Purple to dye a single dress the size of a Roman toga. Other plant-based dyes had been tried but always faded. Perkin immediately realized the value of his discovery, performing experiments on the "fastness" of the dye. Perkin had discovered a durable, inexpensive, yet highly desirable material from rubbish, and quickly applied for a patent. By the time he was nineteen, he opened a dyeworks outside London, massively profiting from his serendipitous finding. Alchemy, apparently, was possible after all.

The real mother lode in Perkin's discovery was not in coloring clothing, but something much, much bigger. Chemistry evolved into an industrial discipline, with chemists scrambling to create other colors from coal tar in hopes of cashing in like Perkin. In a surprising twist, the chemical experimentation led not to new dyes but to new molecules that had biological effects. One of the early products created with the new learning was N-acetyl-p-aminophenol, today known as Tylenol. From the nascent synthetic dyes industry exploded new knowledge about chemical reactions, with enormous advances in medicine, photography, perfumery, food, and explosives.

The improved understanding of chemical structures led to massive growth of European companies with chemical expertise, particularly in Germany, where companies such as BASF, Bayer, Agfa, and Hoechst were founded. The modern pharmaceutical companies were born in short order, starting in the 1880s; some of them (like Merck) had existed for years as apothecary shops, peddling plant extracts, but the new understanding of synthetic chemistry transformed the companies into major industrial chemical research entities. Previous small concern companies like Schering, Burroughs Wellcome, Abbott, Smith Kline, Parke-Davis, Eli Lilly, Squibb, and Upjohn all metamorphosed into giant companies rushing to create new medicines.[12]

The quiescent field of microscopy, where little progress had been achieved in over two hundred years, was poised for an awakening. "Because good fixative, paraffin embedding, microtome and eosin stain were not available prior to the 1860s, pioneer microscopic pathologists most often obtained their specimens for microscopic examination by scraping and teasing out the cut-surface of tissues or by preparing smears from fluids and aspirates."[13] Little wonder that the first breakthrough observations of the 1830s and 1840s would be achieved when sampling blood and skin.

With the German openness to dyestuffs and chemical experimentation, it was natural that microscopists would start to tinker with *tissue* dyes. Scientists were used to altering chemical protocols in order to achieve better color penetration and colorfastness in cloth, and it was only a matter of time until the right recipe would be determined for medical use. There were essentially no worthwhile stains for microscopic slides until the decade after Perkin's discovery, and it was another South American plant that garnered attention. The logwood tree, *Hematoxylon campechianum*, is an indigenous tree of the New World whose roots and trunk exude a ruddy turbid colorant when boiled or steamed,[14] and was used for centuries as a dye for cotton. The Spanish used the dye (as did the Mayans), and also American soldiers during the Civil War.

A century and a half ago, *hematoxylin* was identified as a potent mammalian tissue stain, causing a bland, colorless tissue sample to adopt a deep purple, india ink–like hue. Experimentation with various chemicals added to hematoxylin yielded a combination that readily stained the inner parts of the cell, later revealed to be the nucleus, where the chromatin (DNA and RNA) is housed. A decade later eosin, a reddish-pink dye, was discovered as another dye that readily attached to other cellular structures, yielding a fuchsia shade over the entire representation. While the newly discovered dyes provided much improved visualization of the material, it was like looking at a coloring book pigmented with only one crayon.

Washing the slide material with alcohol and other drying agents led to visual changes in the tissues, and a scientific game of hide-and-seek transpired as the German histologists played with the sequence and timing of chemical exposure. A double-staining technique using two stains in succession, and, finally, the combination of hematoxylin and eosin in 1876 set the standard that is still used everywhere today.[15]

The combination of hematoxylin & eosin (H&E), with almost three million slides per day prepared in the cytopathology labs around the world, must be regarded as one of the most monotonously successful chemical combinations on earth. All the chemical and pharmaceutical advancements achieved since the modernization of medicine have not changed the fact that the two chemicals in H&E staining are perhaps the most reliable molecules in medicine, touching more lives over the last 150 years than almost any drug. The yin-yang of H&E staining meant that various elements in tissue were reliably stained either pink or deep purple, and researchers could now focus their eyes on the individual *cells* that made up organs.

While the birth of industrial chemistry occurred in England, it rapidly found a home in German academia, and the multiple scientific bastions that would prop up medicine in the future—optics, pharmaceutics, engineering, physiology, and radiology—co-evolved with decidedly German sensibilities as well. The Italian leadership in medicine, most recently championed by Morgagni, had resulted in a renaissance of French medicine that turned physicians' attention to the patient and her symptoms. Viennese medicine was at the forefront of the birth of many specialties in the mid–19th century, and Rokitansky, the last great naked-eye pathologist, tutored many accomplished physicians around the globe. But the Germans embraced all the new sciences with such gusto and with a cultural alignment that made full adoption of the scientific leadership mantle unquestioned. The title of ascendency among physicians in the world would pass from Rokitansky to a maniacal worker and savant in Berlin, a man who embraced the microscope, with its dyes and German-made lenses (like Zeiss and Leica), and who established the concept of the *cellular basis of disease*.

There has scarcely been a medical student and young physician who labored harder than Rudolf Virchow. The eager young Virchow was born in Pomerania in 1821 to a farmer and local treasurer and after graduating at the head of his class in 1839 from the local secondary school, enrolled in medical school in Berlin, in a military unit of the University of Berlin. Here, at the Friedrich-Wilhelms Institut, Virchow was tutored by Johannes Müller, "a biologist, comparative anatomist, biochemist, pathologist, psychologist, and master teacher," who trained generations of great German physicians. Müller began his career as a physiologist, focusing on nerve function, the mechanism of retina, and the functions of the sense organs in the ear. As happens in science, the objects of interest become increasingly small, and Müller's early subject matter was at the extreme boundaries of plain-vision investigation.

Müller had legendary energy (perhaps suffering from bipolar disorder, exhibiting bouts of mania and, alternately, severe, incapacitating depression[16]) and tended to attract like-minded and similarly indefatigable pupils. An early student was Theodor Schwann (1810–1882), who became the principal advocate of the *cell theory* newly proposed by his botanist friend Matthias Jacob Schleiden (1804–1881). Together, the works of Schleiden and Schwann in the years 1838 and 1839 set a firm foundation for the new appreciation of the importance of cells in plants and animals, explaining how they grow, function, and interact. Chemistry had the atomic theory; biology now had the cell theory.

Müller rapidly turned to the microscope in 1838, and soon was examining the cellular structure of tumors microscopically. Into this torrent of activity and revolutionary upheaval in 1839 strode the new medical student Rudolf Virchow. It was like two supernovae colliding, and the explosion of insight and output is almost unrivaled in scientific history.

Rudolf Virchow was incredibly intelligent and monstrously energetic. He was fluent in many European languages, and had learned Greek, Latin, Hebrew, and Arabic. Besides his multilingualism, he was an ardent archeology, ethnology, and political science devotee. At age twenty, he wrote his father from Berlin that his aim was to acquire "no less than a universal knowledge of nature from the God-head down to the stone." The brash and ultra-confident German, short and thin with bespectacled dark eyes and a piercing owl's gaze, wrote shortly before medical school graduation, ". . . you misunderstand me if you think my pride is based on my knowledge, the incompleteness of which I can see best: it is based on the consciousness that I want something better and greater, that I feel a more earnest striving for intellectual development than most other people."[17]

Virchow graduated medical school in 1843, initially working at Berlin's Charité Hospital, associating himself with the pathologist Robert Froriep. Within two years of graduation, in 1845, Virchow published a case report of a cook in her mid-fifties who had died in Berlin from an unknown disease. At autopsy, the blood in her organs contained a thick, milky layer, floating like a waxy blob. At first glance, it must have appeared like pus to the twenty-four-year-old physician, but unlike John Hughes Bennett, the Scottish physician who first described the disease four months before Virchow, Virchow did not declare it a "suppuration of blood," or an infection. Smearing the blood on a microscope slide, utilizing the primitive carmine dye to stain the cells, and carefully observing the constituents of the fluid, Virchow was at a loss to explain the phenomenon of the hordes of large round cells (interspersed among the small red blood cells) he was observing, and decided upon simply describing the disease by its visual appearance: *weisses Blut* (white blood). In a subsequent 1856 publication, Virchow adopted the Greek term for white blood, *leukemia*, including the description of two forms of the disease, one in which the spleen is enlarged and the other in which the lymph nodes are infiltrated by the white blood cells.

Virchow published another article in 1846 on the nature of blood clots, proposing theories about the genesis of deep venous thrombosis (large blood clots) and emboli (traveling blood clots) that have proven true all these years later. The twenty-five-year-old decrypted the enigma of embolism, where a large blood clot detaches from a leg or arm vein and travels to the lungs, where it completely obstructs blood flow and causes catastrophic death, a concept no one had ever considered. In the space of a year, Virchow had correctly identified (and even postulated the cause of) two major diseases that had plagued mankind forever. Bolstered by his success, Virchow decided to publish a journal, *The Archive of Pathological Anatomy and Physiology, and Clinical Medicine*. It is still published to this day, as one of the most important journals in the world; it is simply referred to as *Virchows Archiv*.

In the first issue, Virchow outlined his scientific world view in a tour de force statement. He declared, "Pathologic anatomy is the doctrine of deranged structure; pathologic physiology is the doctrine of deranged function . . . [t]he science of pathologic physiology will then gradually fulfill its promise, not as a creation of a few overheated heads, but from the cooperation of many painstaking investigators—a pathologic physiology which will be the stronghold of scientific medicine."

As has been seen repeatedly in this work, the Europe-wide Revolutions of 1848 had broad scientific, political, and artistic implications. Virchow was swept up by his ideals of social medicine, which destabilized his position in Berlin. Finding a new home in nearby Würzburg, Virchow entered the most productive epoch of his life. He tackled the subjects of inflammation, cancer, kidney disease, and the anatomy of the skin, nails, bone, cartilage, and connective tissue.[18] Lacking electricity, photomicroscopy, and projection of images, Virchow invented the "table railroad," a track that passed microscopes from student to student so they could peruse what the master wanted them to examine. He implored his scholars to "see microscopically" and to adopt his view that the cell was the *fundamental unit of life*.

After almost a decade in Würzburg, Virchow returned to Berlin in 1856 with great fanfare and to a purpose-built pathology institute. His time in Würzburg had resulted in several quantum leaps in the understanding of cellular function and behavior. Sometimes adopting the ideas of other German researchers, the influential Virchow made ever-increasing claims about the primacy of the cell, at first (in 1852), declaring that any new cell can only arise from the division of a cell already present; in 1854 he wrote, "There is no life except through direct succession." Finally, in 1855, in *The Archiv*, Virchow powerfully concluded, *Omnis cellula a cellula* (Every cell comes out of a preexisting cell).

It is perhaps impossible to convey the solemnity of the statement, *Omnis cellula a cellula*, other than to compare it to a book that would be published four years later, in 1859, by Charles Darwin—*On the Origin of Species*. When we greet a stranger, we ask, "Where are you from?" It is natural to ask about someone's origin and upbringing. The most insightful and ingenious researchers have always been able to delve more profoundly than their brethren, to see further and connect the philosophical dots. Virchow, like Darwin, combined imagination and years of scientific struggles to formulate an overarching idea about our beginnings. Each of us is a conglomeration of cells, dividing again and again and again, achieving specialization and unique functionality.

Embryologists, in time, would discover that every animal starts its journey as a single cell, multiplying its number of cells through division; the only exception is at the spark of life, when two cells (the egg and the sperm) combine to form one.

The original cells in the morula (from the Latin word for mulberry) are "indeterminate," able to become any cell in any organ of the body. These are the original *stem cells*, almost supernatural in their ability to respond and adapt and transfigure. All our lives our cells are responding and obeying the chemical messages from surrounding cells, committing themselves along a particular cell line, thereby forming an advanced cellular neighborhood and eventually functioning tissues and organs. Gone awry, the abnormal cell loses function, and worse, achieves a diabolical characteristic that not only impedes normal cellular and organ function, but hastens death.

Virchow and his successors fathomed the significance of the cellular basis of life—forever destroying the ancient, mystical speculations about vital spirits, humors, and life forces. The need to restore "an ill-understood balance that had become jangled"[19] was repudiated with the understanding of disease as a set of "disordered biochemical phenomena"[20] that would, in the future, be addressed by therapeutic interventions aimed at the locus of dysfunction.

Miasma, bad air, unbalanced humors, and astrology were swept away by anyone willing to pay attention. Virchow's magnum opus was his textbook *Cellular Pathology*, published in 1858, which demanded a new approach in the "advancement of medical science." This became the playbook for the next century's medical research accomplishments. William H. Welch, the "Dean of American Medicine" at the founding of Johns Hopkins University, ranks Virchow's book alongside the works of Vesalius, Harvey, and Morgagni as "the greatest advance which scientific medicine had made since its beginning."[21]

Perhaps Virchow *did* achieve "no less than a universal knowledge of nature from the God-head down to the stone" that he hoped for as a young man. Sherwin Nuland describes him as "Hippocrates with a microscope." Together with his Teutonic microscopy colleagues (who would use advanced stains like H&E in the 1870s), Virchow established Germany as the medical mecca in the mid- to late–19th century, and the surgeons in Germany and Austria (Langenbeck and Billroth in primacy) shared the limelight as the epicenter of learning.

It has been claimed earlier in this book that for the first 295 generations of modern man's existence, an afflicted individual was always better off "going it alone" rather than seeking care from a healer or physician. It is only in the last five generations that a wise patient could expect improvement in their lot by seeking medical attention. Rudolf Virchow, as much as any physician-scientist, deserves the credit for turning our attention to the cell as the foundational building block of life, the currency used by the universe to absorb nutrients, exchange energy, build tissue, respond to stress, store information, serve as communication centers, and to form gametes (ova and sperm) to create another life. Virchow's record is not unblemished—he denied Darwinism and the germ theory his entire life—but his concept of the cellular basis of disease, the *Archive*, his two thousand authored manuscripts, and his long list of apprentices, enshrined him in the Pantheon of medicine, and more importantly, ushered in a metamorphosis in medicine that cracked open the vault of the truths of the inner workings of all cells, tissues, and organs.

In the space of a century, physicians had awakened to the notion of the organ basis of disease, which rapidly advanced to the cellular basis of disease. This, of course, would further evolve into the genetic basis of disease once the understanding of deoxyribonucleic acid (DNA) was realized. Understanding the cell as the building block of life unfettered physicians from millennia of superstition, and the rise of industrial chemistry would soon result in chemotherapeutics that were efficacious. In the late 19th century, surgeons transmuted from bleeders and abscess drainers into diagnosticians—coconspirators with pathologists in their quest to identify and treat disease. Surgeons had long attempted to shake off the association with barbers, but their search for significance would be achieved not by heroic acts and displays of dexterity, but via a scientific reorientation. It is no accident that the greatest surgeons were cultivated in centers where pathology was most warmly embraced; surgeons have never been "wellness" professionals, but are instead mercenaries called upon in the face of catastrophe and therefore, by necessity, must be nurtured in environments where disease and traumatic injuries are investigated and explained.

The contributions of a group of surgeons in Europe, and for the first time, America, would finally raise the stature of surgeons from the lowliest to the recognized. These pioneering surgeons conducted investigations, used experimental tools (like microscopes), altered techniques, reviewed their outcomes, and, for the first time, started to improve the lot of their fellow man. Amazingly, in the late 19th century, surgeons did the unthinkable. Instead of just operating on people in extremis, at the point of death, surgeons began the practice of elective surgery, paving the way for our modern world, where patients seek operations for conditions that not only are non–life threatening, or even causing great pain, but for conditions that are inconvenient, annoying, or even just aesthetically unpleasing.

SEVEN

Germs

"Dans les champs de l'observation le hasard ne favorise que les esprits prepares."
—Louis Pasteur, 1854

With only a few weeks remaining of my surgical internship, I am counting the days that I am held hostage to the surgeons outside of orthopedics. Having spent grueling months in the SICU, attending to the most seriously ill and frail patients, and having survived the endurance tests of vascular surgery and the transplant service, I can't believe my good fortune of ending my intern year on the Plastic Surgery service.

In Plastics, I gloriously sleep almost every night. Even in a major academic center, much of what we do is elective and aesthetic surgery. How do you spell relief? P-L-A-S-T-I-C-S. The Plastic surgeons are kind, relatable, and patient. They even let me sew a little bit, helping me gain confidence with my burgeoning surgical skills. Granted, I am not closing critical surgical incisions and traumatic wounds of the face, where the reputation of a plastic surgeons is made or broken, but I actually feel a part of a team, and not an ignorant grunt who has nothing to offer.

When you are an intern on call on a busy service (like the transplant team), there is a virtual guarantee that you will be up all night. On the transplant service, even if there are no "harvest runs" occurring, the incessant calls from the surgical floor and outside transplant patients conspire to keep you awake. In the event that a patient has suffered a terminal injury in a nearby hospital, and is being kept alive just long enough to donate their organs, the transplant team springs to action, flying or driving to that hospital to procure the organs. Otherwise, the intern's role on the transplant team is to manage the peri-operative patients, minimizing the chance of organ rejection while trying not to accidentally kill patients. When it comes to intern competence, this is a real threat.

The demands of call on the Plastics service are so minimal that I am allowed to field the phone calls from home, a welcome luxury after eleven months of in-house call, usually taken every other night in thirty-six-hour shifts. Now I am only working twelve-hour days, and only taking call every third night. My work week has been reduced to less than eighty hours per week, and I am loving it.

This being Sunday evening, the only real threat to my night blowing up is a fresh surgical patient suffering a wound dehiscence (skin separation) or infection. We are not on for hand emergencies tonight (it's Orthopedics' turn), and I admit it: I'm almost gloating about the great night of sleep I'm about to enjoy.

My Motorola pager buzzes next to me—it's the ER. Dang it. I dial the main number to the ER on my PalmPilot phone, and after a brief hold, a familiar voice greets me.

"Dr. Schneider, this is Paul from Infectious Disease. We were called to see a forty-four-year-old AIDS patient who I've seen in the past. He's been HIV positive for seven years, but his meds aren't working very well. His cell counts are pretty low, and now he's got this bizarre skin infection." (In 1996, being HIV positive was still a death sentence. The anti-retrovirals were in their infancy, and patients still were dying from AIDS after prolonged illnesses.)

I respond, "What do you guys want us to do?"

Paul says, "I talked to my chief, and he told me to call Plastics, to have a look and see if you should do surgery and cut these fungus balls out of his arms and leg."

"Fungus balls?!?"

"Yeah—we think it's subcutaneous phaeohyphomycosis, where fungi from the environment invade the bodies of immunosuppressed patients, growing into large colonies under the skin. Can you come to the ER and take a look at this guy?"

Scrambling for power, and trying to figure out my role in this strange story, I ask the classic question any intern must ask. "Why Plastics and not another service? Why not general surgery?"

Paul retorts, a little frustrated, "I dunno, maybe they thought you guys would do a better job of sewing the skin."

Dang it. No way out. I gather my ID badge, phone, and keys, changing into my green scrubs for the short drive to the hospital.

On my way to the ER, I call the Plastics fellow who is on call as my supervisor tonight. "Fungus balls?" he fires back at me. Okay, now I don't feel so ignorant. I really have no idea what I'm about to see.

Walking into the emergency room, I make a beeline for bay 15. I am intercepted by the ER nurse, a young rookie like myself, asking if she can be there when I talk to the patient. "Sure, but I have no idea what I'm going to say. I've never heard of this. What's his name?"

"His name is Rick, and he seems pretty nice. But his infection is gross."

Folding back the privacy curtain in bay 15, I see Rick on a gurney, in a hospital gown. Seated next to him is an elderly woman with ashen hair in a ponytail, dressed in a gray Penn State T-shirt, jeans, and worn out New Balance shoes. "Hi Rick, I'm Dr. Schneider. I'm an intern on the Plastic surgery team. The infectious disease doctors have asked us to come talk to you and to see if we can help you. I know things aren't great for you—how do you feel?"

"Not too bad, I guess," Rick replies with a palpable sense of fatigue. It sounds like he is genuinely exhausted and is barely able to suppress a cough. He is thin, has no teeth, and

conspicuously, has grotesque tumor-like nodules the size of small tangerines on his arms and his right hand. I glance back at his face. "Kinda gross, huh?" he responds.

"Well, I've never seen anything like it, but I'm only an intern, so that doesn't mean much," I say. "Anywhere else on your body?"

"There's one on my right leg," Rick concedes, folding the sheet to the side, showing me the protuberance over his right calf. This one looks like it's about to explode, looking more like a huge pimple with a thin sheen of protective skin overlying a congealed glob of pus. It is now becoming very hard not to grimace in disgust. I'm not sure how to examine the patient, and it seems pretty obvious that treatment with drugs is not working here.

I call the Plastics fellow, Ken, explaining Rick's surreal situation. We agree—surgical excision of these fungus colonies is imperative. "Put it on the schedule for tomorrow in the main OR. We'll figure it out then," Ken concludes.

When our team meets, we realize that we are short-staffed for the day. Room 8, the simple procedure room, is available this afternoon. While most of my Plastics cohorts are committed to the outpatient ORs for aesthetic surgery, our chief, Dr. Bonamassa, will be finishing a skin flap operation for a breast cancer patient in the main OR. It is decided: we will do the case in Room 8, overlapping the more complex skin flap operation, and the team has tapped me to do the simple act of incising the skin and extracting the fungal assemblages. Not only is it convenient for the team, it is safer to do an HIV case with only one surgeon, lessening the chances of accidentally lacerating a team member during the case. And for me, the intern, this is a huge leap forward: my first case where I am the only surgeon in the room. A mixture of exhilaration, intimidation, courage, and fear builds as the afternoon case approaches.

Once the patient has undergone induction of anesthesia and placement of the breathing tube, I prepare his arms and right leg for surgery. A couple technicians and a nurse help me prep the skin surfaces with yellow-orange betadine soap, and carefully placing the blue surgical cloth drapes around the limbs, I cover Rick's face and torso, only exposing three limbs. A few minutes of surgical choreography result in overhead lights illuminating Rick's right arm, with Christi, my scrub tech, and I sitting on opposite sides of the limb. A stillness descends upon the room after a flurry of activity, and the moment that I have dreamed about my entire life is poised to happen.

"Knife," I ask of the scrub tech. Holding the scalpel, I glance around the room. There at the windowed door, I see Dr. Bonamassa monitoring my progress. Every surgeon has this moment, the first independent cutting of skin, and our eyes lock for a beat. My boss gives me a subtle nod, which I reflexively reciprocate, and turn my attention back to the arm.

I draw the knife edge across the skin overlying the blob of fungus. The thin flesh edges gape open, revealing the doughy cream-colored homogenous fungus. There is no pus, and although I was prepared for a putrescent stench, the globule is odor-free. Immediately, I notice punctate bleeding along the freshly sliced skin boundaries, and the grave consequences of that HIV-tainted ichor demand a solemnity in my task.

THE INVENTION OF SURGERY 101

Grasping a blunt stainless-steel elevator instrument, I tease the encapsulated fungus out of its saccule. The human body has a knack for materializing a pouch-like home for low-virulence invaders and foreign bodies, walling off intruders into a stalemate chamber that can last for years, even decades. I scrape the last of the material out of the cavity, pondering that it looks like cream of wheat (seemingly everything in the body can be described by food analogies). A bulb syringe full of saline and antibiotics (resembling a turkey baster implement full of dish soap) is used to lavage the pocket, and within a few minutes, the first fungus ball infection is conquered. I am intensely focused on eliminating all seven fungus collections, and not on the HIV virus pulsing through Rick's body and not on the fungus that could theoretically infest my own body. I know the possibility of contracting either pathogen is extremely unlikely, but this mindset of not giving a damn about the risks of surgery is mandatory for those contemplating a career in the operative theater. You simply cannot be disquieted over sleepless nights, noxious fumes, sore feet, X-ray exposure, caustic surgeon-bosses, deadly viruses, and agonized patients. Those medical students who wish to preserve a more normal existence for themselves go into other fields like radiology and dermatology. If you read the previous paragraphs with wonder and curiosity, wishing you could see fungus balls being sliced out of an AIDS patient's limbs . . . maybe you have the makings of a surgeon.

Halfway done excising the fungal colonies, I notice Dr. Bonamassa at the window again, scrutinizing our case. A knowing glance among us confirms all is well. Removing subcutaneous infectious tumors is among the simpler tasks an intern can perform, but I am nonetheless euphoric over my first solo case. Intricate and risky cases will be mine in the future, but for now, I can luxuriate in the fact that I am making a difference in the life of this patient while taking another step in becoming a surgeon.

The innovators and pioneers who change the world are often tortured souls who perceive the truth far earlier than everyone around them. This is certainly the case in the development of medicine and surgery, where the visionaries are often considered malcontents and rabble-rousers, and are commonly shunned for challenging the status quo. This is perhaps best illustrated in the case of Ignác Semmelweis, a Hungarian-born physician who turned Vienna on its head and set in motion a series of discoveries that convinced the scientific world that germs are real.

At a time when Marie Antoinette was fearing for her head in Paris, her brother Joseph II, Emperor of the Holy Roman Empire, was dramatically reforming Austria's legal, educational, and medical systems. As one of the great Enlightenment monarchs of the 18th century, Joseph's contributions would be long-lasting, particularly his massive hospital complex, the Allgemeines Krankenhaus der Stadt Wien, or Vienna General Hospital. The numerous palaces, opera houses, majestic government buildings, statues, and fountains are a testament to the glory of the City of Music on the Danube, but it is the Vienna General Hospital (locally referred to as the "AKH") that is important to students of the history of science.

Emperor Joseph II built a sprawling hospital, fifty feet high, with multiple court-yards and many divisions, separating the hospital by specialty. Today, the buildings are intact but have been transformed into non-medical educational facilities for the University of Vienna. Opened in 1784, the AKH followed the European pattern of the 18th century, where numerous hospitals were built, arising from a consciousness of society's responsibility to the poor. With increased modernization throughout Europe, peasants were flocking to cities, and with the burgeoning Industrial Revolution, the cramped conditions and dangerous working environments made medical care more needed than ever before. The old medical institutions throughout Europe, such as the Hôtel-Dieu in Paris and St. Bartholomew's in London, had cared for the poor in the great cities for centuries, but the population explosion created a demand for more hospitals.

The Vienna General Hospital emerged the same decade as the Industrial Revolution. Medical research continued to be severely limited just as the world was becoming modern, and as cities became more congested and even more susceptible to contagion, our helplessness became more obvious. *Modernity surprisingly worsened disease and emphasized our ignorance.* The history of surgery almost always entails heroic failures, accidental breakthroughs, and unimaginable solutions. This was especially true on the maternity ward of the Vienna General Hospital in the mid-1800s.

A startling transformation began in the 19th century: hospitals stopped being death houses, and became healing institutions, and even a venue to cultivate life. The French Revolution had transformed physicians' ideas about the body, and had ushered in an era of uninhibited physical exam of the female body. With an improved scientific understanding of anatomy and pathology, the mechanics of childbirth became interesting to physicians themselves. Childbirth had only ever been commandeered by midwives, but obstetrics arose as a specialty, challenging the supremacy of midwifery. Nobility and the upper classes began choosing to have physicians (rather than midwives) deliver their babies in Europe and throughout the Continent. But the ultimate shock was this: women (and their babies) were *much more likely* to die if delivery was handled by a doctor.

A confounding duality therefore existed throughout the 19th century, wherein women were (directly or indirectly) pressured to deliver with a physician obstetrician in a hospital, in spite of the fact that the known risk of dying was many times higher than delivery with a midwife. What was the cause of death? Puerperal fever, also known as childbed fever. Typically, it would strike a woman in the hours after delivering, and would start as a lower abdominal pain, with a striking tenderness and swelling of the vaginal tissues. A foul discharge of pus would follow, and within hours, a gaseous distention of the belly and a spiking fever would develop. Most patients would rapidly progress toward shock, with shallow breathing, delirium, and profuse sweating in the hours before death. There was simply no effective treatment for a patient suffering from puerperal fever, and there was no explanation. Why—and how—were doctors making matters *worse*?

Infections were a complete mystery to every generation of physicians, from Hippocrates to court physicians to every emperor and king in Europe in the 19th century. Epidemics had occurred in waves, including the plague, typhoid, yellow fever, malaria, and cholera, but lacking proper science to analyze the means of transmission, and having no way to visualize the culprits, contagions were as scary and insoluble as the demons in a Botticelli or Michelangelo painting. Most theorists pondered the "foul air" associated with an infection, wondering if there was something noxious in the atmosphere (in Italian, bad air is "malaria"). The miasma theory of infection posited that bad air was indeed to blame, and therefore, when puerperal fever was ravaging through a maternity unit, physicians concluded that some phantom agent was responsible. It simply wasn't in the minds of men to conclude that small germs, bacteria, or viruses, were to blame.

◆

Ignác Semmelweis was always an outsider. Born in Budapest, Hungary, in 1818, to a grocer, he would always speak German with a Buda-Swabian accent, reinforcing his role as a Hungarian interloper in Austria. He transferred to Vienna to complete medical school, and after two years of wrangling for a residency position, Semmelweis landed in the newfound department of obstetrics. Biding time, Semmelweis volunteered in the pathology division of Carl von Rokitansky, focusing on autopsies of women who had perished of gynecological diseases and operations. Alongside Morgagni, Louis, and Virchow, Rokitansky was the one of the major physicians who established Austria and the German nations as the new leaders of medicine by adopting the anatomic pathological basis of disease—the recognition that diseases (and the symptoms they generate) are organ-based. Semmelweis absorbed Rokitansky's methods of analysis and observation, and armed with the cognitive tools to untangle the mysteries of disease, he solved the enigma of puerperal fever, paving the way for the eventual understanding of germs.

Arriving at the Vienna General Hospital, Semmelweis would have noted the Wiener Gebärhaus, a maternity wing to accommodate single women who were discreetly admitted through a dedicated private entrance. The "Pregnant Gate" was the entrance from the Rotenhausgasse, a narrow alleyway that today faces Austria's National Bank. In the 1800s, the Pregnant Gate was the covert entrance that was accessed by laboring women, sometimes wearing "a mask or veil, and [were] unrecognizable as they wanted."[1] Once admitted, the women were directed to one of two divisions: the First Division where deliveries were carried out by doctors and medical students, and the Second Division, where midwives and students of midwifery performed the duties.

Assignment was based upon the day of the week, which included weekend admissions to the First Division. With obstetrics gaining a foothold as a separate specialty, a specialized division to handle the deliveries of Vienna's unwed mothers seemed to be a blessing for all parties involved. The newcomer Semmelweis discovered the horrifying reality that women cared for in the First Division were dramatically more likely to perish from puerperal fever than those attended to by midwives in the Second Division. Puerperal fever would strike in the hours following childbirth, initially causing painful swelling and redness of the birth canal, followed by severe, agonizing inflammation of the skin, and eventual systemic infection and lethal sepsis. Death was an excruciating certainty and an almost welcome respite from the ravages of fever.

Semmelweis began investigating the oddly lethal effect of physician care, and as a young trainee, became responsible for the welfare of the women in the First Division. He "sought knowledge in the library, the autopsy room, and at the bedside, and few of his waking hours were spent elsewhere."[2] In his reading, he realized his hospital was not unique: there were publications in the preceding decades that detailed similarly poor outcomes among obstetric physician deliveries. In London, between 1831 and 1843, the mortality rate when delivered at home was ten for every ten thousand mothers, versus London's General Lying-In Hospital where six hundred women per ten thousand died of puerperal fever—a sixtyfold increase.[3] Similar articles from Paris, Dresden, Australia, and America showed the same trend.

Ignác Semmelweis, twenty-nine years old, considered every variable. He contemplated the different techniques of midwives versus physicians, the surroundings, the conditions of the buildings, the exposure of the women to medical students, the way drugs were administered, and the protocol of postpartum care. Semmelweis even altered some of the physicians' practices to match those of the midwives, including altering the ventilation, but with no change. Doctors were still more dangerous to pregnant women than midwives. Semmelweis was "like a drowning man, who grasps at a straw;" nothing was adding up.[4] If it was not the air, nor the bed linens, and not the delivery technique, what could possibly explain the scandalous difference?

As the awful death hastened by puerperal fever was becoming almost routine for Semmelweis, he continued his daily practice of dissecting cadavers in the deadhouse of the Imperial and Royal General Hospital, thanks to the "kindness of Professor Rokitansky, of whose friendship I could boast . . ."[5] Engulfed in disease, death, fever, and confusion, Semmelweis decided to take a break, and departed for Venice for a vacation, hoping to clear his mind and somehow untangle the clues behind the problem that tortured him.

When Semmelweis returned to Vienna, a catastrophic finding awaited him: his close friend, Jakob Kolletschka, a Rokitansky disciple and forensic pathologist, was dead. Kolletschka had been performing an autopsy days before when his finger was accidentally sliced by a student's knife. With little delay, Kolletschka became ill, eventually succumbing to a massive infection. His body was dissected by his grieving coworkers, who encountered pus throughout his abdominal cavity and organs in a pattern all too familiar. Semmelweis was understandably traumatized by the grisly nature of his friend's death, reading and rereading the autopsy transcript, when a shock wave of insight came over him. A decade later he wrote:

> Totally shattered, I brooded over the case with intense emotion until suddenly a thought crossed my mind; at once it became clear to me that childbed fever, the fatal sickness of the newborn and the disease of Professor Kolletschka were one and the same, because they all consist pathologically of the same anatomic changes. If, therefore, in the case of Professor Kolletschka general sepsis arose from the inoculation of cadaver particles, then puerperal fever must originate from the same source. Now it was only necessary to decide from where and by what means the putrid cadaver particles were introduced into the delivery cases. The fact of the matter is that the transmitting source of those cadaver particles was to be found in the hands of the students and attending physicians.[6]

Semmelweis realized that his dedicated practice of daily morning cadaver inspection, in an era of no handwashing and prior to the invention of rubber or latex gloves, was leading to the introduction of "cadaver particles" to his own obstetric patients. He concluded, "puerperal fever was nothing more or less than cadaveric blood poisoning."[7] In a slight twist to the convention of the day, disease was not caused by the smell in the air, but was instead triggered by the particles from the cadavers that generated foul-smelling air. What Athanasius Kircher guessed were "invisible living corpuscula" in 1658[8] and what Leeuwenhoek referred to as "animalcules" he had visualized with his crude microscopes in 1677,[9] were microscopic creatures that had now become Semmelweis's enemy.

Already in the 19th century Westerners were using chloride solutions to rid homes and workplaces of the noxious odors of putrid materials; Semmelweis reasoned that chloride's effectiveness was the destruction of the particles themselves. Within two months of Kolletschka's death, a bowl of *chlorina liquida*, a dilute concentration of the disinfectant, was placed at the entrance to the First Division, with the order that every medical attendant wash his hands. Within months, the puerperal death rate plunged, until it was equal to the midwives' ward, where no handling of the cadavers had ever been performed. The seeds of a revolution were sown and, in Vienna alone, the lives of thousands of women were poised to be saved.

There is a reason you have never heard of Ignác Semmelweis, whose serendipitous reasoning and insight should have made him a household name. With many European countries teetering on revolution in 1848, doctors were distracted and digging their heels into accepted tradition; Semmelweis was not able to convince his colleagues of the rightness of his radical idea. Instead, controversy raged on, and in time, Semmelweis would lose status and eventually, his job. Despite the Viennese medical giants (Rokitansky, Joseph Skoda, and Ferdinand Hebra) backing Semmelweis, he could not break through the obtuse obstetrical leadership, who were too defensive. He retreated back to Budapest, where he languished for years until finally publishing his seminal book on the topic of childbed fever.

Sherwin Nuland has called Semmelweis's book "logorrheic, repetitious, hectoring, accusatory, self-glorifying . . . virtually unreadable."[10] With copies of his book he wrote open letters to his main detractors. To a professor of obstetrics in Vienna, he wrote, "And you Herr Professor, have been a partner in this massacre. The murder must cease, and in order that the murder ceases, I will keep watch . . ." To a professor of obstetrics in Würzburg, he cried, "I declare before God and the world that you are a murderer and the 'History of Childbed Fever' would not be unjust to you if it memorialized you as a medical Nero."[11] He became a raving lunatic and a righteous martyr.

In the end, Semmelweis seems to have lost his mind. Roaming the streets of Budapest, consorting with prostitutes, dressing like a vagrant, and mumbling to himself, driven either by the constant state of stress, organic mental disease, or possibly syphilis, the forty-seven-year-old doctor was undoubtedly going insane. His own wife coaxed him to return to Vienna, with Ferdinand Hebra (considered the father of dermatology) meeting him at the Vienna train station and asking him to visit a private sanitarium. They eventually took Semmelweis to a state-run insane asylum, where he was forcibly restrained and incarcerated. Within two weeks, on August 13, 1865, Semmelweis died, his body being transported to the Allgemeine Krankenhaus, and was autopsied in the very deadhouse and on the same table as Kolletschka. The cause of death? Infection, sepsis, and a large collection of pus in this chest—the selfsame illness that he had decrypted twenty years before. Most scholars believe his lethal infection originated from trauma and open wounds sustained during his apprehension and restraint, not unusual for 19th century "madhouse" patients. His was a sadly ironic end for the man who first showed the world the value of handwashing and who set the stage for antisepsis and broad acceptance of the germ theory.

◆

As Semmelweis was dying in a Vienna insane asylum in 1865, one thousand miles away in Glasgow, Scotland, the English surgeon Joseph Lister was preparing a clinical experiment that was elegantly simple, but profoundly important. An eleven-year-old boy was brought to the Glasgow Royal Infirmary with a fractured tibia, the result of being run over by a horse-drawn cart. His fracture was "compound," or as we say today, "open," with the bone protruding through the skin. Open fractures have always carried an alarmingly high complication rate, including loss of limb and death. When shattered bones shear through the skin, a significant degree of soft tissue destruction occurs, which devitalizes the skin, potentiates bone infection (osteomyelitis), and complicates skin healing. In 1865, almost no one had considered that putrefied post-traumatic wounds were the byproduct of germs; Joseph Lister (independent of Semmelweis) had already contemplated that possibility and was poised to prove it to himself. Therefore, the day before Semmelweis died in Austria, surgical antisepsis was begun in Scotland when Lister, instead of reflexively amputating the boy's leg, cleansed the wounds with carbolic acid and applied a dressing doused with the same. In time, the boy's wounds healed, the bones knit themselves together, and the leg was saved. Perhaps animalcules were real.

Joseph Lister was born in 1827 to Society of Friends (Quaker) parents in a village to the east of London. As was considered characteristic of Quaker families, the Listers were industrious, pious, peaceful, and serious. With no interest in sports, hunting, or frivolity, Quakers focused on religion, business, education, and the life of the mind. Joseph's father was Joseph Jackson Lister, a successful wine merchant who himself made a serious contribution to science as a microscopist.

The elder Lister, a scientific autodidact, had befriended a young London physician, Thomas Hodgkin (also a Quaker), who would describe the eponymous blood disorder in 1832,[12] and the two would collaborate on microscopy for years. Lister, even with very little formal education, solved a 150-year-old problem that had limited the early compound microscope to "little better than a scientific toy,"[13] transforming microscopy into a serious scientific investigative tool. Prior to Lister's innovation, microscopy was limited due to *chromatic aberration*, where there is a dispersion of the light traveling through the microscope's tube. This results in blurred, wavy images that make accurate interpretation impossible. However, Lister's creation corrected the visual distortions, turning the compound microscope into a tool that revolutionized medicine, and fueled his son's curiosity and empowered his investigations. History is replete with families who conceptually enabled their progenies futures. In producing the finest microscopes the world had ever seen,[14] Lister's own father was literally building the optical machines that facilitated his son's vision.

Although Joseph Lister had excelled in his preparatory Quaker schools, he was not eligible to attend Oxford and Cambridge, where subscription to the Thirty-nine Articles of the Anglican faith was out of the question for a young Quaker. In 1844 (at age sixteen), Lister enrolled at the University College in London, an "Oxbridge" for students who were religious or social rank misfits, but who showed great promise. Three years later, Lister began medical school at the University College, where he would graduate with honors in 1852.

Lister had been graced with one of his family's finest microscopes when he matriculated to medical school. While there, the young prodigy presented two papers to the

Hospital Medical Society, presciently turning his focus on "Gangrene" and the "Use of the Microscope in Medicine" at a time when University College offered no formal instruction on either topic. [15] Steve Jobs presenting a speech on the future of personal computers and portable music machines in college could not be more perspicacious.

Lister's first twenty-six years were spent within a ten-mile radius in London, but his unquenched thirst for knowledge propelled him outside the circle of London to Edinburgh, Scotland, to complete his surgical training under James Syme, acknowledged as the best technical surgeon in the British Isles. While Lister was reserved and modest, Syme (then in his fifties) was obstinate and intense, but perhaps the men "saw a submerged part of his own personality in the other, and allowed a secret admiration of his unconscious alter ego to forge a deep friendship." [16] Lister flourished in Edinburgh, furthering the Hunterian tradition of British surgeon-scientists from the antecedent century, writing his father that he couldn't "conceive what a high degree of enjoyment I am from day to day experiencing in this bloody and butchery department of the healing art." [17]

Lister was hired as Syme's house surgeon (forerunner to Halsted's surgical residency), and after a year and a half, became assistant surgeon to the Edinburgh Royal Infirmary in 1855. Edinburgh was at the very forefront of medicine and surgery in the mid–19th century. With Paris, Berlin, and Vienna vying for world supremacy, physicians in Edinburgh first demonstrated chloroform anesthesia in 1847, within months of ether's discovery in Boston. (Late in life, Lister recalled that he witnessed Europe's first operation under ether in December 1846, in London, probably cementing his resolution to become a surgeon.) Surgeons, numbed to the cries and protestations of agonized patients, suddenly found themselves sovereign over the afflicted, thus creating new possibilities for cure. The Oxford English Dictionary defines *frontier* as "the extreme limit of settled land beyond which lies wilderness," and for the first time in mankind's existence, the frontier boundary of awareness had been demolished, galvanizing Lister in Edinburgh and Billroth in Vienna into exploration of the wilderness of the deeper parts of the body.

Fortuitously for all involved, Joseph Lister was spending increasingly more time at the Syme home, befriending the eldest daughter Agnes. In 1856, Joseph and Agnes were married, beginning an almost forty-year relationship, which, although it bore no children, was based on shared scientific interests, with Agnes serving as his most important critic, editor, research assistant, and travel companion. The newlyweds traveled to Paris, Padua, Bologna, and Vienna, visiting the great thinkers and leading hospitals, before settling back in Edinburgh, microscope at the ready. While in Vienna, Lister went to the Allgemeines Krankenhaus, meeting with Semmelweis's former colleagues, but it is unknown if his story or observations were a topic of discussion.

Lister plunged into work in Edinburgh, balancing his young surgical practice, teaching obligations, assistantship to Syme, new marriage, and most critically, his burgeoning research laboratory. For years, his lab was in his own kitchen, with Agnes helping investigate blood coagulation, physiology of nerve and muscle fibers, lymph flow, and inflammation. Lister (and Agnes) gathered animals from local fields, parks, and streams, procuring organs from slaughterhouses, slicing, staining, and microscopically evaluating the structures and guessing about their function. In 1860, the Listers moved westward to Glasgow, Scotland, where, for a decade, Lister's momentous discoveries took place.

Lister fully threw himself into research on inflammation. Logicians had reasoned that contagions result from living organisms, and not just foul-smelling, or noxious, air. Lister was becoming convinced that putrefaction of wounds was due to some invisible thing that resided in the air and made its way into open wounds. Reading in English, as well as the French- and German-language journals, Lister encountered Jacob Henle, the influential professor from Göttingen, Germany, who reasoned that contagions had to be organic, that is to say, of living microscopic material.

Venereal disease was always a topic of interest to germ theorists. The early Renaissance scientists could never deduce the source of plague, typhoid fever, smallpox, or cholera outbreaks. Random persons always seemed to be affected, and prior to the analytical tools of epidemiology, it was simply too difficult to ascertain the germinal source of epidemics. But every European doctor had encountered gonorrhea, herpes, (and since the conquest of the Western Hemisphere) syphilis; everyone knew that virgins never contracted those types of venereal diseases that resulted in sores, scabs, scars, pus, boils, and all sorts of effluvia and detritus on the privy parts of the licentious citizenry (whom Lister described as having an "inclination to venery,"[18] or sexual indulgence). Venereal diseases, therefore, could *not* be spread through the air, and it didn't demand genius insight that (like procreation) it was ejaculate and the fluids shared during intercourse that carried the seeds of infection. The venereal disease epidemics provided further evidence that *contagium animata* were real.

Jacob Henle, writing in his classic 1840 essay, "On Miasmata and Contagia," conducted a series of Einsteinian "thought experiments," using only reasoning to ponder epidemics and pandemics.[19] Because microscopy was still limited, he used his imagination to elucidate the foundational underpinnings of germ theory (later inspiring his star pupil Robert Koch, in Göttingen). He considered cow-pox, where "an atom of pox poison can produce a rash over the entire body."[20] Predating Pasteur, Henle considered fermentation of wine, concluding that it was the "decomposition of an organic fluid by vegetable bodies."[21] In both these instances, the multiplication or amplification of byproducts suggested *living organic beings*. It simply did not make sense that an elemental poison or toxic gas could rip through a community on a more than additive (an exponential) basis. Amplification was occurring, and something living had to be multiplying itself within people's bodies.

Lister was becoming increasingly convinced that a surgical wound oozing with pus was not simply inflamed and "healing," but was instead putrefying, infected, and necrotic. His investigations on wild rabbits, dead cows, and oxen continued, as well as experiments and microscopic evaluations of inflamed frog feet, with Agnes at his side, in their kitchen. Had Lister been a lone genius with a magical microscope, he still needed a gentle nudge to help him take an important conceptual leap. Prior to the invention of the printing press and the formation of peer-reviewed scientific journals (e.g., *Philosophical Transactions of the Royal Society*), isolated savants could never "see further," but with the revolution of information sharing, geniuses could connect. In Glasgow, in 1865, Lister's colleague in the chemistry department recommended that he read the publications of Louis Pasteur regarding fermentation of beer and wine in the *Comptes Rendus de l'Académie des Sciences*. Lister, the English surgeon practicing in Scotland, sat down to read the Parisian chemist's French publications, an act which would change surgery forever.

◆

Louis Pasteur (1822–1895) is widely considered the father of microbiology for his pioneering work on fermentation, microscopy, vaccination, and bacteriology. A chemist (and not a physician), Pasteur was one of the first scientists to put the microscope to work for the good of mankind. An early publication in 1855, a study of the formation of lactic acid in the souring of milk, included the observation of "budding organisms" that he guessed were the active causes of fermentation, similar to the "tiny self-replicating organisms" associated with alcoholic fermentation.[22] At the time of the career-organizing publication, Pasteur was the new dean of sciences at Lille University (on the French-Belgian border), and was approached by a local manufacturer of beet-root alcohol who told the chemist of a mysterious catastrophe that was threatening the local beer and wine industries.

Pasteur, the new man in Lille, listened to the story of the "slimy juice of useless sour ooze"[23] and instead of resorting to chemical experimentation, turned to microscopic examination. Perhaps the brewer had heard of Pasteur's previous souring-milk publication, but there simply did not exist in the world a bacteriologist. Professor Pasteur gathered samples of the spoiled beet-root alcohol, placed droplets on a glass slide, and, adjusting the little mirror of his microscope to pinpoint the sunlight onto the morass, envisaged a peculiar partnership. There in the liquid were tiny yeast bodies, buoyantly adrift potatoes in the slurry, and what had been conjecture a year before, Pasteur now petitioned with confidence: the yeast organisms were the agents of fermentation.

Fermentation of sugars (found in beets, grapes, wheat, potatoes, corn, rice, and even bananas) by naturally occurring yeast has unwittingly been mankind's test kitchen for purification of liquid, making it intoxicatingly potable. Pasteur had unraveled the conundrum and for good measure, made another stunning observation: also present in the soured batches were great numbers of rod-shaped microbes. These, he reasoned, were the germs that were causing the spoilage. What earlier explorers had postulated might be operating organically, Pasteur now was demonstrating: the collaboration between the yeast and the sugars cloaked inside the fruits and grains was fermentation, while the interaction between bacteria and the sugars evidenced rotten slime. Fermentation good, putrefaction bad.

"Dans les champs de l'observation le hasard ne favorise que les esprits préparés" ("Where observation is concerned, chance favors only the prepared mind"), Pasteur famously proclaimed to his students at Lille the year *before* his famous observations about yeast, bacteria, and fermentation. In the late 1850s Pasteur published articles about his experiments on fermentation, and as a chemist, the papers were in scientific (not medical) journals. No wonder it was Lister's chemist colleague who alerted Lister to Pasteur's work.

Lister pored over Pasteur's investigations on spoiled beverages and butter, and set his father's microscopes to work in his kitchen laboratory, repeating the experiments and finding the same microbes in what Pasteur called the "world of the infinitely small."[24] Lister was not the first medical man to recognize Pasteur's sibylline analysis; Thomas Spencer Wells, a future president of the Royal College of Surgeons, proposed that microorganisms in the air caused disease[25] but did not consider practical interventions to combat them. Lister "perceived that Pasteur's work could be applied to surgery, but he took one more giant step: he began the formulation of methods for doing so."[26]

At the International Medical Congress of 1876, held in Philadelphia, Lister persuasively recollected, "When I read Pasteur's original paper, I said to myself, just as we destroy lice in the head of a child, who has pediculi, by poisonous applications which will not injure the scalp, so, I believe, we can use poisons on the wounds to destroy bacteria without injuring the soft tissues of the patient."[27] Lister brainstormed about a chemical agent that could be sprayed into the air (still focusing on the mistaken belief that germs were primarily floating in the air and descending upon surgical sites), and all that was required was Lister holding his nose to the Glasgow air. In maritime regions, ships' timbers were protected with creosote—a coal-tar derivative also used to coat railroad ties. Further distillation of coal-tar yielded *phenol*, an aromatic organic compound that was used to reduce odors of decomposition in sewage, with the serendipitous finding that it mitigated cholera epidemics (well before the bacteriology of cholera was understood). Dumped into rivers, including the Thames during the "Great Stink" of 1858, phenol, or *carbolic acid*, greatly diminished putrid smells around London. The aromatic, semi-fruity odor belied the truth of its chemical, bactericidal actions upon the bacteria teeming in the foul waters.

Carbolic acid, therefore, was a natural implement in the war against surgical infections. So it came to be on August 12, 1865, in the Glasgow Royal Infirmary, James Greenlee—the eleven-year-old boy—was treated by Lister for an open tibia fracture. The young lad's leg was thoroughly washed out (under chloroform anesthesia) with a dilute solution of carbolic acid. At the end of the operation Lister dressed the wound with a mixture of putty and carbolic acid, covered by a sheet of tin foil (to reduce evaporation), and splinted. Four days later, the wound was uncovered. Instead of finding a troubled wound with inflamed skin edges, a discharge of pus, rotting flesh, and rancid smells, the traumatized area was healing well. The leg was redressed in a similar fashion and left on for five days. At the second redressing, a superficial burn to the skin was diagnosed by Lister, so the third dressing included a solution of carbolic acid and olive oil. The wound completely healed, in time, with no further surgery required. What would have been routine amputation was instead uneventful healing, and James was discharged six weeks after injury.

"Antiseptic surgery emerged when Lister realized that the analogy between suppuration and fermentation which Pasteur had drawn could be applied in practice."[28] A series of scientific observations, simple experiments, and clinical applications had been linked to birth the era of "*antisepsis*," the process of eliminating microbial organisms that cause disease.

Immediately thereafter, Lister performed every surgery antiseptically. Lister and Pasteur would become coconspirators against the microbial enemies that caused disease. Crude as the early science was, with no specific identification of bacteria and with no real appreciation of how they grew, thrived, and died, the mere act of assiduously cleansing the skin and the traumatized tissues dramatically improved elective and accident surgery. In time, the entire world would adopt "Listerism," although acceptance came begrudgingly in some nations, even though the proof was right under their noses.

Lister succeeded where Semmelweis had failed. Most scientists and physicians had never heard of Semmelweis and his analysis of puerperal fever. Lister, on the other hand, enacted great change in medicine, setting the stage for a dramatic change in the feasibility of surgery. The industrious Lister always made friends easily; insatiably curious and unusually determined, he balanced uncompromising dedication to work with unselfish, loving consideration for everyone who came into his sphere. These qualities helped him, more than anyone else, convince the world that germs were real. Semmelweis, devoid of charm, became an annoyance, and lost to madness, died in cruel irony the day after Lister's first antiseptic operation. Francis Darwin, son of Charles Darwin, concluded, "In science the credit goes to the man who convinces the world, not to the man to whom the idea first occurs."[29]

In only a few years, Lister's antiseptic method would be examined on a continental scale. The Franco-Prussian War lasted less than a year (July 1870 to May 1871), but provided a unique laboratory for Lister's techniques. The Germans were among the first to adopt Listerism, establishing field hospitals attended by physicians armed with carbolic acid. The Franco-Prussian War was the first war in history fought with accurate guns and cannons—knifepoint and bayonet wounds were rare—with the end result that wounded soldiers "bore the unique scars of long-range rifle shots: diffuse, open wounds with splintered bone fragments and bits of clothing mixed in with the damaged flesh."[30] The injured German soldiers were doused, scrubbed, dabbed, and swabbed with the carbolic poultice, with the end result that, for the first time in warfare history, fewer men died from infections from their wounds than from the trauma itself. Alternatively, the French, who clung to the ancient and unscientific method of globbing on greasy salves to open wounds, beheld a 70 percent mortality rate from infection following amputation.[31] The theater of war had staged its greatest laboratory experiment in Listerism, and only the most obstinate would cling to their outdated black magic.

Of the thousands of medical personnel in the Prussian army, there were two remarkable physicians whose immediate impact was negligible, but whose eventual contributions were enormous. One was Edwin Krebs, a protégé of Rudolf Virchow, and the other was Robert Koch, a young German man from a small village. Krebs, away from the front lines of the Franco-Prussian War and emulating his esteemed professor, decided to sample some tissue from the wounds of deceased soldiers. Utilizing his microscope, Krebs observed a multitude of rod-shaped and spherical bodies among the infected tissues. Similar to Pasteur's observations of spoiled milk and beet-juice alcohol, Krebs made a groundbreaking hypothesis: the odd-appearing bodies were germs that not only were associated with infection and death but were the *causes* of disease and final destruction.

To validate the germ theory, one needed a physician scientist who could meticulously identify a germ, show how it grows, and differentiate it from other germs that caused dissimilar diseases. It may not be obvious to the nontechnically trained reader that each bacterial species (staphylococcus, for example), has an exact appearance under a microscope, precise growth pattern, narrow inhabitable environment, DNA profile, and unambiguous effect on particular plants or animals. The first light of comprehension was beginning to shine upon the Teutonic researchers that the bacterial kingdom could be observed, described, and perhaps, in due time, combatted, and it was Koch who would lead the way.

Shortly after the war, the reticent and cloistered Koch eagerly accepted an appointment to become the local health officer in Wöllstein, eastern Prussia. He relocated there with his wife and young daughter, embracing a chance of being the area's sole physician and perhaps of making an important scientific discovery. Wöllstein was, as the name implies, a hub of sheep farming and wool production, and with wool being the world's leading textile in the 1870s, Wöllstein was a vital, if not urban locale.

Robert Koch hailed from north-central Germany, reared before the unification of German states at the conclusion of the Franco-Prussian War. Unlike other European countries, Germany had no single, dominant intellectual center, so cities like Berlin, Munich, Leipzig, Wittenberg, Würzburg, and Nuremberg all had centuries of proud academic achievement. Koch had attended medical school in Göttingen, fortuitously training under Jacob Henle, a pioneer in microscopy. In fact, his *Handbook of Systematic*

Human Anatomy was the first great work of anatomical description since Vesalius's *Fabrica*, delving into the microscopic anatomy of our organs. What Vesalius had done for human anatomy, illustrating our structure, perceiving the interconnectedness of organs and fibers, and confronting falsehood, Henle did for *microscopic* anatomy, delving deeply into the microstructure of each organ, and bringing to light the infinitesimal elements that constitute the human corpus.

Thomas Goetz observes that Henle and Koch were an ideal match; both were chronically shy, more inclined to "heads-down research than to socializing," and both paid great attention to detail.[32] Decades after Henle's less voluminous, but no less significant *On Miasmata and Contagia*, Herr Professor's ruminations on germs were no doubt reverberating in Koch's ears. His bio-mathematical calculations had led Henle to conclude that little animated "vegetable bodies," or "infusoria," simply had to be at work in the microscopic world. The slight delay in symptoms followed by a precipitous decline in an afflicted individual mirrored a community's collapse in the face of the same disease. Neither the lone patient nor an entire tribe could be felled by simple chemical amalgamation; replicating, organic beings were feeding at the hosts' expense. Fresh from the war, it was up to Koch to prove Henle right.

❖

No longer in Germany, Wöllstein lies in Poland, where it is known as Wolsztyn. It remains a small town, surrounded by a patchwork of fields and pastureland that are reminiscent of rural Americana. The town boasts a few churches and many small shops whose windows are embellished with undecipherable Polish lettering. Within minutes you realize this little village has precious few visitors, and the scant few pilgrims who make their way to this pastoral enclave are here to encounter Koch a century and a half later.

In the center of town, along *Roberta Kocha Road*, stands Robert Koch's house and clinic, which previously had been a hospital for the poor, completed in 1846. The family of three moved here in 1872, occupying the upstairs of the Gothic building with a large bay window and red tile roof. On the lower floor, below the bay window, is a large double door that provides access to a passageway into the house, functioning as a receiving portal to Dr. Koch's downstairs examination room. That arcade was frequented by stricken patients who likely could receive no benefit whatsoever from Koch.

Within a few years, Koch had settled into his routine of seeing patients and tinkering with his beloved microscope. Demanding more, Koch acquired cutting-edge microscopes at a time when he could scarcely afford them. Happily for mankind, Koch lavishly drained his bank account, obsessively examining tissues and experimenting on animals from his backyard collection.

Within a year of his arrival, in 1873, sheep in the area began to die. Worse, local farmers and sheep shearers began to sicken. The malady was all too familiar to the denizens of the area: woolsorters disease (also known as anthrax). Herbivores (e.g., cattle, sheep, horses, goats, antelopes) are most commonly affected, but humans are at risk if exposed to the tissues of affected animals. When sheep and cattle are infected, the clinical course is characterized by sudden onset and a rapidly fatal course, with staggering, dyspnea (labored breathing), trembling, collapse, and even convulsions.[33] Occasionally, bloody discharges from body openings (mouth, nose, anus) occur; in humans, skin lesions with ghastly ulcerations and necrotic, blackened skin lesions often develop. Since ancient times, anthrax had been interpreted as a type of celestial judgment or biblical plague, so great and unexplainable was the terror.

As the outbreak spread to humans that summer of 1874, Koch started to see an influx of patients in his Wöllstein house, likely using folk remedies to care for the scabbed, swollen, and carbuncled victims. Ever the scientist, Koch would collect blood and fluid from his affected patients. In his makeshift lab on the first floor, he would examine his test subjects' samples under his microscope and record his findings in his personal research notebooks. On April 12, 1874, he used the term "bacteria," (following the lead of German zoologist Christian Gottfried Ehrenberg), when he recorded in his notebook, "The bacteria swell up, become shinier, thicker, and much longer."[34] Koch's observations mirrored what a few researchers had seen in the years before, but no scientist had ever taken the *next* step to evaluate if those bacteria, resembling grains of rice lined up end-to-end, were the *cause* of anthrax.

Eighteen months later, a few days before Christmas 1875, a local constable in Wöllstein appeared one evening at Koch's home, likely mounted on horse-drawn wagon, with a carcass of a dead animal whose blood was dark and thick. Terrified that the lifeless beast would resurrect the anthrax plague, the police officer brought the remains to the one person in town who might know what to do next.

Robert Koch surveyed the animal, and immediately knew that his microscope was his best analytical tool. Koch had a hunch the beast was suffering from anthrax, and eagerly extracted blood from the dead animal and inspected it under his microscope. The bearded and bespectacled thirty-two-year-old was delighted to see his slide suffused with the same type of bacteria he had seen the year before. Jacob Henle's dream had been that "It could be empirically proven that [bacteria] are actually the active part (the cause of the disease) if one could isolate the contagious organism and the contagious fluid and observe the powers . . ."[35] reverberating in his ears, Koch had a spark of inspiration.

Walking out of his makeshift laboratory and exiting the rear doors, he entered his garden and pulled a healthy rabbit out of its cage. Controlling for normal, Koch sliced the rabbit's ear, drawing a crimson trickle, and verified microscopically that the rabbit's blood was free of bacteria. He then injected the dead animal's *animalcule*-laden blood into the fit rabbit's ear. Bacterial life cycles were completely unknown at the time—there wasn't even agreement that germs were real, let alone an appreciation of their mode of operation and lethality. Retiring to bed that night, Koch had no idea what the next day (Christmas Eve) would bring.

I believe, given Koch's singlemindedness and obsessive focus, he had trouble getting to sleep that night. He faced a full clinic on the twenty-fourth, tending to his patients in his upstairs exam room overlooking the backyard (where the rabbit was caged). He worked from morning to night. When his work was done he checked on his leporid patient.

The rabbit was dead.

Bringing the rabbit inside, Koch sampled the blood and found the same bacteria he had seen the day before. Recording in his journal, he described the selfsame bacteria in "moderate numbers" in his research notebook. That Christmas Eve, Koch retired to bed, as expectant as any child, planning his next move. Instead of disposing of the dead rabbit's body, he saved it for more experimentation the next day.

Christmas morning, Koch sampled more tissue from the dead animal, no doubt starting to reek. (This was the 1870s, before the invention of electricity, light bulbs, and refrigerators.) Even more bacteria were growing, and Koch was becoming more convinced that these multiplying organisms were the cause of anthrax. Not just recording his observations in his notebooks, Koch began planning experiments to verify his hypothesis that these plain little microscopic bodies were the culprits of the primeval scourge.

Koch was able to pass on the disease from the rabbit to additional pet animals from their household, including pet mice. Each animal perished from anthrax, and each specimen showed the telltale bacteria under his microscope. Koch was the unwitting exemplar of laboratory animal experimentation: his use of rabbits and mice (including his own breeding program of white mice) continues in every university setting in the world, gaining its start in tiny Wöllstein.

Koch experimented further, attempting to culture the anthrax bacteria outside a living host. Undoubtedly, the pioneer was confused about the waves of anthrax plagues that had always occurred. What was the secret of the organism's ability to rapidly propagate after months (or years) of quiescence? Robert Koch was now sailing into uncharted waters. The few visionary scientists who suspected the villainy of the innocuous-appearing germs that existed at the other end of the metal tube of their microscopes had never succeeded in incubating germs.

Convinced that the bacteria were living organisms that needed sustenance, Koch tinkered with body fluids that could provide a growth medium. In a macabre solution, he used the aqueous humor, the fluid contained within eyeballs (procuring them from slaughterhouse cattle). Koch used his dissection tools to gather a small sample of infected mouse spleen, mixing it into the aqueous humor on a thin glass slide. He mated that slide against a stouter glass slide that had a small concavity he had created.

The fluid mixture was held in place by a small ring of petroleum jelly, creating a sealed culture environment.

Koch placed his contrivance on his Zeiss microscope stage and tuning the large brass knurled focusing ring, peered into the fluid, finding nothing. He scrupulously scanned the slide, but couldn't find any bacteria. He set his microscope aside for an hour, anxiously returning to see if any growth had occurred, but still found no change. Two hours later, Koch inspected, but was again disappointed. Then, after a few hours, Koch saw something wonderful: whorled stacks of rice-shaped rods started to appear. In time, the entire world underneath his own eyeball was replete with the anthrax bacteria, which Koch would call *Bacillus anthracis*.

The anthrax colonies started revealing different shapes as they matured. Koch observed that the long rods enlarged in time, and then birthed small round *spores*, which recently had been reported by Ferdinand Cohn, of the University of Breslau (now Wroclaw, Poland). Koch's breakthrough began with subsequent experiments, where he began altering conditions. He had found success with aqueous humor, but now Koch (accidentally?) considered the effects of drying or heating cultured samples. He grew colonies of *Bacillus anthracis,* and then dried them under heat, noticing that the rods and the spores would cease growing. When the rods were placed back into the growth medium, nothing changed—they were apparently inactivated and their contagiousness halted—and it was obvious to Koch that the rods were not the robust form of the contagion. On the other hand, the spores were resilient to temperature and dryness, able to spring to life once placed back into culture, generating new rods and spores capable of producing disease. This, he reasoned, was how anthrax infected herbivores: in spore form it could reside in a field, surviving years in suspended animation, waiting for invigoration in an animal's body. He had solved the life cycle puzzle of anthrax.

The Wöllstein physician and part-time scientist knew his work was important, but isolated in his small Prussian town wasn't sure with whom he should liaise. Koch reached out to Ferdinand Cohn, who invited Koch to Breslau to demonstrate his experiments. Within days, Koch "headed to the train station carrying microscopes, slides, cows' eyes, mouse spleens, and boxes of rabbits, frogs, and mice—many, many mice, both living and dead. Some were flush with anthrax. Rushing through the Wöllstein station to make his train, laden with boxes and trunks, he must have been quite a sight."[36]

When Koch arrived in Breslau, he immediately set up his equipment in Cohn's institute, inoculating animals and culturing eyeball fluid. In the days that followed, scientists visited him, assaying his progress and taking stock of his meticulous techniques. Soon, rods and spores were emanating in media and animals were succumbing to disease. An esteemed professor, Julius Cohnheim, the director of the Institute of Pathology and a former assistant of Virchow's in Berlin, came to survey the progress. Cohnheim "couldn't get over how methodical and thorough this Koch was; he apparently emerged from nowhere but was calmly demonstrating the most deliberate and decisive laboratory techniques Cohnheim had ever seen."[37]

Professor Cohnheim, dazzled, departed from the makeshift experimental setup and rushed across campus to his assistants, telling them to stop what they were doing and to get over to Cohn's lab to witness what the wunderkind was demonstrating. He said, "I regard it as the greatest discovery ever made with bacteria and I believe that this is not the last time that this young Robert Koch will surprise and shame us by the brilliance of his investigations."[38]

Koch soon published his work on anthrax, and in a pulse of experiments and publications over the next several years, demonstrated the presence of bacteria in wounds, thus supporting the concept of Listerism in surgical science. He would continue to refine culture techniques, both in Wöllstein and later Berlin, that are used every day in every hospital and every lab in the world, including the use of agar as a culture medium and the use of round glass plates with side walls, named after his assistant, Julius Petri (the "Petri dish"). Koch worked with the Zeiss company to improve microscopes and with Leica to invent photomicrography, mammoth achievements themselves.

Koch's apotheosis came on March 24, 1882. After establishing himself as one of the great young innovators in all of Europe, the thirty-eight-year-old announced he was giving a talk entitled "On Tuberculosis." Even though the previous decade had seen the fortification of the germ theory, agreement on the origin of tuberculosis (TB) was not yet consistent, even among medicine's greatest geniuses, including Rudolf Virchow. Rumors about the gravity of the lecture had circulated among the cognoscenti in Berlin; the library at the University of Berlin's Physiology Institute that Friday night was packed in anticipation of Koch's demonstration.

Koch started by reviewing the statistics of TB in the 19th century: one in seven of all human beings died from tuberculosis, but, "If one only considers the productive middle-age groups, tuberculosis carries away one third and often more of these."[39] TB was a true, slow-motion world pandemic; his erstwhile audience was mired in that reality, but the more he emphasized its import the more they must have expected a profound announcement. The trouble was, no one had ever visualized the germ that caused the disease.

The secret of its covert cold-blooded efficiency was also the characteristic that made it difficult to image. The tuberculosis bacterium is now known to be a member of the *mycobacterium* genus, with over 150 species (including *M. leprae*, the cause of leprosy) and characterized by a thick cell wall that is waxy and hydrophobic (not attracted by water). This thick cell wall helped the small bacillus (from the Latin word for *stick*, *baculus*) hide in a sea of cells, impervious to the typical dyes used to stain tissues. It was up to Koch and his team to try other chemicals and to alter conditions to draw out his quarry.

Koch and his associates prepared blocks of tuberculous tissues, and using the newly invented microtome, shaved extremely thin layers of samples and mounted them on glass slides. The typical approach was to expose the slide to an alcohol-methylene blue stain at room temperature, but this was ineffective. Through trial and error, Koch experimented with many reagents, settling upon the addition of potassium hydroxide and Bismarck brown, an industrial brown dye to counterstain the surrounding tissues. The chemical sorcery was perfected by simultaneously heating the slides to 40° C, which reduced the reaction times to only an hour.

Almost four hundred years earlier, the first voyagers to the New World had peered through their telescopes to search for new lands as they sailed westward. The excitement of visualizing a speck of land, magnified through the use of glass lenses, could not have been greater than what gripped Robert Koch when he gazed down his German microscope in his Berlin laboratory. Harnessing his new technique, he now was faced with a swath of chestnut-colored tissue, sparsely populated with eye-popping cerulean bacilli, centered within the caseous tubercles. [40]

Standing at the podium at the University of Berlin's Physiology Institute before the impromptu genius society, flanked by microscopes and slide preparations, test tubes and culture dishes, Koch announced that he had visualized the enemy. More significantly, he calmly proclaimed, he had grown TB in culture.

Instead of growing the finicky bacteria in broth culture, Koch used solidified cow or sheep serum, which had been heated and poured onto slanted tubes to increase the surface area for growing colonies. The tubes were then inoculated with small batches of the isolated TB samples, grown at 37° C (human body temperature) and monitored for colony formation. In turn, guinea pigs in his lab were inoculated with the TB cultures, and after ten to fourteen days, were sacrificed for lung tissue microscopic examination.

Koch was able to see the same bacilli in the guinea pig lungs, thus completing the circle of isolating an organism, culturing the bacteria outside an animal, infecting another animal with the germ and producing the same sickness, and finally observing the same organism under the microscope. These rules from Koch, the father of bacteriology, are the gold standard of bacterial investigation, and are simply referred to as "Koch's postulates." These insights started with Henle, but came to full flower with Koch.

In his careful, methodical, if not mundane, manner, Koch had delivered his magnum opus. He detailed how he identified the organism and how he cultured it. He concluded, "All of these facts taken together can lead to only one conclusion. That the bacilli which are present in the tuberculosis substances not only accompany the tuberculosis process but are the cause of it. In the bacilli we have, therefore, the actual infective cause of tuberculosis." [41]

In the history of science, medicine, philosophy, and mathematics, when great proofs are published or presented, there has been a tradition of concluding with the

Latin phrase *quod erat demonstrandum*, abbreviated *QED*, meaning "thus it has been demonstrated." On that Berlin evening in 1882, Koch delivered perhaps the greatest *QED* moment in the history of mankind. After he was finished, Koch quietly shook hands with some of his compatriots, but faced no challenges. His assistants, later to become famous in their own right, recalled, "I hold that evening to be the most important experience of my scientific life," and that Koch's presentation was "pure unadulterated gold."[42]

Koch was awarded the Nobel Prize in 1905 for his TB research. He suffered humiliation and enjoyed exaltation throughout his career, at times finding himself on the wrong side of history. His work on other diseases, such as cholera, helped save millions of lives, and it is a shame that his name is not as well-known as his two most important allies, whose names are immortalized with the monikers "Listerine" and "pasteurization."

Smooth, uniform, and rounded, little germs are microscopically monotonous and unthreatening. While a pus-filled and fetid discharge is revolting, up close, the little bodies swimming in the secretions are humdrum in comportment, not resembling dragons or sea monsters. Ticks, lice, tapeworms, and maggots are disgusting; all bacteria are tediously mundane. Mankind has perennially been terrified of bears, sharks, tigers, wolves, elephants, and especially humans, but the insipid world of bacteria have always (even to this day) led to more worldwide deaths every week than all the combined mammalian and predatory annihilations that have ever occurred. Dreary in appearance, but ruthlessly efficient in operation, bacteria ruled our human existence without any challenge whatsoever until Pasteur, Lister, Koch, and colleagues identified, nurtured, decoded, and short-circuited the kingdom of germs.

Practically, the exposé of germs led to an immediate decrease in the scope of epidemics. TB was curtailed within years through the simple steps of pasteurization of milk and the alteration of public behavior of coughing and spitting. Even before the introduction of antibiotics that were effective against TB, it was in decline. More important, the discovery of germs, the description of their life cycles, and the attenuation of their impact on our lives was the greatest unraveling of any disease mystery that society had ever contemplated. Fathoming germs meant that man could explain illness and, for the first time ever, made it worth an invalid's time to submit to a doctor.

Morgagni, Louis, Rokitansky, and Virchow had made death understandable, but disease could not be explained until germs could be discerned. And for this we have Semmelweis, Lister, Krebs, and Koch to thank.

In Wöllstein, in Koch's upstairs parlor, where he tended to patients (and could gaze out the back windows to his animal cages and horse barn), there is a curious demarcation in the pinewood floor. We know that Koch was so preoccupied with his bacterial research that he partitioned this room with a heavy curtain, so that he could treat his patients upstairs without having to take the only (and very public) stairway to his downstairs simple laboratory. In the middle of the floor is a slightly angled strip of wood running across the grain of the rest of the floorboards. This is the border between his clinical and research space. In this tiny little town, hours and hours from the closest academic bastion, a self-funded and eccentric young voyager laid the foundation for modern medicine.

Perhaps more profoundly, Lister laid the cornerstone for modernity. Lister was the first to propose a therapeutic intervention—and then to evaluate and alter his technique—and achieve a significant therapeutic benefit. (One could argue that anesthesia is the first effective medical ministration, but prior to Listerism, anesthesia did little to improve outcomes.) "It is striking that it is surgery, the least theoretical of the medical disciplines, that was the first to be transformed,"[43] and it was the scientific inclinations of Joseph Lister (later emulated by Langenbeck, Billroth, Frederick Treves, Edoardo Bassini, and William Stewart Halsted), that led the transformation. The new order of things, with German surgeons meticulously executing operations (antiseptically under anesthesia) was soon to replace the archaic practice of speed and dazzling dexterity.[44] It was the most salubrious insurrection ever staged, and it was based on Lister's hunch that Pasteur (and others) were right about germs.

The founders of the germ theory paved the way for antiseptic surgery, first facilitating lifesaving trauma and abdominal operations, and fifty years later (with the advent of antimicrobials), enabling elective surgery with incorporation of implants. The pioneering germ hunters, therefore, provided a fastidiously clean and antiseptic environment that made the implant revolution a possibility.

EIGHT

Antibiotics

Amputating a man's arm is a gut-wrenching and shocking act. Regardless of the clinical justification and no matter the years of practice, severing a limb from a body requires stubborn resolve and intense personal subordination. Perhaps some surgeons grow callous to cutting off a limb—I never have.

I am a surgical intern at Pennsylvania State University and all I want is a couple hours sleep. I figure, if I can lie down now, I'll instantly fall asleep, and will get enough shut-eye before 4:00 A.M. to last me through another day of grunt work. But on this winter night, just as my body twitches and shocks itself to slumber, my pager vibrates me to reality. Like all surgical interns, I am taking call while "in-house," staying in the hospital all night fielding phone calls from the Emergency Room, the hospital floor nurses, and outside patients.

In the darkness, I fumble for my little black Motorola pager on the nightstand next to my head. Checking the four-number code, I dazedly recognize "6550" as one of the extension numbers to the Medical Intermediate Care Unit. We don't get many calls to that number, and I hope that I have been paged incorrectly. Without turning on the lights, I prop up on my left elbow and punch the number on the institutional green AT&T office phone.

A nurse answers my call, informing me of an urgent surgical consult on a seventy-eight-year-old man with elbow pain. She explains that he had been admitted hours before with heart attack–like symptoms, but that all preliminary tests were ruling out an MI (myocardial infarction, or heart attack). Oftentimes an MI patient complains of crushing chest pain, with associated left arm or jaw pain; alert ER personnel hear these complaints and immediately begin testing the patient for a "cardiac event." Although all initial tests were ruling out an MI, the severity of his symptoms warranted a hospital admission. As the hours progressed, his arm pain worsened, and by 2:00 A.M., the medical team was getting nervous. The aged patient was developing blisters and "ecchymoses" (bruises) on his left arm. Instead of an MI, they are now considering some ominous issue with his musculoskeletal system.

I am a know-nothing surgical intern, just months removed from medical school graduation, but I agree to come evaluate the patient as a first-line responder for my surgical team. I sit up in bed, take a deep breath, and slide on my day-worn, slightly smelly socks. After fumbling for my shoes, my thoughts become more organized, and I'm already starting to generate a *"differential diagnosis," the list of possible causes behind this man's presentation. While trying not to wake my bunkmate and fellow intern, I slip out of the night call room and jog up the echoic stairway to the medical floor.*

Briskly walking down the darkened hallway, I arrived in the Medical Unit, which is a beehive of activity. Nurses and aides are darting around, and are oddly relieved to see me. Typically, floor nurses in an academic medical center rightfully have disdain for interns. They arrive every July with their new MD degree, but are as helpless as a newly licensed motorist trying to drive a stick-shift, uphill, for the first time. But these were medical nurses, adept at caring for cardiac patients, but greenhorns themselves when dealing with an odd musculoskeletal patient who would normally be a couple floors down on the orthopedic floor.

A young nurse points to the corner bed, where a seventy-eight-year-old gentleman restlessly lies in his hospital bed, his left arm propped up on pillows. Rapidly, I can see the bruises that the nurse was telling me about, and I can also see that his forearm is swollen. I ask, "Mr. Louis, does your arm hurt?"

This aged man is truly sick, and can only mumble a feeble, "yes." Growing concerned, I approach his bedside, and focus on his arm. There are dark, splotchy patches of bruises the color of grape jelly. I lean over the bed, inspecting the inside aspect of his elbow. There are several raised burgundy-colored blisters above his elbow, and I am starting to feel out of my league. What am I looking at?

I reach for his wrist to lift his arm, and instantly feel the crackle of air under the forearm skin that feels like squeezing a bag full of wet Rice Krispies. My stomach drops, and while I don't have much experience or judgment in the practice of surgery, I know this is gas gangrene, the byproduct of "flesh-eating" bacteria. There are classes of bacteria that are infamous in causing rapid infections that result in the death of the body's soft tissues, so-called "necrotizing fasciitis," with the occasional byproduct of "subcutaneous emphysema," or gas underneath the skin. The physical exam finding of subcutaneous emphysema is frightening, to say the least.

I gently place Mr. Louis's arm back on the pillows, knowing that I am seeing my first case of "nec fasc"—pronounced "neck fash," in common parlance. (This is how residency works—you can read all about subcutaneous emphysema and necrotizing fasciitis, but until you have someone's limb in your hands with crunchy air underneath the skin, you have not been properly initiated. Somehow the numbers work out. Relatively rare, every surgery resident has seen nec fasc.)

I turn to the nurse and say, "necrotizing fasciitis." All conversation stops and everyone freezes.

"Really?" she says.

"Yes. I'm going to call Dr. Moulton, my senior resident."

Connecting with Mark Moulton, I explain the details of the case. Getting to the point, he asks me, "Are we early enough to save his arm or will we have to amputate?" I confess to

Mark that I really don't know, that I don't have any experience. Mark tells me to get the patient rushed to the operating room immediately. We will try and save Mr. Louis's life, if not his arm.

A flurry of phone calls to the operating room and the anesthesia team achieves the impossible, and we are rushing to the OR within half an hour. Life is on the line. The rest of the orthopedic team has made its way to the hospital by 3:00 A.M., and my boss, Dr. Spence Reid, quickly concludes that an amputation is mandatory. In the pre-op holding area we get a portable X-ray that reveals air going all the way to the shoulder. Typical for necrotizing fasciitis, the bacteria are on a warlike march, leaving a plume of air in their wake, and before the bugs get to the chest, daring surgery must be performed. Not only do we need to amputate his entire arm, the collarbone and shoulder blade must also be removed, a so-called "forequarter" amputation (as opposed to hindquarter, or lower limb).

Even before transporting the patient to the OR suites we gave Mr. Louis a large dose of penicillin, but necrotizing fasciitis is notorious for not responding to antibiotics in the emergency setting. Penicillin helps, but surgical magnificence is demanded if the patient is to live another hour.

Once we urgently transfer the patient to the operating room and the anesthesia team intubates him, we rapidly position him on the surgical table. Racing to save his life, he is propped on his side and his entire left side and arm are swathed with greenish-blue surgical drapes. Dr. Reid works very quickly, making a dramatic, football shaped incision around the shoulder blade and chest. Under nonemergency situations, this dissection would likely take ninety minutes, but under the circumstances, the dissection is done at lightning speed, in barely a dozen minutes. The collar bone, the shoulder blade, the entire arm, and all the muscles attached to those bones are rapidly cut away. The nerves emanating from the neck and the large blood vessels emerging from the chest cavity must all be tied off and cut.

As a resident at the beginning of my training, I know I would kill this patient if I attempted to do the operation. I just don't have the skills yet. Dr. Reid is a superb surgeon, a master craftsman with unique understanding, adept hands, supernormal concentration and stamina, and most important right now, heroic courage. Moments like this will kindle all these attributes in me for the rest of my life, and Dr. Reid's greatest gift to me will be the gift of confidence, the ability to take on impossible shoulder and elbow cases in the future. Surgeons are criticized for arrogance and brashness; this critique is probably fair, but at this moment, a fearlessness nurtured from deep self-assurance is mandatory.

A surgeon can perceive if he has outflanked a flesh-eating bacterial infection—there is no crackly air in the layers of soft tissue that he is cutting through. A cocktail of life-supporting medicines continues to be injected into Mr. Louis's IVs as our team completes the final steps of the forequarter amputation. Cutting edge antibiotics, in addition to penicillin, are being pumped into his body even as the team races to detach the limb.

The moment of liberation of the putrefied appendage finally occurs, leaving a gaping wound over the rib cage. There is a simultaneous sense of triumph over the bacterial

horde and an acquiescence to the power of microorganisms as the limb is separated from the
thorax and dropped into a hazardous waste trash bag. Aggressive irrigation with antibiotic-
laden saline is performed, and a palpable optimism flickers to life in our operating room.
  Mr. Louis, although bizarrely disfigured with no arm and no shoulder, will live.

Mr. Louis's life was saved by surgery and by penicillin. I have posed the question many
times to friends and patients: How many years ago was the first dose of penicillin
given? In ancient times, or five hundred years ago, or during the Revolutionary War,
or after the First World War? Few people realize that the first clinical administration
of penicillin in a small English hospital was only seventy-five years ago.

The pioneering work of Pasteur, Lister, and Koch convinced scientists and physicians that germs were real. As Robert Koch microscopically elucidated their life cycles
and interactions with humans, the dark veil of ignorance regarding infectious diseases
was lifted. Semmelweis and Lister, among others, were able to show the advantages of
handwashing and sterilization, and it is not surprising that public health institutions
were created in the years after John Snow helped create epidemiology and Florence
Nightingale influenced hospital design. Although improved sanitation and cleanli-
ness dramatically decreased epidemics, there was still no answer for acute or chronic
infections in individual patients.

The advent of modern chemistry coincided with the triumph of germ theory
during the 1880s, in no small part because manufacturing dyes provided contrast and
color to an otherwise drab and blurry microscopic world. The bourgeoning German
industrial chemical companies began as dye manufacturers, only later turning to
fertilizers, perfumes, photography, and pharmaceuticals. Paul Ehrlich (1854–1915),
a Prussian-Jewish physician-scientist continued the proud German tradition of perfecting the art of histological staining, eventually gaining fame for differentiating the
component cells in peripheral blood.[1] A contemporary of Robert Koch, Ehrlich had
a breakthrough insight when he considered the chemical processes that were occurring during the staining of tissues and bacteria. There was a primitive understanding
that certain dyes had a special affinity for certain cells (and their constitutive parts);
further trial-and-error testing with dyes by the Danish physician Hans Christian
Gram (1853–1938) yielded the most important finding in the history of bacteriological
microscopic analysis—that bacteria could be grouped into two main classes of cells
that either stained purple ("Gram-positive") or red ("Gram-negative") in response to
a series of staining steps with crystal violet and safranin stains.

Paul Ehrlich was intrigued by why different dyes were attracted to particular species
of bacteria, but handicapped by primitive research tools, had no way of formulating
a scientific response. However, demonstrating the type of keen insight that geniuses
possess, Ehrlich skipped several steps ahead and wondered if the dye materials could
be manipulated not to just embellish a slide but to kill bacteria. If a staining material

could be identified that targets and binds with a particular class of bacteria, it made sense to the pioneering scientist that a dye could be used as a weapon.

Ehrlich traveled to London in 1907 to lecture to Britain's Royal Institute of Public Health, delivering a lecture for the ages. He dreamed that one day there could be a "targeted drug, one that would attack a disease-causing microbe without harming the host suffering from the disease."[2] Ehrlich conceived of chemical compounds that would serve as *magic bullets*, just decades after researchers had finally proven the germ theory. Barely fifty years removed from John Snow's revolutionary epidemiological research during the cholera outbreak of 1854, Ehrlich returned to the very London neighborhood that had been (literally) awash in diarrhea, stumping for magic bullets.

By the time Paul Ehrlich had traveled to London, he was already well on his way in the quest for the magic bullet. Modern chemistry was in full bloom, with Dmitri Mendeleev's periodic table coming into focus and a developing appreciation of how atoms bind together to form complex molecules. For an extremely insightful researcher like Ehrlich, the mysteries of simple chemical compounds were beginning to dissolve at the turn of the 20th century, and as one of the fathers of histological staining, it's not a surprise that he turned to azo dyes like methylene blue, congo red, and alizarin yellow in the search for a chemical breakthrough. Since the mid-1880s, Ehrlich had experimented with the azo dyes as potential therapeutic agents, and although he was inadvertently turning his patients' eyes and urine various colors of the rainbow, he and his lab partners were able to show a response to malaria.

Azo dyes—aniline derivatives like the mauveine discovered by William Perkin in 1856—are chemically stable and not very changeable; Ehrlich and his cohorts were hoping to find another substance that acted like a dye (showing a propensity to bind with certain bacteria), but was more chemically unstable and easier to manipulate in the lab. Ehrlich knew of a chemical compound named *atoxyl* that had been shown to kill trypanosomes, single-cell parasites that cause diseases like African sleeping sickness. He was intrigued by atoxyl, particularly once he realized that it was a chemically unstable arsenic-based molecule and not a true aniline dye.

And so the testing began. Ehrlich and his colleagues Alfred Bertheim and Sahachiro Hata began to chemically modify atoxyl in 1907, feverishly altering the composition of the molecule bit by bit. Different versions were further modified, and a numbering system was generated based upon these modifications. The eighteenth version of the fourth compound (number 418) was effective in curing sleeping sickness, but was causing blindness in some of Hata's lab animals and was therefore abandoned. By the summer of 1910, in what can only be described as crude experimental processes, Compound 606 had been created and tested. The sixth version of the sixth compound (606, arsphenamine) showed tremendous success in lab animals with various diseases, including syphilis.[3]

Syphilis likely was not present in Europe before explorers brought it back from the New World in 1495, and it raged for four hundred years across the continent with its slow-motion terror of blisters, aching testicles, sore throat, raised skin rash, and in its final stages, facial deformities and brain infections. With no effective treatment, mankind was defenseless against the corkscrew-shaped bacterium. Until Compound 606.

The German chemical company Hoechst AG, also located in the Frankfurt area, began marketing Compound 606 in 1910 as "Salvarsan." Through trial and error, Paul Ehrlich had created a molecule that was part stain, part poison. The dye portion of arsphenamine would bind to the surface of the syphilis bacterium, whereas the arsenate portion killed it. In so doing, he had developed the world's first synthetic chemotherapeutic agent. For good measure, Ehrlich coined the term "chemotherapy."

Salvarsan rapidly became the most prescribed medicine in the world, leading to hopes that it would have broad application among many different types of bacteria. Unfortunately, Salvarsan, and its improved version, Neosalvarsan, had extremely narrow efficacy across the microbial world. This, paired with its significant side effects, made it a qualified success. More significantly, the development of Salvarsan was a false lead, as all future antibiotics (after the sulfonamides) would be "natural" molecules gleaned from nature—from fungi or bacteria—and not synthetically created from dyes or other simple chemical molecules. When sophisticated chemical engineering is performed by pharmaceutical companies in the search for a new antibiotic, it is upon naturally occurring chemicals already being produced by living organisms.

World War I (1914–1918) introduced horrific methods of combat, and while there were the predictable medical advances achieved from the theater of war, there was a transitory disruption in the German pharmaceutical industrial machine. The German biochemical revolution was fueled by rigorous academic programs at decentralized universities, a cultural identification with industriousness, and the creation of durable funding that was the envy of Germany's European neighbors.[4] There was a grand consolidation among German chemical and dye businesses following the conflagration, setting in motion the powerful chemical, agricultural, and pharmaceutical manufacturing enterprises. Familiar names like Bayer, Agfa, BASF, and Hoechst combined together to form IG Farben in 1925, resulting in the largest chemical company in the world.[5] As will be seen, the German chemical corporations involvement in World War II was much more diabolical and vastly more damaging.

In the years leading up to World War II, the Teutonic drive for innovation in chemistry had led to great breakthroughs in fertilizer development, which even today accounts for half of the world's crop production.[6] Assembly-line manufacturing, pioneered by Henry Ford, was fundamental to the next wave of the Industrial Revolution in the early 20th century, but instead of making vehicles, the German research machine would use mass production organization to tackle scientific challenges with brute force. The testing of prospective chemical compounds was formalized on a grand scale, exposing huge numbers of potential drugs to various bacteria in what was described as an "endless combination game [utilizing] scientific mass labor."[7]

Paul Ehrlich, the father of histological staining, immunology (he was the first to grasp antibodies), and chemotherapy, died in 1915, just as World War I was exploding. Wartime disruptions and the vacuum left after his visionary leadership led to a lull in chemotherapy discovery. The formation of IG Farben in 1925 and the arrival of Gerhard Domagk (1895–1964) in 1927 to Bayer set the stage for a muscular approach in the quest for a true antibacterial medicine. "If Ehrlich had tested dozens of different recipes in order to find the antisyphilis treatment, Bayer would try hundreds. Or thousands."[8] In a foreshadowing of the petrochemical polymer industry, Bayer chemists began producing thousands of chemical compounds from coal tar, the thick liquid that is a by-product of the production of coke and coal gas from coal.

Domagk, as a pathologist and bacteriologist, had gained a specialized understanding of the microbial enemy (including being a wounded soldier in World War I), and was critical in constructing the experimental framework, having identified a particularly virulent strain of *Streptococci* (Gram-positive cocci which links in twisted chains). *Streptococcus*, the pathogen famous for throat infections, pneumonia, meningitis, and necrotizing fasciitis, was an ideal test bacterium, not only because it was common, but because it killed laboratory animals so terrifyingly efficiently. Domagk, like his famous German predecessor Robert Koch, intentionally infected laboratory white mice with his test bacteria. Thousands of diseased mice died over the first few years of the project, helplessly succumbing to *Strep* despite being injected with myriad coal-tar derivatives from the Bayer chemists.

Trudging along, as science demands, the scientists continued tinkering with the azo dyes, chemically modifying the compounds with the addition of chlorine atoms, then arsenic, then iodine. Years of failure and almost no hope demanded a resiliency that was perhaps battle born, but a breakthrough did finally occur in 1932, when the team began linking the azo dyes with a sulfa-based molecule. The protocol that he had practiced for years yielded a monotonous outcome: injecting live *Strep* cultures into the abdomen of a mouse would result in death within a day or two. But in late 1932, outside Düsseldorf, Germany, twelve mice were administered a new drug—an azo dye amalgamated with sulfanilamide—shortly after being injected with the deadly bacteria. Concurrently, fourteen mice were injected with the same bacteria but were not given any medicine. All fourteen of these control animals were dead within days, while all twelve that had received the new compound, KL-730, had lived. The Bayer scientists had stubbornly forged ahead as the carcasses of rodents piled up, but in 1932, the world's first antibacterial magic bullet had finally been crafted.

Bayer knew that their new medicine, KL-730, which they would name "Prontosil," was effective against bacteria because of the unique marriage between the azo dye and sulfanilamide. Except that it wasn't. What the Germans had never performed was an isolated test of sulfanilamide alone. A group of French scientists at the Institut Pasteur in Paris repeated an experiment with various sulfanilamides on a group of forty mice, including a treatment group with sulfanilamide alone and no azo dye.

After a few days, the Parisian team evaluated the response among the test animals. Almost all of the mice died who were treated with newer azo-sulfanilamide combinations, but all of the mice lived who were treated with Prontosil, Rubiazol, *and sulfanilamide alone.* The Bayer scientists had assiduously labored to protect their patent rights over Prontosil, sure that it represented a bonanza, but they had never considered that sulfanilamide alone might be the subjugator. At about the same time that the Institut Pasteur scientists made their discovery, the Bayer group was unearthing the same sobering fact. While it was a tremendous moment for mankind, it was a financial catastrophe for Bayer; the sulfanilamide molecule had been discovered (and patented) in 1908 by Viennese chemist Paul Gelmo, and was now in the public domain. The financial goldmine had evaporated before their eyes.

Bayer did profit from sulfanilamide. They marketed it around the world as Prontosil, even after realizing that sulfanilamide alone was the effective agent, without the need for the azo dye. (It also explains why Prontosil was only effective in vivo and not in vitro. In a test tube full of bacteria, Prontosil posed no risk. Only animals have the enzyme that separates the dye from sulfanilamide. If testing had only occurred in test tubes, and not animals, Prontosil would have appeared as a failure, and it was this and other drugs that educated the early pharmaceutical manufacturers that "pro-drugs" were genuine. At times, pro-drugs are ideal—a pro-drug is intentionally manufactured so it can survive digestion, turning into the active metabolite once in the bloodstream.)

Prontosil and other forms of sulfanilamide hit the world market in 1935, immediately making an impact. "Virtually overnight, mortality from childbed fever [*Strep pyogenes*] fell from 20 to 30% to 4.7%."[9] Physicians across the United States and Europe embraced the new drug, but the American public became intimately acquainted with the new sulfa drug in 1936 when Franklin Delano Roosevelt Jr., while a student at Harvard College, contracted a life-threatening *streptococcal* throat infection. Physicians in Boston administered the new magic bullet, saving his life, and in the process, helped propel America into the modern age. The *New York Times* trumpeted the news on its front cover, helping ignite a "sulfa craze" across the country, even leading to patients asking their physicians for the new wonder drug by name (a first). Even at the outset of the antibiotic revolution, overprescribing was a temptation.

The European quest for synthetic chemotherapeutic molecules was in full launch mode as the world tilted toward a second Great War. Chemists were obsessed with a haphazard survey of chemicals, believing that the new man-made particles could outsmart the bacterial enemy. While the modern pharmaceutical industry has created, de novo, chemicals that lower blood pressure, increase blood flow, and alter cholesterol levels, the source of antibiotics would be from mother nature, not from the minds of scientists. Unbeknownst to the chemists, several years before sulfanilamide was given to a human, an accidental discovery in London had already opened the vistas of future medical care.

Alexander Fleming was a young Scottish physician working at St. Mary's Hospital in London, and although he was trained as a physician and surgeon, his talents in laboratory research had led him to an eventual career as a bacteriologist. Small and slight, Fleming had joined the inoculation department at St. Mary's in 1906, soon turning his attention to Paul Ehrlich's Salvarsan.

Bacterial researchers have always followed the pioneering example of Robert Koch, studying the lives and sensitivities of microbes by growing colonies of bacteria in Petri dishes in a nurturing environment. Fleming and his colleagues focused on important pathologic bacteria like *staphylococcus* and *streptococcus*, culturing the bacteria and evaluating the conditions that altered colony formation. In 1922, Fleming and a lab assistant were cleaning up Petri dishes that had been seeded with bacterial colonies when they noticed an odd pattern. Typically, in a Petri dish of bacterial colonies, there is widespread, even growth of bacteria across the dish; instead of seeing such growth, Fleming noticed that there were blank areas of no bacterial colonies. In a victory for everyone who has suffered from the common cold and a drippy nose, Fleming recalled that nasal mucous from his own nose had dripped onto the culture dish days earlier, and he rapidly surmised that his own nasal drippings had somehow hindered the growth of bacteria. The shy and reticent researcher concluded that there must be a substance in the nasal discharge that had inhibitory powers, naming it *lysozyme*. For the first time in world history, a purely organic substance had been characterized as having antibacterial properties.

Lysozyme became a fascination for Fleming, albeit a research dead-end. In time, researchers were able to show how lysozymes function to weaken the cell walls of bacteria, but more important, the recognition of a molecule that inhibited, or killed, microbes prepared Fleming's mind for his revolutionary observation in 1928.

As summer turned to fall in 1928, Alexander Fleming returned to London from a holiday by the sea. When he arrived at his petite laboratory at St. Mary's Hospital (preserved today as a memorial to the man and his momentous discovery that September 3), a jumbled stack of Petri dishes was on a tabletop, including a dish that had fallen off its perch and lost its lid. The story goes that he glanced at the Petri dish and quickly did a double take—dozens of round spots of *staphylococci* carpeted the dish but their spread was limited by a large island of white mold on one side of the dish. Recognizing a pattern similar to what he had seen five years earlier, the blotch of mold had a surrounding beltway, a demilitarized zone of sorts, where there were no bacterial colonies and no fungus.

Fleming muttered softly to himself, "That's odd."

For thousands of years, humans had unwittingly harnessed mold to make wine and beer and bacteria to make cheese. Fewer than one hundred years before Fleming's discovery, Louis Pasteur had solved the riddle of fermentation, and less than half a century before, Koch had demonstrated that bacteria were real. Fleming had already concluded five years earlier that lysozymes from human fluids retained antibacterial properties, and now, perched in his little lab above Praed Street, began conceptualizing that the mold itself was making a substance that was deadly to the *staphylococcus*.

The name of the mold? *Penicillium*. (Read that carefully. It doesn't say "Penicillin.")

The *Penicillium* mold was likely a contaminate in the building or from the air from an open window. There has been much conjecture about the source of the mold—was it from a nearby lab, was its presence a hallmark of sloppiness of research, did it taint the bacterial culture because Fleming's assistant was slovenly?—but in the final analysis, *Penicillium* is a common mold that has been making its own special chemical as a defense, likely for millions of years. How it got into that lab is not important, but the fact that Fleming paused to consider its actions *is* significant.

Correctly ascertaining that *Penicillium* was producing a substance that inhibited bacterial encroachment, Fleming and his assistant, Stuart Craddock, (initially) became obsessed with farming *Penicillium* and harvesting the resultant "mold juice." Fleming then tested this concentrate on other bacterial samples and found that it was effective against *staphylococci* and *streptococci*, finally settling on the name "penicillin" as the name of the substance that would make him world famous. In March 1929, Fleming published an article titled, "On the Antibacterial Action of Cultures of a Penicillium, with Special Reference to Their Use in the Isolation of B. *Influenzae*." This predates, by several years, the German discovery of sulfanilamides, but Fleming and his team lost out on the designation as providers of the first antibiotic because they could never adequately cultivate the finicky mold in sufficient quantities to make it clinically significant.

In fact, *Penicillium* was so persnickety that Fleming gave up. It is confusing today to reconcile Fleming's abdication on mastering the development of (arguably) the most significant drug ever discovered, but the lack of sophisticated research tools, lab space, manpower, and most important, intense drive to corral the fungus meant that it would be up to another team, more than a decade later, to harness the power of *Penicillium*. Amazingly, Alexander Fleming walked away from *Penicillium* and never published on it again.

Eight years passed after Alexander Fleming's publication with no success, by Fleming or any other researcher, in cultivating *Penicillium* and producing penicillin. While several scientists had been inspired by Fleming's 1929 article, none could overcome the same technical challenges in understanding its actions, including George Dreyer of the Oxford University Dunn School of Pathology. "The Dunn" had been founded in 1922 with a £100,000 gift from Sir William Dunn, a Scottish merchant banker and politician who had made a fortune in South Africa. The institution would become world famous for disease process and bacteriological research, and by the time the building was completed in 1927, an impressive roster of resourceful minds was being assembled.

Two industrious, ingenious, fatherless, and indomitable researchers arrived at Oxford in the mid-1930s, one from Australia and the other from Germany. Together, they would tame *Penicillium*, perfect the production of penicillin, and conspire with researchers around the globe to make the breakthrough medicine available just when the world was bent on collapse.

Howard Florey initially voyaged to Oxford as a Rhodes Scholar having just graduated from medical school in Adelaide. His father had died a few years earlier, and the ambitious young Australian made the first of many career moves when he began a three-year study program in Pathology. Florey was awarded numerous scholarships during his academic matriculation; in addition to the Rhodes scholarship he was awarded the Rockefeller Foundation fellowship, which led to intermittent research trips to New York, Chicago, and Philadelphia during his graduate work. Brief stints in Copenhagen, Vienna, and Madrid, combined with an eventual doctorate from Cambridge in 1927, provided him with an unmatched educational background. In 1935, he was named the second director of the Dunn School of Pathology, turning his attention to bacterial gut impermeability and investigating whether or not lysozymes were involved in the protection of the gastrointestinal tract against bacteria. Florey was honing in on an area of expertise, having demonstrated prodigious drive, intelligence, and leadership skills; all he needed was a comrade who had similar soaring ambition and talents.

Ernst Chain was born in Berlin in 1906 to Russian-Jewish immigrant parents. Like Florey, Chain's father died while he was still in school (in Chain's case, when he was thirteen). Similar to Florey's athletic achievements (he excelled in tennis, cricket, and football), Chain was a piano virtuoso who gave concerts on several continents. Chain graduated from Friedrich Wilhelm University (now Humboldt University of Berlin) and the Institute of Pathology at Berlin's Charité Hospital in 1930. In photos, Chain bears a striking resemblance to Albert Einstein, and the young pathologist, looking in every way like a true genius (he was), began work in the chemical pathology lab at University College Hospital in April 1930. A few years later Chain landed a research position in Cambridge, and after a couple years there, was offered the job of biochemist at the Dunn School, working under Howard Florey.

Howard Florey had succeeded in hiring a world-class-trained (read: German chemist) scientist who could help him investigate the biological aspects of infection and immunity. He could not have found a better colleague—Chain later wrote that his "principal motivating principle . . . was always to look for an interesting biological phenomenon which could be explained on a chemical or biochemical basis, and attempting to isolate the active substances responsible for the phenomenon and/or studying their mode of action."[10]

Those who work in a research laboratory understand that research meetings, usually held weekly, are the lifeblood of the investigative effort. During that meeting, the lab director will ask for updates on specific experiments, and will invite commentary from various members of the team as to the meaning of the results. Unanticipated results are a major area of focus, because they represent either possible failure or potential new avenues of examination. Another occasional agenda item in the weekly meeting is the consideration of a completely new area of investigation, usually based upon a newly published article or podium presentation. A fresh research prospect is scintillating to a lab that craves a breakthrough, and sometimes the best spark for a new idea is to dig up old research publications and dust off an inadequately explored concept.

The stories about the penicillin pioneers seem a bit apocryphal at times, but in a well-remembered afternoon tea discussion among lab workers at the Dunn, Florey and Chain discussed the dead-end paper of Fleming from 1927. While no research team had achieved success in investigating the byproducts of *Penicillium*, it had been contemplated by Florey's predecessor, who had frozen away samples of the *Penicillium* and other microorganisms as potential sources of antibacterial substances. So in 1937, one year after FDR Jr. had received his lifesaving sulfa medicine, Florey and Chain began the nearly impossible task of efficiently growing *Penicillium* and producing penicillin. The gauntlet was laid down to the team: while it must be impossible to achieve, an inspired effort was demanded if they hoped to perceive the mechanism of defense of the fluffy-white mold. Impromptu or not, it was a lab meeting for the ages.

Norman Heatley was a contemporary of Ernst Chains in Cambridge, and although he was a PhD scientist, his exceptional skill was in building laboratory equipment out of random parts and castoffs, a second coming of James Hooke. With the lab budget so severely limited ("to call the Dunn experimental program impoverished is to flatter"[11]), Heatley was essential. No one in the world knew how to successfully grow *Penicillium,* and it was going to take creativity, stubbornness, keen insight, and luck. Heatley—humble, elegant, tall, and slender—did not waste time in deciphering the ideal conditions for *Penicillium* cultivation.

The life cycle of *Penicillium* came into clear view in 1939. The *Penicillium* culture would grow a thin white carpet over the agar, and as it matured, the branchlike mycelia would grow and generate penicillin-rich droplets that yellowed as they dried. These droplets could be harvested with a pipette, but harvesting too early limited the yield; waiting too long oversaturated the fungus, and further growth was squelched.

The mold grew adequately on agar, but precious little "mold juice" was produced without extra nutritive substances. Heatley turned to different growing containers and altered the temperature. Fertilization with nitrates, salts, sugars, glycerol, and meat extracts, combined with enriching the air with oxygen and CO_2, was performed. Brewer's yeast was added, and when one reads of Heatley's maneuverings you are left questioning whether he felt more like a chef, horticulturist, brewer, or scientist.

Adding urgency to the effort was Hitler's invasion of Poland on September 1, 1939. Chain could no longer travel back to Germany, and, unable to rescue his mother and sister, both perished in Nazi concentration camps. Great Britain and France declared war on Germany within days of the invasion, further heightening the urgency of the penicillin production and testing. Refinement of the growing process and the expansion of penicillin production continued into 1940, but no testing of the finished product had been performed.

On March 19, 1940, the first suitable batch of penicillin was finally processed and tested for stability. Chain, as the expert chemist, set to work on determining what type of molecule penicillin was. Primitive testing conditions notwithstanding, to Chain's great surprise, the molecule was not a protein, but what it *was* was not immediately obvious. The first step in clinical analysis was to inject two mice in the abdomen with the entire amount of collected penicillin. To the team's great relief, the mice tolerated the injections without incident and more amazingly, excreted it in their urine, unaltered.

Production of penicillin continued, and within two months, on May 25, 1940, an experiment with groups of infected mice was carried out at Oxford's Dunn School. Eight mice were infected with *streptococcus*, with four treated with a series of penicillin injections and four untreated as controls. By the next morning, all four of the untreated mice were dead, while all four of the mice who received penicillin were alive and well. The magic bullet had been derived from nature and none too soon. The following day, the evacuation of Dunkirk began, and it didn't take too much imagination to consider how a war effort could be facilitated with an antibiotic that, for the first time, was widely tolerated and highly effective.

Penicillin production would continue to be a great logistical challenge for the Oxford team, particularly considering the lack of material support the English enjoyed as the Nazi noose tightened around the British Isles. By the onset of 1941, the primitive apparatus at the Dunn School had ramped up creation of the antibiotic to levels sufficient to test on the first human guinea pigs.

In the fall of 1940, an Oxford policeman, Albert Alexander, scratched his face with a thorn from one of his rose bushes. Simple cleansing of the wound proved useless, and a secondary infection with Gram-positive bacteria developed in his face and scalp. As the English winter, with low cast gray clouds and short days of light, dragged on, Albert's infection spread to his torso, arms, lungs, and left eye. Treatment with sulfa medicines was ineffective. Abscesses with oozing pus had cropped up all over his body, and surgery to remove his left eye was mandated. After months of suffering, and with death imminent, Mr. Alexander became the first person in the world to receive penicillin for an infection on February 12, 1941.

The intravenous injection of penicillin was started in the morning, with doses given every three hours. By the next day, the patient's face was no longer swollen, his fever had normalized. The gush of pus almost immediately had slowed down, leaving everyone ecstatic and the policeman able to eat. It must have seemed like a miracle, but the sobering reality of their production failings tempered the sense of triumph, particularly when a second patient, a fifteen-year-old boy named Arthur Jones, contracted a life-threatening infection after a hip operation. Alexander had been given a five-day course of penicillin, essentially exhausting the stockpile held by Florey and Chain; another ten days passed with his condition remaining stable.

Both patients had received injections around the clock; dire shortages of the medicine mandated that Alexander's urine be collected and reprocessed. A bicycle brigade between the Radcliffe Infirmary (now the site of the Radcliffe Humanities, a teaching space that houses faculty offices, a library, and classrooms, located on Woodstock Road) and the Dunn laboratory maintained the lifeline for the first recipients. Arthur Jones was given penicillin recovered from Alexander's urine and the newly harvested penicillin from Heatley's manufacturing contraptions.[12] After a month's struggle, Albert Alexander succumbed to his ancient enemy, but not without demonstrating a profound response to penicillin. On the other hand, young Arthur Jones lived.

Florey and Chain rightly concluded that their little molecule might be a stupendous breakthrough; the type of discovery that earns fame and trips to Stockholm. But a more pressing demand was to improve upon large-scale manufacturing. The British Commonwealth was incapable of meeting the demand; Germany, Japan, and Italy were enemies.

Alternatively, the United States was rapidly becoming the world's lone superpower, even before World War II. Manufacturing giants in America had transformed the relatively young country into a GDP colossus, and while the United States was relatively inexperienced in chemistry and science (compared to Germany), "the sophistication and productivity of America's agricultural sector was like nothing else in the world."[13] So, in June 1941, Florey and Heatley flew from England to Lisbon, Portugal, and after three days' layover, boarded a Pan American Boeing *Dixie Clipper* for the transatlantic flight to LaGuardia Airport's Marine Air Terminal, landing on July 2.

The voyage to the United States, in retrospect, was an out-and-out triumph. The young pharmaceutical companies in the greater New York City area, like Pfizer, Squibb, and Merck, were eager to meet with Florey and Heatley, having read the August 1940 publication, "Penicillin as a Chemotherapeutic Agent," in the *Lancet*. But the greatest collaborative relationship would be with the scientists at the United States Department of Agriculture (USDA) research labs.

The USDA labs were (and still are) tasked with improving agricultural production while ensuring the safety and wholesomeness of crops, meat, poultry, and eggs. The Northern Lab of the USDA is based in Peoria, Illinois, and prior to the arrival of Florey and Heatley, had received dozens of samples of *Penicillium* mold from around the world. Although the Oxford team was an assemblage of masterminds, none were mycologists (fungus scientists). The challenge of identifying the most potent strain of *Penicillium* and the most efficient means of production of penicillin was entirely in the wheelhouse of the Peoria USDA lab, and in a matter of months, production of penicillin had improved one thousand fold. After rounds of testing, a strain of *Penicillium* was isolated that became the "ancestral source for virtually all of the world's penicillin,"[14] originating from a cantaloupe from a local Peoria market.[15]

The USDA lab succeeded in discovering an ideal strain of *Penicillium* and also in improving fermentation. They prevailed over the "better seed, better soil, better cultivation and harvesting"[16] contest—and it would fall upon the pharmaceutical companies in the United States and England to utilize those techniques to meet the demand. To entice drug companies to participate in the penicillin production challenge, the US government developed an unprecedented system of financial support and patent protection for the burgeoning corporations that established firm foundation to their explosive growth in the immediate postwar period. The wartime Office of Scientific Research and Development (OSRD) and Committee on Medical Research (CMR) undertook a comprehensive program to confront the medical problems of the war, in essence, weaponizing American science against the Axis powers. In the 1950s, a dizzying confluence of government funding, research sophistication, new hospital construction, and surgical know-how launched the modern medicine revolution; all of these developments can, in part, be blamed on penicillin's mandated industrial cultivation.

The American production of penicillin expanded exponentially from 1942 to 1945, but oddly, German development of antibiotics was virtually nonexistent. Decades before, Lister's carbolic acid treatment of wounds (and the German adoption of that method) had changed the balance of power in the Franco-Prussian War. German soldiers survived their battle wounds; French soldiers didn't. However, as war raged on during World War II, tens of thousands of German soldiers died from septic wounds, while American manufacturing ramped up penicillin production in preparation for D-Day.

Why did the Germans spend so little time and money on antibiotic development? Were they not the greatest chemists in the world?

Part of the answer lies in the need for fuel. Outside of Romania, there wasn't a decent-sized oil field between the Atlantic and the Urals, mandating that the Nazi state exhaust their scientific resources in the development of synthetic oil and synthetic rubber manufacturing. The Germans clung to a partially effective class of drugs, the sulfa-based antibacterials, and spent the rest of their assets and energies on propping up their war machine.

Another major reason for the Nazi scientific failure was the reversal of the educational system that made it the envy of the world in the first place. The scientific autonomy that had been earned over the course of decades suddenly vanished under

Nazi control, while "American scientists, universities, and the medical profession . . . performed under minimal control primarily in independent institutions, rather than in government laboratories."[17] In addition, the loss of so many gifted Jewish scientists, either through murder or defection, weakened the talent pool of the formerly proud German institutions.

They have never recovered their world leadership in chemistry and biology.

On December 10, 1945, with Europe in tatters after the conclusion of World War II, Alexander Fleming, Ernst Chain, and Howard Florey were awarded the Nobel Prize in Physiology or Medicine for their discovery and development of penicillin. Although sulfanilamide was antibacterial, it was not a molecule made by an organism. Sulfa drugs are therefore not "antibiotics," as coined by scientist Selman Waksman, which he defined as "a chemical substance produced by microorganisms."[18] One year before the Nobel ceremony in Stockholm, a tuberculosis patient at the Mineral Springs Sanatorium (near the Rochester, Minnesota, site of the Mayo Clinic), received the first dose of streptomycin, forever changing the treatment of tuberculosis, the development of antibiotics, and the world.

Streptomycin was discovered by Selman Waksman and Albert Schatz, soil scientists who specialized in the study of *actinomycetes*, a sub-order of bacteria that make the soil their home and have mold-like branching filaments. Additionally, *actinomycetes* were presumed to secrete antibiotic-like molecules, due to their ability to fight off other bacteria in their loamy world. Waksman and his colleagues at Rutgers University had spent the 1920s and 1930s collecting soil samples and testing the thousands of bacteria that were dredged up. A teaspoon of soil might contain billions of bacteria—competing for scarce resources and evolving molecular weapons to defend themselves from other bacteria and various members of the plant and animal kingdoms.

The first compounds isolated from soil-borne bacteria, in 1939 by French-American René Dubos (who later won the 1969 Pulitzer Prize for his book *So Human an Animal*), were effective against other bacteria, but were also toxic to mammalian cells. It would become obvious (in time) that the most effective antibiotics would target structures and machinery peculiar to bacteria, sparing animal tissues. Although the 1939 substances were clinical failures, they did inspire the microbe hunters to continue the quest. The overwhelming challenge seemed beyond scale: investigate thousands and thousands of bacterial species and somewhat arbitrarily test for antibacterial properties.

Despite the daunting task, Waksman and his followers developed a protocol for isolating a bacterium that produced a minimally toxic, yet clinically effective drug. Years later, he said, "We isolated one hundred thousand strains of *streptomycetes* (as *actinomycetes* were then known). Ten thousand were active on agar media, one thousand were active in broth culture, one hundred were active in animals, ten had activity against experimental TB, and one turned out to produce streptomycin."[19] While these numbers were rough approximations, they do capture the concentric circles of possible antidotes to disease; more intriguingly, although Waksman is the genius who pioneered antibiotic research (and rightly earned the Nobel Prize in 1952), the quote is very likely inaccurate, as it seems that Schatz himself carried out the critical research on streptomycin in isolation, from June to October 1943.

The Merck-funded research at Rutgers came to the attention of Mayo Clinic researchers William Feldman and Corwin Hinshaw. The Mayo Clinic had transformed from a father-and-sons practice in the tiny town of Rochester in the 1880s into one of the world's greatest research institutions. This was accomplished by embracing Listerism, implementing modern cellular pathology, nurturing collaboration among scientists and physicians, and the unselfish reorganization into a not-for-profit charity.

At the Mayo Clinic, William Feldman was a world-class veterinary pathologist and Corwin Hinshaw was a physician with an interest in bacteriology. Feldman and Hinshaw were obsessed with lung disease and in particular tuberculosis, the deadliest infectious disease in history. TB had killed one-seventh of all human beings—roughly fifteen billion people. The cure that had eluded Robert Koch became the obsession for Feldman and Hinshaw. The two Mayo researchers communicated with Waksman after reading his initial 1941 streptothricin paper[20] (which proved to be an experimental disaster after it was realized that streptothricin was dangerously toxic to kidneys), hoping to collaborate on the testing of any future antibiotic discoveries.

Albert Schatz's discovery of streptomycin is the story of stubborn dedication and personal subordination. While working in seclusion in a basement lab in Waksman's research building at Rutgers, scouring soil samples for a bacterium that could defeat TB—specifically against the most dangerous strain that his Mayo colleagues could provide—Schatz isolated two variants of *Streptomycetes griseus*. One sample was from a heavily manured field soil, and the other was swabbed from the throat of a chicken. Both samples of *Streptomycetes griseus* were antagonistic to TB in vitro, but in vivo testing would clarify if streptomycin was an effective *and* safe antibiotic.

Feldman and Hinshaw were among the first to receive an advance copy of Schatz's and Waksman's famous 1944 paper trumpeting the arrival of streptomycin,[21] and by April 1944, the two Mayo researchers began testing streptomycin in guinea pigs infected with a variety of diseases: bubonic plague, tularemia, shigellosis, and TB. By late June 1944 it was obvious that streptomycin was a miracle drug; it was curing every guinea pig of every disease, *including TB*. Additional testing was performed in the following months, and by the fall of 1944, Hinshaw was prepared to administer the first dose of streptomycin to a human. On November 15, 1944, Patricia Thomas became the first patient to receive the wonder drug. Severely infected with TB, and with no hope of survival, Patricia received five courses of streptomycin over the next five months, with dosages based upon a patchwork of early science and guesswork. Not only did Patricia Thomas live, she married and had three children, living another twenty-two years.

To really determine if streptomycin worked as well as was initially believed, a groundbreaking analysis was needed. While there had been simple trials comparing diets and primitive drugs (dating all the way back to the story of Daniel in the Hebrew Bible), including the important 1793 experiment of Scottish surgeon James Lind, in which a controlled trial of citrus fruit administration was shown to be effective in the prevention of scurvy, a true randomized controlled trial had never been performed prior to 1948.

"Alternate allocation" trials, in which every other patient is given an alternating experimental remedy, are prone to error because clinicians cannot undo selection bias when assigning patients to treatment arms, no matter how stringent the assignment of patients is. The British epidemiologist-statistician Austin Bradford Hill realized the shortcomings of previous experimental designs, concluding that the only sound way of evaluating a drug would be to blind *both* the clinicians and the patients. A trial for streptomycin was designed in early 1947 in which a triple-blinded design (patients, treating clinicians, and evaluators) would all be ignorant to whether patients received the actual antibiotic or placebo medicine.

With the war having just ended and British funds at paltry levels, there simply weren't the resources to treat a large number of patients. In fact, with supplies of streptomycin almost nonexistent and without much research financing, a randomized trial of patients getting no medicine was not only scientifically intriguing, it was necessary. Hill later wrote, ". . . in this situation it would not be immoral to do a trial—it would be immoral not to, since the opportunity would never come again."[22] The world's first randomized controlled trial, intentional but fortuitous, was concluded six months later, and the results were undeniable. Of the fifty-five patients who received streptomycin, only four had died (and twenty-eight improved); in the control group of fifty-two patients who received only placebo drug, fourteen had died. (A follow-up study demonstrated a reversal of the trend, and researchers would later conclude that resistance was building to streptomycin. Later studies showed improved outcomes if aspirin was concomitantly given, thus bolstering the case for streptomycin.)

Streptomycin became an immense success story but not without its controversies, including who should receive the credit for its development. The most important wrinkle in the story of streptomycin is that it paved the way for future chemotherapeutic development—both for antibiotics and anticancer medicines. Most antibiotics are derived from bacteria from the soil (not industrial dyes and chemicals), and increasingly, from bizarre places in the world, including the ocean depths and from the air. But scientists learned that grinding hard work and a little good luck achieved something that physicians and scientists would have thought impossible a mere fifty years earlier: the control, if not cure, of TB, and the ability to address almost any infection that may arise . . . at least until drug resistance and bacterial evolution outfox modern intellectuals. Cockroaches have nothing on bacteria.

The ability to identify, stain, culture, and test bacteria helped lead to the formation of a new industrial behemoth of pharmaceuticals as we know them today. It is stunning to realize that "penicillin, streptomycin, the versions of tetracycline, chloramphenicol, and erythromycin had all been introduced between 1941 and 1948,"[23] simply by working the soil. I've asked dozens of patients how drug companies bring new antibiotics to the market, and most draw a blank expression and say something like, "Don't they design them in their corporate drug offices?" The fact is that pharmaceutical scientists rely on billions of years of evolution among the tiniest inhabitants of our world, deciphering which molecules have novel methods of defense and confrontation, and utilize these newcomers in the battle against our attackers.

Given the stakes, both financial and of legacy, one would think that hundreds of drug companies would have discovered and modified thousands of antibiotics over the last seventy-five years, but from "1938 to 2013, only 155 antibacterial compounds received FDA approval. Because of resistance, toxicity, and replacement by a newer-generation derivative, only ninety-six antibiotics remain available today."[24] The partnership among microbiologists, chemists, statisticians, physicians, and businessmen has yielded a relatively limited palette of weapons to prevent and fight infections, but if modern man is incompletely garrisoned, we can glory in the fact that, over the last several generations, mankind is no longer vulnerable to the caprices of microbes. The antibiotic revolution, building on the breakthroughs of the understanding of the organ and cellular basis of disease, and the founding of bacteriology meant for the first time *ever* that it was worth going to a doctor when you were sick.

Pathetic treatments with Dr. Rush's heavy-metal laxatives (so-called "thunderbolts"), snake oil patent medicines, arsenic poisoning, toxic industrial solvents, noxious animal feces, and lethal plant material were in their twilight by mid-century. (Though never gone—witness the never-say-die cottage industry of home remedies, alternative medicine "experts," and television infomercials for cures the "medical establishment doesn't want you to know about.") As flimsy medical interventions crumbled under their own weight, the reputation of doctors began to grow, and as Paul Starr has so thoroughly described, the "social transformation of American medicine" meant that Americans were able to couple their new postwar prosperity with a renewed interest in health, with a shift from the (necessary) preoccupation with infectious diseases to a fixation on the betterment of chronic illnesses, like cancer, heart disease, neurosis, and arthritis.[25]

Infant mortality plummeted, life expectancy doubled, dread diseases were alleviated, and, on occasion, cancer was able to be cured in the decade of the possible in the 1950s. Perhaps the antibiotic revolution's greatest contribution in the transformation of the philosophical outlook of Westerners was not a loss of fear of infection but an openness by physicians (and their patients) that antimicrobial medicines were making it safe to implant foreign materials into human beings. "In 1950, about 230,000 physicians were practicing in the United States, and the overwhelming majority had left medical school well before the first antibiotics appeared."[26] Yet, it was those physicians who pioneered the use of implants as their scientific partners simultaneously innovated implant materials, like alloys, plastics, and transistors.

A certain clairvoyance was shared among the healers and dreamers—an idea that started four hundred years ago—an inclination that a line and race of inventions could be synthesized for implantation into the fabric of the human body, under the auspices of antibiotics.

NINE

Anesthesia

"The medical profession can be the greatest factor for good in America. The greatest asset of a nation is the health of its people . . . the real job of the medical profession is the extension of knowledge of what the medicine of today is doing and can do in the future, and this must be done by collective effort."

—William James Mayo[1]

"Suffering so great as I underwent cannot be expressed in words . . . but the blank whirlwind of emotion, the horror of great darkness, and the sense of desertion by God and man, which swept through my mind, and overwhelmed my heart, I can never forget."

—J. Ashhurst[2]

How did you get here?

Why are you here?

These existential questions mirror the first question we ask a new acquaintance, "Where are you from?" We want to understand our origins and the meaning of our existence. And the search for meaning and the reasons for our existence are never more imperative than when we are suffering.

Even before scientists and physicians understood the intricacies of the human body and the concepts of disease and healing, there was a necessary preoccupation with the alleviation of pain. If medicine had only evolved to eliminate pain, it would be a triumph. Of course, the advancement of medicine and surgery has been much more than the simple eradication of pain—it is the reversal of degradation, the conquering of microbial infection, the management of trauma, the pharmacological alteration of systemic disease, the halting of cancer, the reconstruction of damaged body parts, and the enhancement of biological tissues. But, in the beginning, the instinctual desperation to lessen pain (even when there was no understanding of the human machine) drove early civilized man to the heavens—and to the earth.

Feeling better, experiencing euphoria, or sensing nothing at all became possible with empirical experimentation with botanicals, starting in ancient times. Comprehensive control of wakefulness, or even the ability to induce profound sleep was the Gordian knot of surgery.

The cross-reactivity among plants and animals in our chemical world is startling. Why does ethanol, produced via plant fermentation, have such powerful effects in our minds? How has Mother Nature dictated that the poisonous chemicals in spiders, snakes, and flying insects have their evil potentialities in our bodies? What is the explanation for the efficacy of the simple chemical element, lithium, in the treatment of bipolar disorder and depression? And why do some plants fabricate chemical compounds that have no internal use and seem to have been tapped by the gods to make wondrous substances that only work on the minds and hearts of men?

The foxglove plant, an ornamental plant with vividly colored, elongated bell-shaped flowers, produces *digitalis*, a molecule that is powerfully used to treat cardiac arrhythmias (like atrial fibrillation) and congestive heart failure. In the foxglove, the digitalis base molecule has no physiologic role, and only functions as a pigment, and more importantly, as a toxic deterrent to pesky herbivorous woodland animals. Why would nature fashion a small molecule in a decorative plant that makes our heart contract more dynamically and rhythmically? It's an enigma, and nature is replete with oddly symbiotic chemicals.

Ancient shamans had no understanding of chemistry and pharmacology, or even the concept of disease treatment. Greek and Roman healers had imagination and curiosity, but a dearth of effective medicines; antiquity's tool chest was mostly empty, and the few remedies that existed were minimally effective and dangerously toxic. The most important characteristic of any intervention, therefore, was that there *was an effect*. The cause-and-effect that is most obvious is rapid altered mental ability, like drunkenness or hallucination. Even the most incurious primitive man could link the stupor or agitation following the ingestion of alcohol or psychedelic mushrooms.

The inkling that plants had biological powers opened the door to hypothesizing about how the body works. The Galenic concept of disease treatment in the 2nd century was based upon the theory of *opposites*. Life was a constant balancing act—good health was evidence of good balance, but imbalance in life was a harbinger of impending disease. Using the foundation of Hippocratic humoral theory, if a patient was plagued with an abundance of phlegm (one of the four main humors of the body), the counterbalancing therapeutic move would be to apply desiccating heat. This is evident in the life of Achilles, who by legend, was nursed on bile (instead of milk), thus giving him a "bilious" and bellicose character.[3]

Paracelsus (1493–1541), and later Samuel Hahnemann (1755–1843), proposed a completely different concept regarding the systemic balances of the human body. For philosophers like Paracelsus, who contemplated the inner workings of our mortal frame without the use of the microscope and before the advent of organ physiology, our bodies were a confusing matrix. His breakthrough concept of *"similia similibus curantor,"* now appreciated as pseudoscience, held that a substance that causes the symptoms of disease in a healthy person would cure symptoms in a sick patient. For example, if one were suffering from diarrhea, the homeopathic intervention would be to take a laxative, forcing even more diarrhea. Unbelievably, there are still subscribers to this crackpot mentality.

The homeopathic mindset, and later the modern allopathic outlook, riveted the gaze of medical sages on the effects (and side effects) of botanical potions. Even in the epoch before modern biochemistry illuminated the true-life molecular world, the methodical analysis of plant and earth materials resulted in a catalogue of useful formulations. As scientists demystified air, organized the periodic table, determined the laws of chemical reactions, and finally laid to rest the pseudoscience of alchemy, the potential of refined and purified drugs came into focus.

More important, an obsession with the unfathomable became a possibility: on-command sleep—the purview of gods—and the ability to regulate consciousness at the wave of the hand.

The Greek god of sleep, Hypnos (Somnus to the Romans), had many sons, but his son Morpheus was a god of dreams, delivering messages and prophecies from the gods to mortals through the medium of dreams. With winged power, Morpheus floated into the dreams of heroes and kings, taking the shape of any human form at will, mimicking "their gait, their face, their moods," according to Ovid.[4] "In the arms of Morpheus" was to be asleep, consorting with the gods.

The poppy plant was tended by the Sumerians at least five thousand years ago, and it has likely been in continuous cultivation in the broad Middle East ever since. While many poppy varieties are prized throughout the world for their striking flowers, it is the *Papaver somniferum* that is the source of opium and other alkaloids that transformed medicine, ignited wars, and propped up dictatorships.

The source of opium is the "tears of the poppy," the milky-white latex that is contained within the walls of the poppy fruit, or seed pod. Opium farmers have, for millennia, knifed the outer hull of the green seed pod with parallel scratches, eliciting a milky exudate. The "tears" are allowed to dry overnight, darkening, and collected by scraping the raised, beaded prize. This tarry treasure has been pilfered from floral hosts in its crude form for millennia, ingested by bronze-age farmers for its gut-soothing and sleep-inducing powers, and only later distilled into its constitutive composites, awaiting the invention of the hypodermic needle.

Thomas Sydenham (1624–1689) introduced *laudanum*, a concoction of sherry and opium, in 1680, and it was medicine's great salve for several hundred years.

With the advent of modern chemistry in the early 19th century, Friedrich Wilhelm Sertürner, a German chemist, perfected the purification of morphine from opium in 1804, becoming the first scientist to refine an individual drug molecule from a donor plant.[5] Industrial production of morphine by the German chemical giants, and later by American companies in the 1830s, meant that self-medicated Westerners could "narcotize" (*Narke*, Greek for *stupor*) themselves with over-the-counter morphine and codeine, even into the 20th century. Nothing could be more sublime than to be carried away on Morpheus's wings when you were in agony, *in extremis*.

A medicine typically becomes active in the body only when it enters the bloodstream; physicians must therefore be cognizant about how a drug is absorbed, chemically changed, and metabolized in the human body. While most medications are absorbed, in varying degrees, in the gut, their use as a medical mind-controlling substance is unreliable. Therefore, the depth and timing of somnolence following the ingestion of laudanum was always untrustworthy, and the pioneers of surgery were saddled with squirming, tormented patients.

The route of a drug into the bloodstream was the riddle our scientific ancestors solved using the evolving field of chemistry, and bit by bit, the puzzle pieces of our biochemical world amalgamated into a coherent whole. However, before the invention of the hollow (hypodermic) needle by Alexander Wood in 1855,[6] there were no medicines that could be injected intravenously to provide surgical-grade sleep. While the isolation of morphine in 1804 was a breakthrough, the search for another class of drugs was no laughing matter.

Actually, it was.

Joseph Priestley (1733–1804) was the one man most responsible for introducing modern chemical and physical studies of gases, even though he could boast of no formal scientific education.[7] While life on earth is implausible without water, oxygen is *the* active ingredient in our planet's atmosphere. Oxygen's presence was hidden from our perceptions until Priestley's experiments uncloaked its existence. Early in his learning, the autodidact Priestley had concluded that air was a "simple elementary substance, indestructible and unalterable."[8] In essence, Priestley was conforming to the ancient Greek conception of air, earth, fire, and water as the four elemental components of creation. He was breathlessly supposed to believe that air was homogenous with no component parts, but his simple experiments would unravel the mystery of air.

Priestley came of age at a time of roiling philosophical and religious turmoil, compelling solitary scriptural study, and a maverick theological outlook. His natural skeptical nature was fomented by his immersion in the Calvinist, and later, Presbyterian teaching. Like Joseph Lister a century later, Priestley was not allowed to attend Oxford or Cambridge as a nonmember of the Church of England. Before long, Priestley was "encouraged to study for the ministry, and study, as it turned out, was something he did very well."[9] He taught himself "Latin, Greek, French, Italian, German, and a smattering of Middle Eastern languages, along with mathematics and philosophy."[10] Instead of seminary, Priestley attended Daventry Academy, a school for Dissenters, and there seems little doubt that his start as an outsider prepared his mind for thinking differently.

Priestley supported himself as a tutor to wealthy families in the decades following his formal education. Like Robert Boyle a century before, Priestley spent a great deal of time thinking about the clockwork universe, tinkering with electricity, plants, minerals, and air. He made regular trips to London, interacting with the cognoscenti, including the American genius Benjamin Franklin, who would become a lifelong friend. Priestley became a member of the Royal Society in 1766 following a series of experiments on electricity, and this exposed him to Society members' presentations, their experimental methods and the tools of the trade of scientific investigation.

How does air work? What is it made of? If you were a hobbyist scientist in the 18th century, how would you deconstruct the meaning of your respirations? Consider your breath—why does air rush into your chest during your animal desire to breathe? Dear Reader—follow my orders as an experiment—draw in a large breath, fill your lungs to the maximum, and hold it. Before reading the preceding sentences, you were completely unaware of your respirations, but I want you to stop breathing and think about air. Do it now.

When you can't possibly hold that big gulp of air any more, blow it all out. I mean *all* your air, comprehensively emptying your lungs. Now stop breathing. How long can you last? Twenty seconds? Two minutes? Why must you breathe? What is actually being accomplished during your inhalations and exhalations? What is the composition of air and why must we feast on it?

Prior to the Royal Society's exploration of the structure of air, there was almost no way to think about it as anything but a homogenous, invisible element of uncertain function. Ancient Greek theories about the vitality of air were no longer satisfactory—the new age of experimentation mandated that air itself be tested. The only way to evaluate air was to isolate it, and the apparatus of choice among scientific detectives was a bell jar made of clear glass; the airtight vestibule was often positioned over a shallow pool of water or mercury, trapping the air inside for experimentation. Some of the earliest experiments by the newfangled scientists of the Royal Society involved the glass bell jar and a vacuum pump. If a small bird was placed in an airtight jar and the air was evacuated with a hand crank vacuum, the unfortunate creature would almost immediately keel over, lifeless. Even without the use of a vacuum, the importance of air was revealed by the burning of a candle in a bell jar occupied by a rodent. As soon as the candle flame extinguished, the rodent itself became limp. These experiments were performed in an era before oxygen was recognized. How to explain a burning candle or a breathing animal?

The theory among intellectuals in the early 18th century was that both animals and inanimate objects contained *phlogiston* (Greek: "burned"), and when creatures exhaled or when things burned, phlogiston was released into the air. When the air got too saturated with phlogiston, the theory went, the fire would spontaneously go out or an animal would collapse inside a bell jar. It seemed like a tidy explanation, except it was dramatically, and laughably, wrong.

Time for a quick quiz. In 1648, the Dutch physician Jan van Helmont published the results of a clever experiment aimed at exploring the nature of the growth of plants and trees. Some years earlier, van Helmont carefully collected, dried, and weighed two hundred pounds of soil, and then placed it in a large pot. He then selected a young willow sapling which weighed five pounds, planted it, and over the next five years, fastidiously watered the tree in its pot. At the conclusion of the experiment, he cautiously removed the tree with its roots, and found that it had gained 164 pounds. Van Helmont then dried the soil and weighed it, discovering that in those five years it had only lost two ounces of weight. His conclusion? Plants gained their weight from water alone.

Question: Do you agree with Jan van Helmont? (Answer in one moment.)

In 1754, Joseph Black (1728–1799) presented his thesis for the doctorate of medicine to the University of Edinburgh. As a trained physician, Black conducted experiments (in both Glasgow and Edinburgh) on the properties of kidney stones, testing the possibility of dissolving the stones in acid. Black collected both kidney stones and gallstones, and then plopped them in various acids. To his surprise, certain stones (such as limestone), when dunked in acid, produced a fizzing effervescence of gas bubbles. He termed this gas "fixed air," surmising that it had been fixed as a constituent of the solid material. Later experiments with fixed air showed that it could not sustain a flame or keep an animal alive; however, plants thrived in the presence of fixed air.

Joseph Black's findings led directly to other foundational studies by the legends of chemistry over the next few decades, and it has been argued that his thesis was "a brilliant model, perhaps the first successful model, of quantitative chemical investigation, as well as a classic exemplar of experimental science worthy of comparison with Newton's *Opticks*."[11] And it started with the young Scotsman leaning over a small flask of acid, questioning why champagne-like bubbles would be emitted from a stone submerged in liquid acid.

Thinking back on Jan van Helmont: is it possible that a tree gains weight not from water, but from the air?

The notion that air was made of molecular building blocks was appealing, but the monumental challenge of untangling the enigma was daunting. Other British chemists produced hydrogen gas (Henry Cavendish, 1766) and nitrogen gas (Daniel Rutherford, 1772) using the bell jar. Those early chemists collected information about the constituent gases in air principally by heating or burning objects trapped in a jar, or by observing how plants and animals responded to each other in the chamber, or by burning objects in the compartment. The explosion in the interest in gases was no doubt fueled by the invention of the steam engine and the burgeoning understanding in combustion. The mounting inquest among the curious was, what fuels combustion?

Joseph Priestley was rising in prestige among the learned English, and as British authority was crumbling in the American colonies, he accepted a position as a tutor and savant at Bowood House, the estate of the Earl of Shelburne, one hundred miles due west of London. In 1773, Priestley became the librarian and intellectual companion to the Earl and his family, and more important, became caretaker of experimental equipment that would lead to one of the most important discoveries in the history of thought.

In the late summer of 1774, Joseph Priestley focused his considerable attention on a curious red material called mercury calx, a crimson powder that resembles paprika. When heated, mercury calx turns into liquid mercury, the legendary quicksilver that had intrigued alchemists for centuries. But even savants like Isaac Newton had squandered the opportunity to learn the secret that was hidden within mercury's powdery granules.

When a scientist applies a flame to a crucible of mercury calx powder, an intense fire is generated above the substance as it turns into mercury. Priestley decided to try and turn the powder into liquid mercury in his glass bell testing chamber to investigate the "air" that somehow caused such a brilliant flame upon the embryonic quicksilver. But how to liquify the powder inside the glass chamber? Cleverly, Priestley used a twelve-inch "burning lens," a magnifying glass that concentrated the power of the sunlight into a small patch of calx, like so many mischievous middle-schoolers tormenting ants. Holding the lens on the outside of the jar and melting the powder into liquid mercury, Priestley was able to enrich the air inside his bell jar, permitting a candle inside the vestibule to burn brighter or for a mouse to live much longer while entombed inside the airtight trap.

The enriched air byproduct was similar to Joseph Black's "fixed air," in that some type of gas was emitted during a chemical reaction, whether it occurred during the melting of a powdery solid, or during the exposure of a stone to acid. Sitting in his small experimental room in the Bowood estate (preserved to this day), Priestley pondered the significance of the air. "The feeling of it in my lungs," Priestley wrote, "was not sensibly different from that of common air, but I fancied that my breast felt peculiarly light and easy for some time afterward. Who can tell but that in time, this pure air may become a fashionable article in luxury. Hitherto only two mice and myself have had the privilege of breathing it."[12] Priestley was, of course, breathing super-concentrated oxygen. Mercury calx is actually mercuric oxide, and when it is heated, O_2 is given off into the air.

Joseph Black's fixed air? It is carbon dioxide (CO_2). And why do plants inside a glass bell jar thrive when exposed to fixed air? Because they are able, through photosynthesis (also discovered by Joseph Priestley), to cleave the carbon and give off the oxygen molecule. That, by the way, is the answer to Jan van Helmont's willow quiz. The willow tree grew larger, not by the soil, and *not by the water*, but by the air around it supplying CO_2. All of the trees and plants in our world grow larger by converting CO_2 into the carbonaceous stuff of their frames.

The pioneers of chemistry—Black, Cavendish, Rutherford, Priestley, and Lavoisier—foundationally discovered the gases that comprise the air we breathe, paving the way for modern chemistry and setting the stage for the invention of anesthesia. Humphry Davy (1778–1829) was an Englishman who was apprenticed to a surgeon as a teenager following the death of his father. In an odd coincidence, the surgical profession can be glad that Davy was uninterested in surgery, instead partaking in home experiments with chemicals and flame. He made the acquaintance of several legendary scientific pioneers while still a youth, and was proffered an invitation to the Pneumatic Institution, a research facility in Bristol purpose-built to investigate the medical powers of gases.

Joseph Priestley had been the first, in 1772, to synthesize nitrous oxide—and the first to note its mind-altering powers. But Humphry Davy assiduously tested nitrous oxide, and with the help of James Watt, built testing chambers and refined the production of the gas. Most important, Davy was the first to consider the medical use of nitrous oxide, commenting in 1800, "As nitrous oxide in its extensive operation appears capable of destroying pain, it may probably be used with advantage during surgical operations in which no great effusion of blood takes place."[13]

The component parts of our worldly atmosphere had been elucidated by the turn of the 19th century, and as chemistry turned from a qualitative discipline to a quantitative science, the jigsaw puzzle of our atomic world was organized. John Dalton (1766–1844) is the Englishman who formalized the *atomic theory*, the concept of a law of proportions wherein atoms (chemical elements which cannot be further subdivided) combine together to form chemical compounds. Dalton observed that these combinations are always in specific ratios, and by codifying the recipe book of chemical formulations, he pulled back the veil of our universe's structure. Every chemistry student warming reagents over a Bunsen burner, and for that matter, every chef combining ingredients like baking soda ($NaHCO_3$) into a bread dough recipe, is recognizing Dalton's insight into the proportionality of our world. (Incidentally, baking soda is used in baking because the chemical decomposition of baking soda results in the liberation of water and carbon dioxide (CO_2), and this gas release results in the expansion of the batter, making it lighter and fluffier. Written chemically, the reaction is: $2NaHCO_3 \rightarrow Na_2CO_3 + H_2O + CO_2$.)

It is prudent to consider the rise of chemistry in the 19th century since its advance improved the world so dramatically and so rapidly. The ability to synthesize drugs, refine petroleum, improve alloys, produce fertilizer, and manufacture synthetic fibers has changed the way we eat, travel, medicate ourselves, store food, and make the clothing and materials that are touching you right now. And one of the most important ways medicine was changed in the mid-1800s was the ability for amateur alchemists to make their own potions that had genuine and powerful effects on the human body.

To genuinely be convinced of the effectiveness of a medicine, the time interval from exposure to the predicted effect should be as limited as possible. When we watch someone get poisoned in the premiere episode of a streaming video series, it works best when the wine goblet is still in the prince's hand when he keels over in front of the assembled crowd. Such a show would never work if the poison of choice required hours, or days to deliver the fatal blow. Men have been getting drunk since before modern times, and no one has to be convinced that too much drink leads to debasement, dizziness, and blacking out. But it does take some matter of time to achieve that level of drunkenness; it would be more impressive to reach a level of stupefaction within minutes, or even seconds, when exposed to a new substance. That is what happened in the 1830s, just at the beginning of the chemistry revolution.

Ether frolics and laughing gas parties became all the rage along the East Coast as America was celebrating its semicentennial. Slipshod wannabe chemists had learned how to make ether gas and nitrous oxide, and like itinerant evangelists, traveled from town to town to make a show of citizens becoming intoxicated on the newfangled fumes. Imagine it's 1839. You are a middle-aged woman in Philadelphia, attending a public spectacle, witnessing your husband on a stage, stumbling and incoherent after a single exposure to a liquid-filled sponge, with not a care in the world when the exhibitionist harms his body. Or, consider the young medical student next to the woman, a twenty-five-year-old from rural Georgia, who watches in stunned amazement as the man on stage becomes insensate. What might be possible in the medical realm with this new concoction?

Crawford Long was the medical student in the crowd in Philadelphia in the late 1830s. He was from Danielsville, a small town ninety miles from Atlanta, the son of a prosperous merchant. After college in Athens, Georgia, Long attended medical school at the oldest medical school in America, the University of Pennsylvania, graduating in 1839. The school was founded by physicians and surgeons who had trained in Europe, and Long was fortunate to receive the best education possible, although in 1839 this meant no understanding of germs, cancer, the cellular basis of disease, and a medical world free of anesthesia and antibiotics. Long did learn a proper scientific approach, even if his medical mentors were practicing a form of medicine that was broadly feeble.

Crawford Long left Philadelphia and completed an internship in New York City by 1841. Bellevue Hospital, the oldest public hospital in the nation (founded in 1736), was a place for primitive, hurried surgery when Long was in the city; in the 1830s there was absolutely no sub-specialization in surgery—because nothing worked. However, the benefit of training in urban hospitals, teeming with patients and overflowing with diseases, has always been vital. In medicine and surgery, volume of experience is critical, and during Long's education, with medical journals in their infancy, nothing could substitute for sheer numbers of diseased and injured patients. Of course, medicine in mid–19th century America was completely ineffectual, but the new art of diagnosis (as pioneered by Morgagni and Rokitansky) at least allowed doctors to make educated guesses about what was killing their patients.

There can be no doubt that the avalanche of suffering patients in New York, coupled with Long's indoctrination to the mysterious authority of laughing gas and ether, triggered a hypothesis about their ability to empower insensibility. There is no evidence that Long discussed the potential use of gases in providing anesthesia while in New York, but curiously, after training in two of the most important medical meccas in America, he retreated to another tiny Georgia town and just one year later, made history.

Within a year of Crawford Long arriving in Jefferson (a neighboring town to Danielsville), history was made, even though it would take years for the world to realize what had happened. On March 30, 1842, Crawford Long provided (what would become known as) anesthesia to a young man with a tumor growing on the back of his neck. The shoebox two-story, redbrick building where the breakthrough occurred is still preserved on College Street in Jefferson, where the tall, slender, bearded Southerner administered ether gas to young James Venable. Like all operations before the advent of antibiotics, the tumor resection procedure was one of extraction, not implantation. It goes without saying that it was performed near a window, illuminated with sunlight, the incandescent lightbulb still decades away.

Initially, Crawford Long did not think of reporting his revolutionary technique. Dozens of the most important moments in science and medicine were achieved in obscurity in small hamlets by lone geniuses; oddly, the urge to proclaim the discovery to contemporaries oftentimes is nonexistent. Several years would pass until Long saw a report in the *Boston Medical and Surgical Journal* (forerunner to the *New England Journal of Medicine*) on December 9, 1846, by John Collins Warren, reporting on the fantastical achievement on the use of ether in the prevention of pain during surgery. One can only imagine Long's horror, being scooped by a group of Boston physicians and surgeons. He simultaneously felt shock at someone else solving the pain conundrum with the selfsame ether and not a little bit of dismay for not capitalizing on his priority. These dismal feelings would germinate and mature, particularly as the triumvirate of pioneers in Boston battled with each other, clamoring for prestige and financial windfall.

When you disembark from Boston's Red Line train at the "T stop" at the Charles/ MGH station, you are confronted with the stark contrast of the gleaming metal and glass of the above-ground station and the intimidating stone edifice of the Liberty Hotel, formerly the Suffolk County Jail. As you walk east along Cambridge Street, you encounter the Museum of Medical History and Innovation (obviously, one of my favorite places in Boston). Turning left, and heading north up Grove Street, you are flanked by huge, urban parking structures. An angled cul-de-sac of medical buildings awaits you, one block in the distance, and at the end of the canyon of a mishmash of redbrick rectangular structures and silver metal and glass clinic medical buildings, lies a rounded white brick tower, The Massachusetts General Hospital (MGH). Actually, all these buildings are part of the MGH campus, including many more that you cannot see.

Instead of walking toward the MGH tower, if you wander east along Parkman Street, past the Wang Ambulatory Care Center, you will encounter a tree-lined lane to your left. It's another gorge of medical buildings, but in an unexpected parklike setting, and as you walk along the sidewalk, you strain your eyes to see the structure that is ensconced behind the trees, up a slight slope. After trekking about four hundred feet, winding along the sidewalk, the structure comes into view. *This* is the *original* Massachusetts General Hospital, now two hundred years old. It is now referred to as the Bulfinch Building (after the architect, Charles Bulfinch) and largely provides administrative offices for the campus.

Gazing at the white granite Bulfinch Building and letting your eyes track upward, higher than the three levels of windows and above the triangular pediment, is a square-shaped central tower which supports a large dome, capped with a small cupola, both clothed with copper, now with a verdigris patina. This regal, classical revival-style hospital seems out of place; the surrounding clinical edifices are swarming with patients, residents, nurses, and attending doctors, but the Bulfinch Building stands in somber silence. Thankfully, as the campus expanded, the grand old lady has remained untouched, and standing in front of the building, no time has passed since 1846, when history was made up in the dome.

The first surgical amphitheater in the Western Hemisphere was at the University of Pennsylvania, opening in 1804. Like all European and American surgical amphitheaters built in the 19th century, the theaters in Philadelphia and Boston were on the top floor, with large windows and skylights. Candles provided some light, but in the era before electricity, sunlight was the greatest illumination surgeons could hope for. The Massachusetts General Hospital was completed in 1823, and while an architect could never be faulted for placing a surgical amphitheater nearest to the sun, only a quarter century would pass until the practical positioning of a theater atop a hospital also became a symbolic representation of the revolutionary change in surgery.

A "grand exhibition of the effects produced by inhaling nitrous oxide" was advertised in the Connecticut's *Hartford Courant* on December 10, 1844, attracting the lay public and professionals alike. Specifically, a dentist in Hartford named Horace Wells attended the demonstration with his wife, and was astonished at laughing gas's ability to mask pain among the volunteers, including a young man who suffered a bloody injury to his leg. In a matter of weeks, Dr. Wells was extracting teeth with patients under the influence of laughing gas, later writing, "I was so much elated with this discovery, that I started immediately for Boston, resolving to give it into the hands of proper persons . . ."[14]

Horace Wells made the one-hundred-mile trip to Boston, relying upon a former partner to make an introduction to the medical leaders of the city. Wells had previously been in a dental partnership in Hartford with William Thomas Green Morton (1819–1868), and while it was short-lived (1842–1843), they apparently had parted under amicable terms. Morton arranged a meeting with John Collins Warren, the head of surgery at MGH in February 1845, just two months after Wells had started using laughing gas to perform dental procedures.

The excitement of going to Boston, mingling with the top surgical figures in America, and demonstrating his technique to the medical students at Harvard, quickly faded to embarrassment and shame. Wells had been invited to administer nitrous oxide to a patient who needed a limb amputation, but after vainly waiting days for the patient to agree (who can blame the patient?), it was proposed that Wells oversee the administration of gas for a tooth extraction. In some ways this was likely a relief for Wells, and by treading familiar territory of dental anesthesia as opposed to major limb surgery, he had to feel sanguine on that late winter day.

Instead, catastrophe occurred. Wells later recalled, "Accordingly a large number of students, with several physicians, met to see the operation performed—one of their number to be the patient. Unfortunately for the experiment, the gas bag was by mistake withdrawn much too soon, and he was but partially under its influence when the tooth was extracted. He testified that he experienced some pain, but not as much as usually attends the operation. As there were no other patient [sic] present, that the experiment might be repeated, and as several expressed their opinion that it was a humbug affair (which is in fact all the thanks that I got for this gratuitous service), I accordingly left the next morning for home."[15]

Humiliated, Horace Wells left Boston, but not before meeting with William Morton and Morton's Harvard chemistry professor, Charles Jackson. While Morton's demonstration had been a failure, the meeting among the three would prove to be the genesis behind the use of ether. Morton was a native of Massachusetts, and had earlier attended the Baltimore College of Dental Surgery, the first dental school in the world (prior to this, dentistry was similar to early surgery, where the apprenticeship model was used). It is debated if Morton actually graduated from dental school,[16, 17] but he returned to New England and briefly partnered with Wells.

Morton practiced a more aggressive form of dentistry, likely influenced by his dental surgery professors in Baltimore. This resulted in him struggling for patients, both in Hartford and later in Boston, when his patients suffered more pain from his more invasive techniques. Ambitious, industrious, and battling for recognition, Morton entered Harvard's medical school in 1844, enrolling in the chemistry classes of Charles Jackson.

Charles Jackson (1805–1880) was another Massachusetts native, coming of age in a combustible era of intellectual breakthroughs in geology, chemistry, medicine, and mineralogy. Jackson graduated from both Harvard College and Harvard Medical School, and while he initially stayed in Boston to teach and research the exciting new field of chemistry, his career and varied interests led him across Europe and America. His life would be characterized by clever insights and vast curiosities, but also with a bizarre tendency to claim priority in invention, asserting his primacy in the development of the telegraph, guncotton, and most important, ether anesthesia. Like his eventual rivals Wells and Morton, his death was ignominious, but there is no doubt that he played a pivotal role in educating Morton about the properties of ether and its possible use as an anesthetic.

In the months following Horace Wells's failed laughing gas demonstration before the Harvard medical students and faculty, Morton began a clandestine investigation into ether. Initially, it was for his own practice, having a "direct pecuniary motive, bearing almost daily upon him, to alleviate or annihilate pain under his operations."[18] Animal experimentation gave way to testing ether vapors on two young dental assistants, and by the summer of 1846, Morton was becoming convinced of the utility of using ether for dental surgery and general surgery as well.

William Morton choreographed his scientific moves with careful business acumen, consulting again with Jackson about the ether preparation and conferring with a patent commissioner about the potentialities of his claiming primacy in the use of ether. Coinciding with his commercial exchanges, Morton finally attempted a tooth extraction using inhaled gas from an ether-soaked handkerchief on September 30, 1846. The procedure was accomplished on a somnolent, pain-free male; instead of the usual tortuous affair, Dr. Morton was able to extract the tooth without fanfare, emboldening him to approach the most respected surgeon in America.

Given Morton's experience of a single patient, his boldness in approaching John Collins Warren, the professor of surgery at Harvard, is curious. Morton was well aware of the previous nitrous oxide misadventure just twenty months earlier, and one must wonder about his "considerable self-confidence or considerable foolhardiness, or perhaps a good measure of each. It certainly flew in the face of scientific restraint, in view of the fact the twenty-seven-year-old dentist's experience with ether was minimal, his knowledge of its hazards was nil, and he had not yet even bothered to do anything about perfecting an apparatus by which the gas could be administered to a human subject."[19]

Morton, however, charged ahead.

John Collins Warren (1778–1856) was the second professor of surgery at Harvard (his father, John Warren, was the first). Although his father had founded the Harvard Medical School, John Collins Warren received his medical education in Europe, training under Sir Astley Cooper in London and Baron Guillaume Dupuytren in Paris, eventually obtaining his medical degree from Edinburgh University.

When Warren returned to Boston, he became one of the "fathers of American medicine," establishing the journal that would become the *New England Journal of Medicine*, cofounding the Massachusetts General Hospital and the American Medical Association, becoming the first dean of the Harvard Medical School, and ascending to the title of professor of surgery for over thirty years. Revered, Dr. Warren was austere and highly skilled—his "flint-faced, grizzled appearance belied the fact that in a lifetime of surgery he had never been able to inure himself to the horrors of the operating theater."[20]

Despite Morton's optimism, he must have been surprised with Warren's willingness to experiment upon one of his surgical patients. William Morton feverishly collaborated with an instrument maker, fashioning a glass apparatus with two openings and a central vessel that contained an ether-soaked sea sponge. With only a few days' notice, Warren invited Morton to come to MGH with his device for a procedure on a young man with a vascular tumor below his jaw.

On Friday, October 16, 1846, Gilbert Abbot, a young TB patient, was readied for surgery in the amphitheater at MGH. John Collins Warren awaited Morton, almost losing patience and proceeding in the usual barbaric manner. With Morton making the finishing touches on the device, he appeared fifteen minutes late on the appointed morning. The long, narrow stone steps all the way to the top of the hospital demanded one last challenge for the young medical student, and there seems little doubt that he arrived, gasping for breath before a large audience of skeptics whose appetites were whetted for another debacle, another humbug remedy.

Morton quickly went to work, readying the contraption and positioning the mouthpiece in front of the nervous patient. He ordered Abbot to breathe, and within minutes, he was asleep. On that late New England fall morning, with daylight flooding through the glass panes above and Harvard students leaning in from the steep and narrow rows, Morton nodded to the regal surgeon, "Sir, your patient is ready."

Five weeks later, Warren's report on that glorious morning was published in the *Boston Medical and Surgical Journal*. Dr. Warren wrote that after three minutes the patient "sank into a state of insensibility. I immediately made an incision about three inches long through the skin of the neck, and began a dissection among important nerves and blood-vessels without any expression of pain on the part of the patient . . . being asked immediately afterward whether he had suffered much, he said that he had felt as if his neck had been scratched."[21]

While it may be apocryphal, Warren is believed to have finished his operation, calmly looked up at the silent crowd and stated, "Gentlemen, this is no humbug."

Warren's article outlined the additional cases performed over the following three weeks under the control of ether anesthesia. He concluded that there was a "decided mitigation of pain," and that the medicine's effects upon the body "soon pass off without leaving any distinct traces behind them." His last sentence of the publication is wonderful: "Let me conclude by congratulating my professional brethren on the acquisition of a mode of mitigating human suffering, which may become a valuable agent in the hands of careful and well-instructed practitioners, even if it should not prove of such general application as the imagination of sanguine persons would lead them to anticipate."

This was the publication that Crawford Long read down in Jefferson, Georgia, now four and a half years after his initial use of ether anesthesia. Thus, his panic.

In the end, Crawford Long did earn the accolades he deserved. One of the Emory University hospitals in Atlanta is named after him, and his statue is in the United States capitol building. But he never got to revel in the notoriety that would one day be bestowed upon him.

The story is much worse for Wells, Jackson, and Morton.

Horace Wells rapidly declined in the months after Morton's successful demonstration. He moved often, failed in his dental practices, and eventually became addicted to ether and chloroform. He committed suicide (just fifteen months after Morton's demonstration) in a prison cell in New York City while under the influence of chloroform.

Charles Jackson continued his lifelong quest for fame and recognition, but instead died impoverished and insane.

William Morton, even hungrier for financial reward and esteem, was crippled by regret and resentment. He lost patent battles and the respect of his colleagues, and was censured by the American Medical Association on the grounds of "unworthy conduct."

At the age of forty-eight, after losing yet another legal battle over recognition of his "invention" of ether anesthesia, Morton killed himself during a summer heat wave in New York City. "On an impulse, he decided to take his wife Elizabeth on a cooling buggy-ride through Central Park. Without warning, he suddenly jerked the horse to a stop, leaped out of the wagon, and plunged his head into the tepid water of the lake. Obviously disturbed, he was urged back into the buggy, but had driven only a short while longer when he precipitously vaulted from the rig, threw his body over a nearby fence, and fell to the ground on the other side, unconscious," dying from a cerebral hemorrhage. [22]

Europe had been at the forefront of medical innovation for centuries, but for the first time, American physicians and scientists had made a major contribution. Within weeks, news reached London that anesthesia had finally been accomplished. For those in London, less than seventy-five years since the Boston Tea Party, a revolution so close to the Boston harbor must have seemed highly unlikely.

Robert Liston, the celebrated professor of surgery at University College, London, scrambled to test the efficacy of ether on one of his patients. On December 21, 1846, Liston performed an above-knee amputation, the first surgical operation in Europe under ether anesthesia. The operation was a success, and instead of butchery, the witnesses were treated to a painless and efficient exposition. Later, Liston performed an excision of the nail of the great toe, a surprisingly painful operation that had always distressed patients and surgeons alike. Before operating, Liston told the assembled surgical students, "We are going to try a Yankee dodge today, gentlemen, for making men insensible."

When the toenail removal operation was successful (with a nineteen-year-old Joseph Lister in attendance), there was a similar refrain, among the operators and spectators alike, that ether anesthesia was no "Yankee dodge." Inhalant anesthesia was so obviously a breakthrough that anyone who witnessed it immediately knew there was no going back. Within months, the ether transformation spread across Europe, in an instant changing the way surgeons treated their patients.

James Young Simpson (1811–1870), a pioneering Scottish obstetrician who trained and practiced in Edinburgh, learned of Liston's ether anesthesia performances and wasting no time, immediately traveled to London and met with Liston and watched several operations. In a matter of weeks, Simpson was experimenting with ether on obstetric patients, a true testament to the rough and tumble days of medicine where patients were entirely at the mercy of their doctors. Before there were guinea pigs and lab rats, humans were the chief experimental subjects.

Simpson began a program of auto-experimentation when he considered that ether's serious drawback was its high flammability, a significant feature in light of Simpson's use of ether in Edinburgh homes, where candlelight, gaslight, and coal fires were ubiquitous. An explosion would be lethal for everyone in the vicinity, and given Edinburgh's crowded 19th-century tenement buildings, a detonation would be cataclysmic. Simpson and his scientific associates assayed and evaluated any chemical they could get their hands on, stumbling upon chloroform in November 1847, on the advice of a chemist friend.

Assembled at Simpson's house in Edinburgh, a group of men and women experimented with the effects of chloroform, reminiscent of the frolics that led to the use of ether in America. The investigative protocol was simple: Simpson and his physician friends held tumblers in their hands, and after pouring the test liquid in their glass vessels, inhaled the vapors to determine if any effect was achieved. On November 4, 1847, Simpson recalled a small bottle of a heavy liquid that he had earlier doubted would lead to success, "and with each tumbler newly charged, the inhalers resumed their vocation. Immediately an unwonted hilarity seized the party—they became brightened, very happy, and very loquacious—expatiating on the delicious aromas of the new fluid . . . but suddenly there was talk of sounds being heard like those of a cotton mill louder and louder; a moment more and then all was quiet—and the crash!"[23]

All who had inhaled the chloroform vapors completely lost consciousness, awakening some time later with everyone on the floor, contorted in the positions in which they had collapsed. The entire supply of chloroform was duly exhausted that night, but not before Simpson's niece took her turn, and after drawing in the chloroform vapors, exclaimed, "I'm an angel! Oh, I'm an angel!"

What sounds like a modern drug party was a quasi-scientific investigation into a better anesthetic, and Simpson's exploration led to his use of chloroform in obstetrics and eventually, to its being the preferred surgical anesthetic in Europe for decades to come.

John Snow is both the world's first epidemiologist (for his work on cholera) and the world's first full-time anesthesiologist. He was adept at using a glass inhaler for the administration of ether anesthesia, and eschewing the relatively dangerous technique of simple chloroform-soaked handkerchief administration, developed an inhaler for the safer administration of Simpson's drug. Unlike the American ether pioneers, Simpson and Snow would both be heralded as heroic innovators, leading fruitful careers, with Snow providing chloroform anesthesia for Queen Victoria's deliveries in 1853 and 1857.[24]

By the 1860s, there was widespread use of chloroform and ether across America and Europe. We have an inaccurate view of anesthesia during the American Civil War—many of us picture gruesome amputation scenes like that in *Gone with the Wind*, where men are begging for mercy, screaming, "Leave me alone . . . I can't stand it! Don't cut, don't cut, don't . . . please!" In actuality, chloroform was used by both sides during the Civil War, although supplies of the anesthetic would have been unpredictable at times.

An interesting twist of fate during the Civil War was that William Morton provided anesthesia for the Union troops and Crawford Long served as an anesthetist for Confederate troops. Even in war, these two antagonists were on opposite sides. It was not the first (nor the last) time that wartime medicine has pitted medical greats against each other, indirect combatants serving their fellow men.

It is a wonder that the inhaled vapors of nitrous oxide, ether, and chloroform have the effects they do. It is easier to ponder why an opiate medicine triggers a feeling of euphoria and tranquility when you consider that an opiate is a mere substitute for our own, endogenous endorphin. The mechanism of inhalant anesthetics is tougher to consider, as they do not mimic one of our own chemical molecules. Interestingly, the chemical behavior of all anesthetics is only recently becoming better understood and mostly relate to alterations in excitation and inhibition pathways, both in the brain and in the spinal cord.

The newer anesthetic agents (like isoflurane, desflurane, and sevoflurane) have rapid onset and reversal, thus making anesthesia much safer, faster, and with fewer systemic side effects than their predecessors. As important as they were, ether and chloroform have become historical relics.

It should be obvious by now that the history of surgery is only recently a history of cutting, dissecting, sewing, reconstructing, and implanting. Only in the last 150 years have surgeons been capable of making positive, and now miraculous transformations for mankind. The revolutions in printing and peer-review publishing made sharing of breakthroughs possible, and the advances in chemistry, stoichiometry, and the understanding of the behavior of gases set the stage for the upheaval of anesthesia.

Nothing about modern surgery is possible if our forefathers had not decrypted the complexity of our chemical world and gained dominance over consciousness. One concern of finding (even primitive) life during an interplanetary space voyage is finding life that has dissimilar molecular building blocks and differently evolved chemical receptors, leaving us vulnerable to life forms that we cannot combat. What if man *was* an original animal on planet earth, not having evolved over hundreds of millions of years and not responsive, or susceptible to the chemicals in our world? It would be possible that we would have no authority over sensations and perceptions, and no governance over pain and awareness. Thankfully, we are a byproduct of every living thing, sharing chemical structures, molecular receptors, and among mammals, anatomical features and organ systems. This makes anesthesia possible, which makes surgery real.

On the west side of the Boston Public Garden (the setting of the beloved children's book *Make Way for Ducklings*), in view of the lagoon's swan boats and Commonwealth Avenue stands the Ether Monument, a forty-foot granite sculpture memorializing the first public demonstration of ether anesthesia in 1846. On one side of the monument is the inscription: "To commemorate the discovery that the inhaling of ether causes insensibility to pain. First proved to the world at the Mass General Hospital in Boston, October AD MDCCCXLVI." The wording is brilliant. Setting aside the controversy of Long, Morton, Jackson, and Wells, the clear and concise verbiage emphasizes the most important point—that in Boston, on October 16, 1846, men proved to the world that ether causes insensibility to pain, true anesthesia. Thousands of years of tinkering with flowers, herbs, and alcohol had resulted in genuine impotence in the face of pain. And while it is now obvious that Crawford Long was the true pioneer in ether anesthesia, it was on that day in Boston that man realized that he had finally conquered pain and established dominion over consciousness. In all of human history, it is one of our most miraculous moments, and for anyone who has languished in agony and submitted herself to the knife of a surgeon, that moment in the MGH ether dome is transcendent, glorious, and the stuff of the dreams of the gods.

TEN

Elective Surgery

Almost every patient mentions how cold it is in the operating room. We keep the room around 60° F, in part to keep comfortable in our layers of clothes and gowns and in recognition that keeping a chillier room helps keep infection rates lower. Lisa is no exception. As she's being wheeled in on the gurney, she's groggily awake and mutters something about the cold.

Before rolling her into the operating room, Dr. Cohen, my anesthesia guide today, performed a nerve block injection on Lisa. He used an ultrasound wand to guide the trajectory of the needle via computer screen as it delicately plunged through the skin and explored the depths of tissues around the neck. Once the shadow of the needle tip approached the brachial plexus (the bundle of nerves that connect brain and limb), he injected a syringe-full of numbing medicine around the nerves. The procedure rendered her arm simultaneously numb and paralyzed. On cue, our little team trekked toward the operative theater.

Underneath her translucent blue surgical bonnet, Lisa's bright auburn hair is visible, tucked safely away in our clean environment. She's another East Coast transplant who moved west to Colorado in middle age. In the last several years her left shoulder has become unbearably painful, and after failing injections and physical therapy, she has opted to undergo shoulder replacement surgery. In the preoperative staging area, Lisa nervously asked about how much pain she'd be in tonight, but that's a faraway concern now. Her sedative medicine has rendered her punch-drunk and unaware. Her sister confided that they watched a shoulder replacement on YouTube last night, only barely enduring the odious presentation.

While in high school, I saw a television special about Dr. Frank Jobe, the longtime team physician for the Los Angeles Dodgers, who pioneered so many shoulder and elbow operations. That day, I decided to be an orthopedic surgeon and never wavered in that pursuit. Following medical school in the Midwest and surgical residency in Pennsylvania, I was accepted at the sports medicine fellowship overseen by Dr. Jobe, the same man who had inspired me almost twenty years before. I have traveled with the Lakers, ran onto the field of the LA Coliseum with the USC Trojans, lived in a bungalow on Jackie Robinson Way during spring training, and listened to the plaintive cries (bad back, bad feet) of a now middle-aged Mike Eruzione in the dressing room of the US Olympic ice hockey team in their only reunion since 1980 Lake Placid gold. Those experiences, teamed with guidance from world-class joint doctors, turned me into the surgeon I am today.

All of us in the operating room wear scrubs, caps, and masks, but only the surgical tech dresses in her sterile gown as she prepares the instrument table. My physician's assistant Ashley and I help Lisa scoot over onto the narrow surgical table.

Dr. Cohen administers more anesthesia through a combination of intravenous sedatives and inhaled gases. Our main goal is pain-free surgery, but we also want the patient motionless so we don't have to worry about injuring a nerve or vessel with a sudden jolt. Cohen renders Lisa unconscious. As the gas mask is held firmly over her mouth, milky white Propofol pushes into the IV. In the few seconds it takes to travel to the heart and circulate into the brain, Lisa plunges into the marvel of senselessness.

Now that Lisa is completely anesthetized, we position her body carefully on the table. Great care is required: while her body is completely alive, it is incapable of self-protection and as vulnerable as a newborn.

We position her to minimize bed sores or nerve injury, and then I adjust the height of the table so that the surgical area matches the height of my elbows. This minimizes shoulder strain and allows for up-close visualization. During shoulder replacement surgery, I stand. Once we're all positioned, the circulating nurse paints the surgical area with skin prep chemicals, including alcohol and other bactericides.

While the nurse paints the patient's shoulder, and after scrubbing our hands, I perform an intricate dance with the scrub tech, who already has her gown and gloves on. While facing her, she slides my gown over my outstretched arms without touching the unsterile parts of my body or clothes. She snaps latex gloves over my freshly doused hands. From behind me a nurse ties my gown and I twist 360 degrees to cocoon myself in a sterile microenvironment.

After all the busyness of collecting Lisa from the preoperative holding area, dealing with her sister's anxiety, hustling to finish a basketball player's physical so he can play this weekend, phoning my office assistant to hear the results of a rugby player's MRI, and readying the patient for surgery, it's time to cut. Dr. Cohen and I catch each other's gaze, connecting in a deep sense of trust over the dominion of life. His somber eyes and a gentle nod of assent confirm that we are ready to journey on.

The scalpel is made of two parts, the handle and the blade. The scalpel handle is flat and made of stainless steel, which allows for repeated washing, sterilizing, and packaging. A scalpel handle can last for years, even decades. The scalpel blade, however, only lasts one case. Instead of stainless steel, the blade is made of carbon steel, which is much sharper. In many operations, a blade loses its requisite sharpness in the middle of surgery and must be discarded.

As a surgeon, you always ask for instruments without turning toward the scrub tech or without taking your eyes away from the surgical field. An experienced tech knows, within a few tools, the instrument the surgeon will ask for next, along with the proper orientation to place the tool into the surgeon's hand. To the close observer, surgery looks as well choreographed as oceangoing sailors maneuvering their craft with minimal conversation and maximal coordination.

My gaze is now firmly fixed on the purple line I've drawn over the front of Lisa's shoulder. All the acts of positioning, scrubbing, gowning, prepping, and draping have come down to this. It is time to make an incision. It is the moment of truth beyond which there is no turning back. "Scalpel."

There are two ways of holding a scalpel. In one position, you hold the stainless steel handle like a pencil; in the other, you hold it like a conductor holds a baton. The former technique makes the majority of incisions, while the latter is employed during larger incisions. When I perform the incision on Lisa, I steady my hand and wrist, and use the muscles around my elbow and shoulder to direct the blade across the skin.

With the scalpel in my right hand, I hold my gloved left thumb and index finger on either side of the incision site, stretching the skin tight. Nothing else in the world exists, no other thoughts, no realities, no controversies. Nothing funny, nothing sad, nothing interesting. I'm in a vacuum where everything further than ten inches away vanishes.

My father was a Marine Corps sniper in Korea before becoming a veterinarian. Curious about his military days, I watched a documentary on sniper training, wherein elite marksmen like him were taught to deeply inhale and exhale before squeezing the trigger. This same technique helps me steady my hand during precise movements; I use it every day in the operative theater.

After my brief breathing exercise, I move the scalpel blade to the uppermost part of my planned incision. The razor-sharp blade is touching the skin and yet no penetration of the metal edge has occurred until I downwardly angle the instrument toward the elbow. New surgeons notoriously misjudge the correct pressure required to properly cut through the skin. Typically, rookie cutters barely scrape the skin, and their supervisors joke about paperclips inflicting greater damage. However, too great a pressure will plunge the knife deeply into the wound, causing potentially catastrophic damage to deep nerves or arteries.

Cutting skin feels like slicing into a fresh peach. As I draw the knife along the skin, yellow fat billows up from the wound. As we age, our skin becomes thinner, so I account for this as my scalpel blade progresses across Lisa's shoulder. The tiny vessels along her skin's edge emit droplets of bright red blood where they have been transected; these must be cauterized, or heat-sealed, with an electrothermal device called a "Bovie."

A perfectly executed cut penetrates only the dermis, leaving further dissection to scissors and electrocautery. Having made the initial surgical incision, I hand over the scalpel and retire the blade. The skin layers contain dangerous bacteria (even after the most fastidious surgical prep), which contaminate the "skin knife." After mere seconds of use, that scalpel blade is done forever. Lisa doesn't move and doesn't recognize the violation of her body's boundaries.

The remainder of the shoulder replacement operation involves exploring deeper and deeper layers of tissue. This kind of deep investigation of the corpus was unimaginable at the same time that photography, the telegraph, the steam engine, and perforated toilet paper were being invented. Every great surgeon requires an innate sense of three-dimensional space buttressed by years of anatomic learning. The location of every muscle and minute blood vessel and nerve is surprisingly predictable; a gifted surgeon comprehends these laminations with rapid, precise dissection. An aging joint, like an ancient tree tentacled onto a rocky cliff face, develops thickened stratums of bony spurs, loose cartilage bodies, and overgrown, inelastic ligaments. As I delve deeper toward the diseased shoulder joint, I have to abandon the scalpel for stouter instruments.

Other specialists deal with soft tissues like brains and bowels. Orthopedic surgeons deal with bones, ligaments, muscles, and joints. Our tool kit includes hardware: metal saws, chisels, drills, and hammers. After dissecting deeply through the soft tissue layers around the humeral head, I expose the arthritic joint, cut off the top of the bone with a metal battery-powered saw, and place successively larger metal stems down the hollow humeral canal. Once prepared, I can insert the final total shoulder implants. At some stages, every operation is like a project, and requires the use of brute force to pound, shape, smooth, and extract tissues and body parts.

After implanting the new shoulder replacement components, I reverse course and backtrack my way out of the shoulder joint. The last step is always skin closure. Back in the days before antibiotics, early surgical pioneers used silk or catgut sutures to bring tissue edges together. Those materials were dissolvable, which created an immune reaction that opened the door to infection, and commonly death. We now use "inert" or low-inflammatory suture materials or metal staples that hold skin together.

As my surgical team applies the final dressings, Dr. Cohen also reverses his steps. He stops the short-acting inhalant anesthetics that were vaporized into the breathing tube and discontinues the intravenous drugs that kept Lisa sedated. Gone are the days where drugs, such as ether, took days to wear off: today's designer molecules wear off in minutes. Lisa begins to move her body and fight against the tube in her mouth. When it is safe, we remove the breathing tube and shuttle her back onto the gurney.

Our little contingent rolls Lisa down the hallway, now progressing toward the recovery room. Patients often fear saying something stupid or embarrassing in the operating room. Most do not. There tends to be only incomprehensible mumbling in the moments surrounding the induction and reversal of anesthesia. However, there is a disarming sentiment most patients have in the first minutes of wakefulness: loss of awareness in the passage of time. With me at Lisa's side helping to guide the gurney, Dr. Cohen asks her if she's doing all right.

"Are we almost ready to go?" she asks.

In 1877, in stately Breslau, Germany, Robert Koch first met a young American named William Henry Welch. Just twenty-seven years of age, Welch had graduated from Yale and Physicians & Surgeons (the medical school affiliated with Columbia University in New York City), and like so many new American graduates, went to Continental Europe for his "grand tour." Stops in Strasbourg, Leipzig, and Breslau had exposed Welch to the world's most sophisticated microscopists, pathologists, and bacteriologists. The eager Welch, encouraged by his serendipitous timing, realized that this new field of pathology would now be his life. Being tutored by the fathers of the new disciplines, like Koch, Friedrich von Recklinghausen, Ludwig, Wagner, and Cohnheim would position Welch for a preeminent role in American medicine.

While Welch was in Leipzig, he met with Dr. John Shaw Billings, the Army colonel who was responsible for building the Library of the Surgeon General (now the National Library of Medicine) and who had been hired by Daniel Coit Gilman, the new president of Johns Hopkins University, to help design the new hospital at the fledging institution, and was tasked with recruiting promising physicians to Baltimore.

Armed with an enormous endowment from the wealthy industrialist Johns Hopkins (a bachelor Quaker), the university was unlike anything ever built. The hospital (and its associated medical school) would be patterned after the German laboratory-centric model and the clinically-based British model.[1] Gilman's and Billings's dreams of a scientific hospital staffed by full-time professors was revolutionary, and demanded innovative doctors who would commit their lives to changing the way hospitals operated. Welch and Billings met in 1877 to drink beer at Leipzig's Auerbachs Keller, a legendary wine bar and restaurant (frequented by Goethe), and the prospect of Welch becoming the linchpin hire at Hopkins was tempting to both visionaries.

In time, William Henry Welch would become the founding physician at Johns Hopkins University. First, however, he would return to New York City, where he oversaw the building of the United States' first pathology laboratory, at Bellevue Hospital. This new discipline required the latest microscopes and their associated tools, chemicals, and supplies, as well as organized morgues and structured protocols to imitate the very latest in German pathology.

Welch, the luminary in his field in the city, was a celebrated educator among the ambitious medical students looking to supplement their learning. Welch's intelligence and unsurpassed training greatly accentuated his social standing, but it was his congeniality, borne from a household of generations of Connecticut country physicians, that endeared him to students and patients. Welch's father was "close to the people not only as their medical adviser but also as their true friend and counselor."[2] The short and pudgy Welch was tapped as a member of Skull and Bones at Yale University (class of 1870), and would forever enjoy the company of colleagues at clubs, dining halls, and home parlors.

In September 1880, William Stewart Halsted, a commanding and vivacious New Yorker returned home from his own European postgraduate tour of duty, ready to claw and climb and outwork every other surgeon in the city. Where Welch had been tutored by the finest pathologists on the Continent, this former Yale football player had been inculcated by the leading surgeons in Vienna and was an enlightened Listerian. Halsted, son of a successful merchant, would lead one of the most remarkable lives in American history, starting with a stint as a surgeon at Roosevelt Hospital and as an anatomy demonstrator at Physicians & Surgeons.

It took little time for Dr. Halsted to make an impact in New York. His enthusiasm, expertise, and panache made an immediate impression; his groundbreaking methods and sheer brilliance fortified his status as one of the medical men of the future, and his zeal led to appointments at multiple hospitals around Manhattan in an era where horse-drawn carriages were the mode of transportation. Halsted's steel-blue eyes, cosmopolitan manners, and impeccable wardrobe, together with his tony Madison Square address, cemented his reputation as a cultivated physician in a specialty that only recently had risen above the level of derision.

One of Halsted's first innovations was to organize an informal evening of medical didactics for the P&S students. Held several times a week, the "quiz" was typically held in his home office in his 25th Street residence,[3] and the teaching duties were shared among the young stars of the medical community. The two most popular instructors were Halsted and his good friend William Welch, two years the surgeon's senior. The two trailblazers were both Yale and P&S graduates, gregarious and gifted educators who inspired devotion to this new form of medicine.

After accepting the offer to become the founding physician at Johns Hopkins University, Welch departed for another expedition to Europe in 1884. He had spent seven years as a pioneering physician at Bellevue, but his Gotham contributions would be dwarfed by his future work at Hopkins. Granted an eighteen-month sabbatical to revisit the leading medical centers, Welch was also leaving his close friend, Halsted, who was transforming medicine metropolis-wide as a surgeon, anatomy prosector, quizmaster, and experimental scientist.

By 1884, Halsted was on staff at five hospitals (including Presbyterian, the New York Hospital, and the prestigious Bellevue Hospital), but his dream of a "modern" operating room with comprehensive antiseptic facilities was unmet. Raising money from friends and family, Halsted organized the construction of the most state-of-the-art operating room in the country, an elaborate standalone tent replete with maple floors, skylights, running water, gas for lighting, and steriliza-tion facilities. In 1885, this was likely the most advanced operative theater in the Western Hemisphere.

Halsted's operating room bivouac at Bellevue Hospital was a utilitarian version of the operative theaters he had seen in Austria and Germany, while hovering over the shoulders of luminaries in Vienna, Leipzig, Halle, and Kiel. His most famous exemplar surgeon was the sensitive and melancholy physician-poet Theodor Billroth, the self-styled "sentimental North Sea herring."[4] For twenty-five years, Billroth was professor of surgery at the University of Vienna following his training under Bernhard von Langenbeck in Berlin (1853–60).

Langenbeck contributed significantly to the development of surgery at a time of great philosophical upheaval. He received postgraduate training in surgery in London in the 1830s (predating anesthesia by more than a decade), serving under Astley Cooper and Benjamin Brodie; thus linking the German school of surgeons to John Hunter (1728–1793).

Langenbeck's clinical career was interrupted several times by war, including the Schleswig-Holstein Wars (1848–52 and 1864), the Austrian War of 1866, and the Franco-Prussian War of 1870.[5] Battlefield medicine has never changed so dramatically in any twenty-year period; in those two decades, anesthesia was discovered and antiseptic surgical treatment was introduced. In the Franco-Prussian War, German acceptance of the antiseptic technique was instrumental in vastly superior surgical outcomes among Prussian forces (cared for by Langenbeck and his associates) compared to the old-fashioned treatment rendered by the French doctors. German and Austrian physicians therefore became among the earliest and most ardent adopters of Listerism.

Bernhard von Langenbeck was heralded as a humanitarian in the treatment of both allies and adversaries, and as a founding member of both the German Red Cross and the Geneva International Convention, concluded, "a wounded enemy is no more an enemy, but a comrade needing help."[6] The seemingly endless wars of Prussia and Germany in the 19th and 20th centuries would demand many contributions from the German surgeons, and in a sick twist of fate, German surgeons were still pioneering surgery during World War II, with injured American soldiers returning home with innovative orthopedic implants net yet imagined in the United States.

Langenbeck's other major contribution was his apprentices themselves; he is credited with training nearly every prominent surgical operator of his time, including Billroth, Emil Theodor Kocher, and Friedrich Trendelenburg. His idea of organized training following medical school, wherein the young pupil would live at the hospital and gradually assume greater responsibility over the course of years, has earned him the sobriquet as the "father of surgical residency."

If John Hunter is the father of scientific surgery, then Langenbeck can rightly be described as the progenitor of modern antiseptic battlefield surgery, physician battlefield neutrality, and surgical residency. Langenbeck was at his pinnacle when antisepsis and anesthesia converged, releasing surgery from its "constraining medieval chrysalis."[7] Theodor Billroth's Berlin tutelage under Langenbeck witnessed two of the most powerful surgeons ever to coexist, with Billroth advancing as his most important protégé.

After a transitory stint in Zurich, Switzerland, Theodor Billroth permanently settled in Vienna, becoming the most influential surgeon in the world for a quarter of a century. From 1867 till the 1890s, Billroth's surgical amphitheater at the University of Vienna was the center of the surgical universe, where he pioneered surgical techniques, tutored numerus graduates from Europe and America, refined his version of surgical residency, published numerous articles and refined his classic textbook, promulgated the process of surgical audits (predecessor of surgical outcomes), and inspired generations of surgical leaders. All the while, he fostered intimate relationships with musicians and composers, including his close friend Johannes Brahms.

Professor Theodor Billroth was uniquely poised to drive the final stake through the vampire heart of ancient, nonsensical humoral theory and quackery. The amalgamation of chemistry, microscopy, bacteriology, embryology, physiology, and diagnostics heralded a stunning transitional moment in medicine, with Billroth the unquestioned dean of surgeons. "It was a yeasty time for researchers, and the atmosphere of the German hospitals was a ferment of possibilities."[8] Bloodletting, cupping, purging, leeching, and poisoning were being replaced with careful German laboratory studies and scientific interventions based on organ and cell function. It was the German understanding of disease that enlightened investigators about normal structure and function.

Billroth spent long hours dissecting cadavers and planning on surgical interventions. He was able to pioneer abdominal surgery with careful preparation and strict adherence to meticulous antiseptic technique. Animal experimentation and cadaveric-rehearsed surgery emboldened the Viennese professor; perhaps the abdomen *could* be entered. Nothing short of a "godlike creative spirit," as Mukherjee calls it, would suffice when it came to intestinal surgery. Vienna has a centuries' old reputation for virtuoso performances; with Imperial spirit, maestro Billroth would take his place for master class performances in the greatest theater in the City of Music: the Allgemeines Krankenhaus operative theater.

In 1872, Billroth resected a portion of the esophagus and joined the ends together. In 1873, he performed the first complete excision of a larynx. Even more amazing, he became the first surgeon to excise a rectal cancer, and by 1876, he had performed thirty-three such operations.[9]

What seems commonplace today (abdominal surgery) is nothing short of a stupendous magic act, in reality.

First, surgery on any part of the bowel is fraught with danger, particularly the lowermost portion of the bowels: the colon and rectum. Conceptually, the gastrointestinal tract connects the mouth to the anus in a continuous tube that averages about thirty feet in humans. It is comprised of the esophagus, stomach, small intestine (duodenum, jejunum, and ileum), large intestine, rectum, and anus. The "tube" of the gastrointestinal tract is comprised of many soft-tissue layers that are waterproof and bacteria-proof, so long as the integrity of the layers is maintained. The curvy, writhing, self-contained conduit has attachments to the deep portion of the abdominal cavity that suspend the intestines and connect the blood vessels that nourish the guts, and absorb the nutrients being processed. Importantly, outside the alimentary cylinder, the abdominal (or "peritoneal") cavity is perfectly sterile and lethally vulnerable to infection. While the stomach and small intestine are relatively "clean," the large intestine and rectum are teeming with bacteria; while these bacteria often maintain a symbiotic relationship with the host (us), they represent a grave danger if the contents of the tube cross the layers of the large intestine and spill into the peritoneal cavity.

The second half of the scientific sleight of hand is the possibility of any two surgeon-connected tissue ends healing together. We take this proposition for granted, but how is it that we presume that the distinct edges will harmonize, nurture each other, and synthesize a bond that is functional, waterproof, and resilient? This is surgery in a nutshell: connecting, stapling, sewing, screwing, splinting, and gluing entities together, and soliciting the body to microscopically, even at the molecular level, lay down connective tissues to supplement those artificial connections, and in time, supersede the temporary scaffold-works.

Billroth, in heroic quantum surges, was making proper diagnoses (remember, they had no MRI machines, CT scanners, ultrasound devices, or X-rays), achieving anesthesia under the most primitive of conditions, and performing surgery *with no electrical lighting*. Hence, the positioning of operative theaters on the top floor of hospitals to harness skylights above; the gods of surgery beckoned the sun to shine upon them. Billroth was avoiding infections with early antiseptic techniques and with limited sterilization of instruments, cutting and manipulating flesh with bare hands, and sewing together intestinal tissue with primeval catgut and silk ligatures. Somehow, hubris be damned, Professor Billroth was achieving significant success, and the world was noticing.

It is estimated that 40 to 50 percent of the leading physicians of the United States between 1850 and 1890 studied in Germany and Austria. No fewer than "ten thousand Americans took some kind of formal medical study at Vienna between 1870 and 1914."[10] They were coming for the new emphasis on laboratory medicine, and Billroth exemplified the German experimental physiology and pathology and the ability to translate that into meaningful clinical interventions, for the first time in human history.

Into that world had come William Stewart Halsted, from 1878 to 1880. He had absorbed the mindset, the techniques and protocols, and had observed what tools and machines were required. He would replicate the structures (organizational and physical edifices) that Billroth had erected, down to the building of a tent on Bellevue grounds. His rapacious animal spirit, initially under Teutonic tutelage and now unbridled in New York City, was on the prowl. Halsted demanded tailored suits, Parisian-made Charvet shirts, French laundering (sending his dirty shirts by steamship to Paris, and weeks later receiving back laundered garments), dandy hats, ties, and eyeglasses. The same sartorial impulses and obsessions were unleashed in his medical practice all over the city at all times of the day and night.

Henry Welch left New York in March 1884, stranding Halsted in the New World. Welch's arrival in Europe coincided with a pharmacological coup; the German pharmaceutical company Merck had isolated and purified an alkaloid from the leaves of the *Erythroxylon* plant that was indigenous to the eastern slopes of the Andes mountains. There was a primordial Incan tradition of chewing and sucking the leaves to achieve an energizing mood alteration, but the transoceanic voyage had always neutralized and weakened the plant's powers once back in European capital cities. Scientists at Merck were able to cultivate the *Erythroxylon coca* plant, and using the newly refined chemistry, were able to isolate the active chemical compound, an alkaloid they labeled "cocaine."

Alkaloids are a diverse group of simple chemical compounds with a dizzying array of conformations, and to science novices, it is surprising that bacteria, fungi, plants, and animals all make these "lock and key" molecules that dock into certain cell receptors to effect a change. Curiously, our mammalian brains have cellular receptors that anticipate interactions with molecules from the coca plant from Colombia, opium poppy seeds from Afghanistan, coffee beans from Ethiopia, and marijuana from Mexico. These alkaloids have wide-ranging pharmacological activities, including psychotropic, antiarrhythmic, anticancer, antimalarial, antibacterial, and vasodilatory, among others. Scientists believe these interactions (which are often toxic or lethal) occur because of evolutionary selection pressure that favors the development of an alkaloid by one species as it interacts with another.

The question must be asked; why would the coca plant synthesize cocaine? Chemists have discovered that cocaine functions as a pesticide, powerfully inhibiting neurotransmitters (the chemicals that nerves use to interact with other nerves) in the brains of insects that would otherwise threaten the coca plant. In essence, the honeybee is offered a tantalizing "bolus of blow" that tempts them to stay and continue to fertilize the plant. A bee that is high is a useful foot soldier in the game of fertilization. Perhaps we shouldn't be too surprised to learn that we have thousands of the same molecular receptors in our bodies that function across the spectrum in the plant and animal kingdoms.

Although the chemical structure of cocaine was not accurately described until 1898 (by Richard Willstätter, a future Nobel laureate), the isolation of cocaine had been achieved by German pharmacists in 1859.[11] It seems like an obvious first move to place cocaine in the mouths of research subjects (read: medical students), replicating the Incan model. Numbness of the oral surfaces was invariably noted, although many young men also behaved bizarrely. In Würzburg, and later Vienna, pharmacists noted positive effects upon persons with melancholy dispositions and among Bavarian soldiers who were under severe physical duress.

Vienna was alive with speculation about the possibilities with the new drug. The Viennese neurologist Sigmund Freud (1856–1939) heard about the new drug, considering it "magical." In 1884 he wrote his fiancée, "I take very small doses of it regularly against depression and against indigestion, and with the most brilliant success."[12] At the Vienna Clinic of Ophthalmology (mere steps away from Billroth's surgical amphitheater), a junior intern, Carl Koller, had been experimenting with drugs to anesthetize the eye, including morphine, sulfate, chloral, and bromide. Koller's mind was prepared.

Koller had been assisting in experiments on cocaine, and decided to place the substance in his own mouth. Appreciating the effects upon his own oral mucosa, Koller knew the next step was an animal experiment. In 1884, with a colleague holding down a large frog, Koller prepared a solution by mixing cocaine powder into distilled water. He placed a drop of the solution into one of the frog's protruding eyes. After waiting a few seconds, Koller touched the eye, testing the reflex. Initially there was no drug effect, but after a minute, "came the great historic moment . . . the frog permitted his cornea to be touched and even injured without a trace of reflex action."[13] After testing rabbits and then a dog, the young interns turned to each other. The solution was trickled into their own eyes, and then with pin in hand, they touched their self-same eyes with the head of a pin. His assistant later recalled, "Almost simultaneously we could joyously assure ourselves, I can't feel anything . . . with that the discovery was completed. I rejoice that I was the first to congratulate Dr. Koller as a benefactor of mankind."[14]

Soon the cocaine solution was used in actual eye surgery with great success. The German Ophthalmological Society Conference was held a few days later in Heidelberg, and Koller, seeking the claim of priority, sent a colleague to present their new discovery. Most medical meetings, then, as now, are characterized by presentations that are moderately mundane and rarely scintillating. Occasionally, a paper is given, and reverentially, the room of professionals marvels at the breakthrough. For good measure, eye surgery was performed the next day in front of the conference audience. An American, Henry Noyes (1832–1902), was in the room that September day in 1884, and rushed home to publish a note on the use of cocaine to achieve local anesthesia. In the October issue of the *New York Medical Record*, Noyes described the use of cocaine, but concluded, "It remains, however, to investigate all the characteristics of this substance, and we may yet find there is a *shadow side* as well as a brilliant side in the discovery."[15]

William Stewart Halsted read the 1884 report in the *Medical Record*, and immediately pondered how to further harness cocaine. Halsted's practical savoir faire, honed over the years, led him to believe that he could use the cocaine solution in a novel way. Instead of dripping it into eyes or swishing it in mouths, Halsted perceived the real potential lay with the use of the newly invented hypodermic needle. As a master anatomist, with incredible knowledge of nerves, where they traveled and what they innervated, Halsted at once conceived the notion of *regional anesthesia*.

This author, as a young medical student in his first anatomy lab, was unsure how large nerves are in the human body. "Are they even visible with the naked eye?" I asked myself. To my great surprise, the peripheral nerves are very large, and as they course down the arm or leg, are as big as a pencil before they branch out into tiny tendrils to where they terminate in muscles or the skin. Within each nerve there are imperceptible nerve fibers that are alternately carrying signals down from the brain or upward toward the brain. *Motor* nerve fibers carry the electrical signal down the spinal cord, along the peripheral nerve, and connect to the muscles they command. Conversely, *sensory* nerve fibers within the peripheral nerve carry the electrical signals from the skin, bones, and soft tissues, communicating messages of pain, touch, sensation, vibration back to the "central processing unit," the brain.

Halsted was sailing in completely uncharted waters. The only way to test his hypothesis of regional anesthesia was to start injecting subjects, and what would be impossible now, he approached the best guinea pigs he could find: his medical students at the quiz sessions. By comparison, Koch seems positively cosmopolitan by experimenting on his daughter's pet rabbit.

Within two weeks of Noyes's publication, Halsted had secured a 4 percent cocaine solution from Parke-Davis and Company, and began injecting students in his Madison Square home office. The injecting parties that ensued must have been sensational. Halsted, elaborate metal-and-glass syringe in hand (no modern sensibilities of "single-use" needles), parading around the parlor and plunging the needle into arms and legs. Occasionally thrusting deeply into a nerve, with an electrical blast of pain down a limb, Halsted would have been as frightened as his vassal. However, when the medicine was deposited adjacent to the nerve, numbness ensued down the limb in an anatomic distribution. Within days, it was obvious that regional anesthesia was not just a conceptual dream, it was a reality.

Many of the students experienced a rush of energy, with occasional nausea, flushing, palpitations, and dizziness. By altering the concentration of cocaine, symptoms were mitigated. Soon, operations were being performed on real patients on a regular basis at Roosevelt, and dental procedures were also accomplished by Halsted's dentist friends.

Today, names like lidocaine, Novocain, and xylocaine are familiar to the reader, but few appreciate their close kinship with cocaine. While the former group of medicines are safely, and innocently, used in clinical settings around the world, cocaine is a different beast altogether. As Noyes had predicted in the first American paper dealing with cocaine, there is a *shadow side* to the remedy.

In New York, by the fall of 1885, signs of trouble were emerging. The students and surgical apprentices who had started to use cocaine snuff and even inject the concoction in social settings were now mired in tribulation. "The students began to drop from sight. The doctors' behavior grew increasingly erratic. They slept less, talked endlessly and excitedly, and eventually performed less surgery and ignored their duties."[16] What started as good-intentioned, quasi-scientific experimentation had led to chemical dependency.

They had become cocaine addicts.

Within a year of the first experimental injections, Halsted, his colleagues, and his students had ceded control of their lives to cocaine. Halsted himself was beginning to miss morning sessions at his Roosevelt Hospital clinic, his erratic behavior resembling that of the junkies that were starting to materialize around the city. His coworkers witnessed his spasmodic gesticulations, nervous tics, impatience, and perspiration-filled encounters. A pharmacologically stimulated comrade is a wild, fiendish alternate of himself, frightening and unnerving.

Halsted made a triumphant return trip to Vienna in the fall of 1885, continuing his cocaine binge while abroad. He demonstrated his regional anesthesia technique to surgeons and dentists, and connected with old friends. There is no record of him meeting with Freud in Vienna, but one can imagine what that interaction might have entailed. Gerald Imber, in *Genius on the Edge*, writes that Halsted had lost command of his life, and those closest to him feared that he would be lost forever.

William Stewart Halsted was back in New York by January 1886, further spiraling out of control, lying, and obfuscating. Welch, his erstwhile friend, now in Baltimore to start the work of establishing Hopkins as an elite institution, was alerted by a mutual colleague about Halsted's deterioration. "Once modest and self-effacing, he was now abrupt, spoke incessantly, and cared little for the response of those he was speaking with." [17] Welch formulated an intervention with two other physician friends, the four professionals meeting in an office to save the young surgeon from cocaine damnation.

The one physician, probably the sole human being, who could confront Halsted was the collegial and brilliant Welch, the scion of Connecticut physicians and Yale Skull and Bonesman, gifted in camaraderie and reasoning. The bachelor Welch directly confronted Halsted about his drug abuse, but didn't stop at the chastisement; he proposed a solution: a lengthy, rejuvenating sea voyage, with the combined benefits of fresh sea air and eventual, forced drug weaning and disentanglement of cocaine's grip. Halsted consented, and by February 1886, Welch had chartered the schooner *Bristol*, bound for the Windward Islands in the southern Caribbean.

The arrangement between Welch and Halsted was for the elder physician to obtain a very large quantity of cocaine and to serve as the custodian and administrator of the medication. The treatment plan stipulated that Welch would gradually decrease the daily dosage until Halsted had become completely weaned off cocaine, and by the time the four-thousand-mile round trip voyage was complete, Halsted would be cured.

On the outward passage, the *Bristol* was a vessel of the gods, a veritable Pegasus of the seas. Welch, the Greek literature expert (and once aspiring college professor), must have been recollecting Odysseus and his evasion of the Sirens. The goddess Circe had warned Odysseus that the Sirens were, in reality, murderous monsters disguising themselves as enchanting women with supernaturally enticing voices. Famously, Odysseus stopped his men's ears with wax to deafen and inoculate them from temptation; but so that he "may have the pleasure of listening," Odysseus kept his ears unplugged. Lashed to the mast so that he could not escape, Odysseus was tantalized beyond his power to resist, straining so vigorously that the bonds cut deeply into his flesh.

The journey degenerated as they arrived in the Caribbean, with Halsted bedeviled by the dwindling doses. The friendship between Welch and Halsted came under extreme duress, and late one night, Halsted broke into the captain's medicine chest

Statue *La Nature se dévoilant à la science* (*Nature is revealed through science*), at the Université Paris 5 René Descartes in Paris. *Photo by David Schneider.*

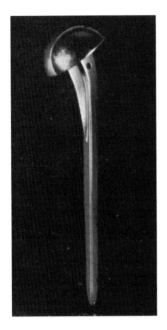

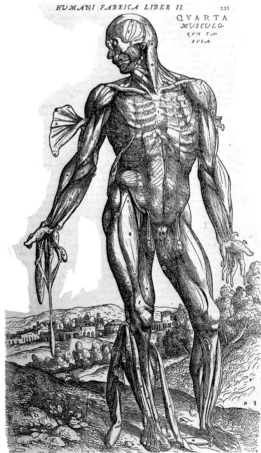

HUMANI FABRICA LIBER II.   221

QVARTA
MUSCULO-
RVM TA-
BVLA.

TOP LEFT: The picture of the future: Dr. Charles Neer's first shoulder implant, presented in *The American Journal of Surgery*, March, 1973., volume 85, issue 3, p. 258. TOP RIGHT: Charles S. Neer II, MD, The father of modern shoulder surgery. *Courtesy of Columbia University.* BOTTOM RIGHT: Muscle man from Andreas Vesalius's *De humani corporis fabrica*, the iconoclastic 1543 human anatomy book that helped launch the Renaissance. *Fourth Muscle Man by Vesalius, Wellcome Collection.*

Title page from *De humani corporis fabrica* with Andreas Vesalius at the center of the engraving. *Title Page by Vesalius, Wellcome Collection.*

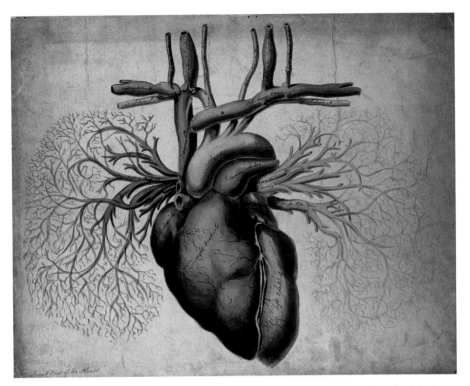

ABOVE: Evolving view of the heart with unoxygenated "blue" blood in the right side of the heart and oxygenated "crimson" blood in the left side of the heart and aorta. *Front View of the Heart, with Arteries and Blood Vessels Attached, watercolor, 18—(?), Wellcome Collection.* BELOW: William Harvey's breakthrough discovery of the valves in the veins of the arms, leading to his discovery of circulation. *The Function of the Valves in the Veins, Wellcome Collection.*

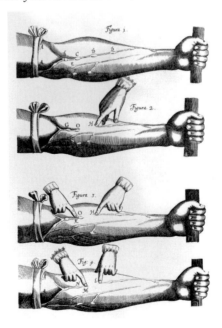

RIGHT: William and John Hunter's eighteenth-century engraving of a nine-month fetus inside the placenta and uterus of its deceased mother, leading to the discovery of uteroplacental nutrition and gas exchange. *The Anatomy of the Human Gravid Uterus by William Hunter, Wellcome Collection.*

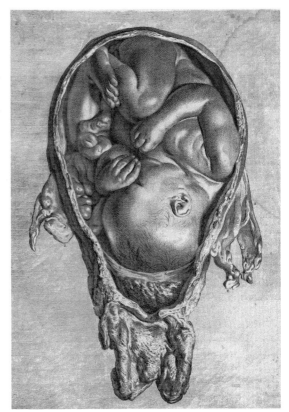

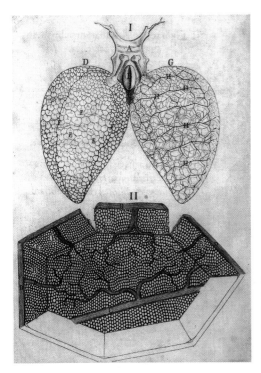

LEFT: Malpighi's drawing of the microscopic architecture of the capillaries of the lungs of a frog, confirming Harvey's conception of circulation. *De Pulmonibus Observation by Marcello Malpighi, Wellcome Collection.*

TOP: St. George's Hospital and Wellington Arch at Hyde Park Corner. St. George's Hospital was the hospital of John Hunter and Henry Gray, and is now the tony Lanesborough, London's finest hotel. *St. George's Hospital and the Constitution Arch, Hyde Park Corner, engraving, Wellcome Collection.* CENTER: John Hunter FRS (1728– 1793), the father of scientific surgery, despite an almost complete lack of formal education. An autodidact and fearless investigator, and perhaps the inspiration behind Dr. Dolittle and Dr. Jekyll and Mr. Hyde. *From The Life of John Hunter: John Hunter, Wellcome Collection.* BOTTOM: Rokitansky's Pathology Institute in Vienna, with the inscription, INDAGANDIS SEDIBUS ET CAUSIS MORBORUM, an obvious nod to Morgagni's revolutionary book, meaning "Investigation of the seats and causes of disease." *Photo by David Schneider.*

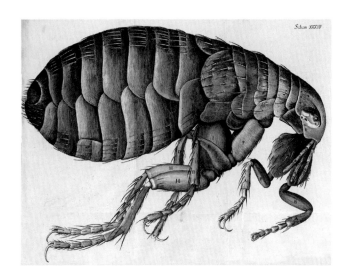

ABOVE: Robert Hooke's revolutionary book *Micrographia* showed the microscopic world in incredible detail. The flea was depicted here, at a time when the linkage between the flea and the plague was unknown. Hooke was playing with fire. *From Micrographia, flea by Robert Hooke, Wellcome Collection.* BELOW LEFT: The Royal Society's first major publication, *Micrographia* was the world's first scientific best seller. This is the engraving that led to the term "cell," for life's basic functioning organ. *Robert Hooke, Micrographia, the cells of the cork plant, Wellcome Collection.* BELOW RIGHT: Hooke's microscope. *From Engraving of a Microscope in Micrographia, 1665 by Robert Hooke, Wellcome Collection.*

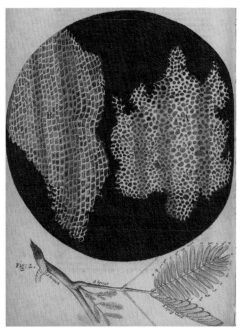

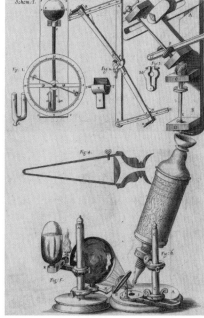

TOP: Typical hematoxylin and eosin slide of human tissue, in this case, the retina, with the nuclei (containing DNA) stained a deep purple and the surrounding tissue a light crimson red. Staining brought the microscopic world out of a dreary nonsensical whitewash into a focused picture of reality. *From Wikimedia Commons.* BOTTOM LEFT: Rudolf Virchow (1821–1902), father of modern pathology, prolific author, and author of the incredible phrase, *Omnis cellula e cellula* ("all cells (come) from cells"). *Portrait of R.L.K. Virchow, Wellcome Collection.* BOTTOM RIGHT: Carl von Rokitansky (1804–1878), one of the chief luminaries of the 19th-century Viennese medical enlightenment, performing 30,000 autopsies over the course of his career. *Portrait of K. F. von Rokitansky by Dauthage, 1853. Wellcome Collection.*

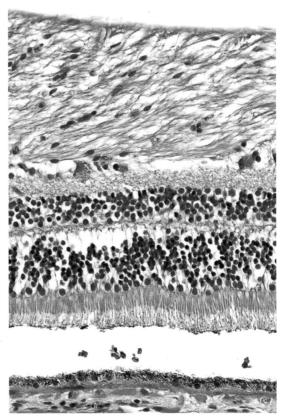

THE ROYAL SOCIETY'S HOUSE IN CRANE COURT (*see page* 104).

ABOVE: The Royal Society's headquarters at Crane Court, London, where Edmund Halley met with fellow cognoscenti to view the solar eclipse. The small alleyway still exists, but like much of central London, the buildings to the left and at the end of the court were lost during World War II bombing. *From Royal Society, Crane Court, off Fleet Street, London: the courtyard, wood engraving after [W. H.], 1877, Wellcome Collection.* LEFT: Isaac Newton FRS (1642–1726), perhaps the greatest genius in the history of thought. Invented calculus, and discovered the laws of gravitation and the properties of light. Became president of the Royal Society and laid the groundwork for vast fields of study to this day. *Sir Isaac Newton. Mezzotint by J. MacArdell after E. Seeman, 1726, Wellcome Collection.*

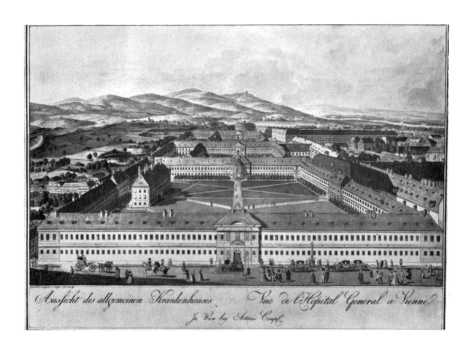

ABOVE: *The Allgemeines Krankenhaus der Stadt Wien*, or AKH, the Vienna General Hospital, where multiple specialties in medicine were invented. *From Wikimedia Commons.* BELOW LEFT: Ignaz Semmelweis (1818–1865), the tortured, unappreciated Hungarian–Austrian physician who unsuccessfully tried to convince doctors to wash their hands. *Portrait of Ignaz Philipp Semmelweis, Wellcome Collection.* BELOW RIGHT: Joseph Lister (1827–1912), perhaps the most important surgeon in world history, the physician scientist who introduced sterile technique and antiseptic surgery. *Portrait of Joseph Lister, 1st Baron Lister, Wellcome Collection.*

ABOVE LEFT: Louis Pasteur (1822–1895), the lone non-physician who ushered in the germ theory, triggering the metamorphosis of surgery into a productive science. *Louis Pasteur, microbiologist and chemist, Wellcome Collection.* ABOVE RIGHT: Robert Koch (1843–1910), pioneering physician who unlocked the secrets of culturing and identifying bacteria. *From Wikimedia Commons.* BELOW LEFT: Paul Ehrlich (1854–1915) and Sahachiro Hata. Ehrlich won the Nobel Prize for his contributions to immunology, and was critical in developing the first effective drug for syphilis, Salvarsan, the precursor to antibiotics. *Paul Ehrlich and Sahachiro Hata, Wellcome Collection.* BELOW RIGHT: Alexander Fleming (1881–1955), discoverer of penicillin, but was unable to master the growth of penicillium, and forfeited the conquest. Co-recipient of the Nobel Prize with Florey and Chain in 1945. *Sir Alexander Fleming. Wellcome Collection.*

Papaver Somniferum.

Poppy – White or Opium.

The opium poppy (*Papaver somniferum*). *Opium poppy* (Papaver somniferum): *white flowers, seed capsule and seeds, coloured zincograph, ca. 1853, after M. Burnett, Wellcome Collection.*

TOP: Humphry Davy (1778–1829), pioneering chemist, the father of electrochemistry, and the discoverer of nitrous oxide, laughing gas. *Portrait of Sir Humphry Davy. Mezzotint by C. Turner, 1835, after H. Howard, Wellcome Collection.* CENTER: Crawford Long (1815–1878), Georgia physician and the first to use ether anesthesia. *Portrait of Crawford Long, Wellcome Collection.* BOTTOM: Massachusetts General Hospital at the very beginning, with the Ether Dome at the top of the hospital. *Image of Massachusetts General Hospital, Boston in 1846–7, Wellcome Collection.*

ABOVE LEFT: Horace Wells (1815–1848), Boston dentist who pioneered the use of ether in dentistry. *Portrait of Horace Wells, photograph of reproduction of stipple engraving, Wellcome Collection.* ABOVE RIGHT: William T. G. Morton (1819–1868), American dentist who first demonstrated ether anesthesia on October 16, 1846, at Massachusetts General Hospital. *From The U. S. National Library of Medicine, digital collection.* BELOW LEFT: Charles Thomas Jackson (1805–1880), Harvard physician who was critical in the development of ether anesthesia. *From Wikimedia Commons.* BELOW RIGHT: John Collins Warren (1778–1856), founder (and dean) of Harvard Medical School and chief of surgery. At the conclusion of the first public demonstration of ether anesthesia, declared, "Gentlemen, this is no humbug." *John Collins Warren, line engraving by S. A. Schooff, 1890, after G. Stuart, 1807. Wellcome Collection.*

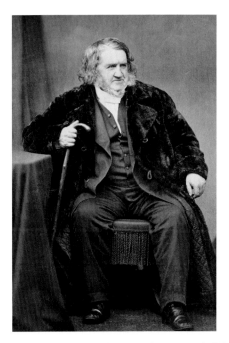

ABOVE LEFT: James Simpson (1811–1870), Edinburgh physician and the first to use chloroform anesthesia. *Portrait of Sir James Young Simpson, photograph, Wellcome Collection.* ABOVE RIGHT: James Snow (1813–1858), father of epidemiology and England's first anesthesiologist. *Portrait by John Snow, 1856, Wellcome Collection.* BELOW LEFT: Bernhard von Langenbeck (1810–1887), legendary German surgeon, founder of surgical residency, and mentor to generations of important surgeons. *Portrait of Bernard Rudolph Conrad von Langenbeck, photogravure after August Hirschwald (?) or H. Riffarth (?), Wellcome Collection.* BELOW RIGHT: Theodor Billroth (1829–1894), world's first abdominal surgeon, courageous and occasionally successful in a time that perilous abdominal surgery was performed without antibiotics. *Photograph of Christian Albert Theodor Billroth by F. Luckhardt, Wellcome Collection.*

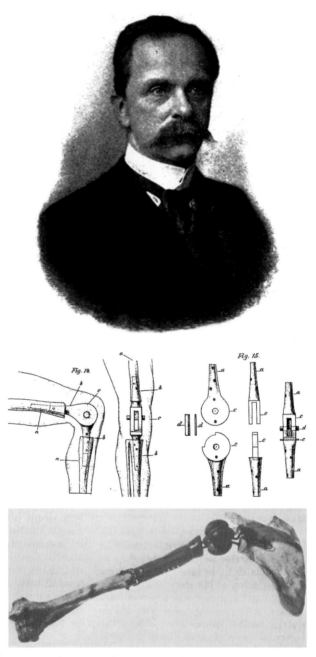

TOP: Themistocles Gluck (1853–1942), Romanian-German surgeon who performed first joint replacement in the world in 1890, with endoprostheses made of ivory. *From Wikimedia Commons.* CENTER: Diagram of Gluck's ivory implants (1890). *From Gluck, Themistocles (Themistokles): Referat über die durch das moderne chirurgische Experiment gewonnen positive Resultate, betreffend die Naht und den Ersatz von Defecten (pp. 101–11; pp.316-82, Taf.3: 20 Ab.).* BOTTOM: The shoulder implant of Pean, located at the Smithsonian Institution. *From Artificial shoulder joint by Pean (1893): The facts of an exceptional intervention and the prosthetic method. Clin Orthop Rel Res June 1978, vol. 133, pp. 215–8.*

RIGHT: Smith-Petersen-type cup arthroplasty that set the stage for all modern joint replacement. *Smith-Peterson-type [sic] acetabular cup for hip replacement surgery, Wellcome Collection.* BOTTOM: Smith-Petersen mold arthroplasties in the first decade. *From Smith-Petersen M.N., Arthroplasty of the Hip, J Bone Joint Surg, 1939.*

FIG. 1

Evolution of the mold:
1923: Glass.
1925: Viscaloid.
1933: Glass (pyrex).

FIG. 2

Evolution of the mold:
1937: Bakelite.
1938: Unsuccessful and successful vitallium molds.

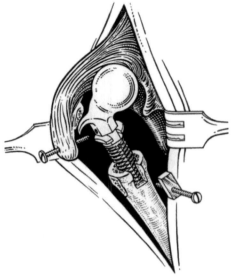

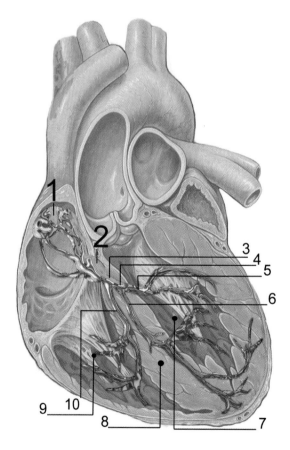

TOP LEFT: McBride's doorknob prosthesis, the harbinger of all future intra-medullary implants. *From McBride, Earl D., A Femoral Head prosthesis for the hip joint J Bone Joint Surgery 1952, Oct; 34 (4): 989.* TOP RIGHT: Dalkon Shield, the disastrous intrauterine implant. *Photo by Jamie Chung; IUD courtesy of Dittrick Medical History Center and Museum/Case Western Reserve University.* BOTTOM: Electrical conduction pathway in the human heart. *From Patrick J. Lynch, medical illustrator; C. Carl Jaffe, M.D., cardiologist, Creative Commons.*

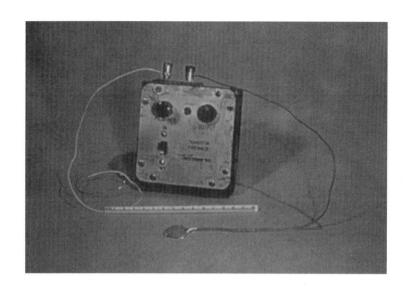

ABOVE: Earl Bakken's first pacemaker. *From Medtronic.* BELOW: Henry VIII granting a royal charter to the Company of Barber-Surgeons. *King Henry VIII Granting a Royal Charter to the Barber-Surgeons Company, wood engraving by H.D. Linton after H. Holbein.*

TOP LEFT: William Worrall Mayo (1819–1911), head of the Mayo patriarchy, father of William and Charlie. *William Worrall Mayo, photogravure with line engraving, Wellcome Collection.* TOP RIGHT: William James Mayo (1861–1939), elder son of W. W. Mayo, revolutionary surgeon despite practicing in the tiny town of Rochester, Minnesota. *Photograph of William James Mayo, Wellcome Collection.* BOTTOM LEFT: Charles Mayo (1865–1939), one of the founders of the Mayo Clinic and renowned operator of unbelievable surgical volume. *Portrait of Charles Mayo, Wellcome Collection.*

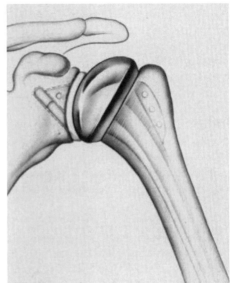

TOP: The total shoulder arthroplasty of Charles S. Neer, MD *From Journal of Bone and Joint Surgery, 1974; 56: 1-13.* CENTER: Broca's Area of the brain. *From Wikimedia Commons.* BOTTOM: Otto Deiters (1834–1863), pioneer in microscopy of nerves. *From Wikimedia Commons.*

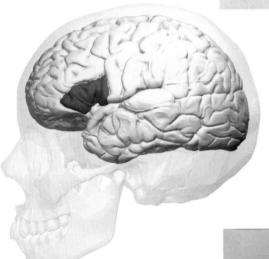

TOP: Camillo Golgi (1843–1926), neuroscience pioneer who invented staining techniques that brought the microscopic architecture into focus. *Portrait of Camillo Golgi. Wellcome Collection.* BOTTOM: The neuron of Otto Deiters. *From Deiters, O. F. K. 1865 Untersuchungen über Gehirn und Rückenmark des Menschen und der Saügethiere. Braunschweig, Germany: Vieweg pp. 1–318.*

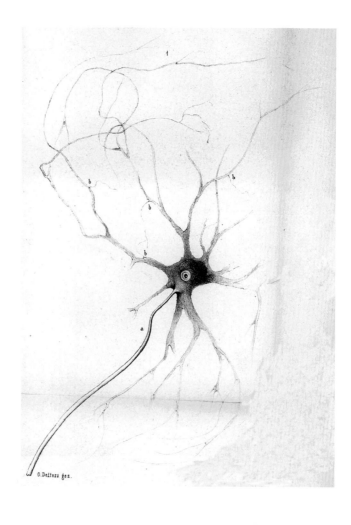

O.Deiters gez.

ABOVE: Santiago Ramon y Cajal (1852–1934), the father of neuroscience. *From Universidad de Ciencias Medicas de la Habana.* BELOW LEFT: Drawing by Ramon y Cajal of the medulla oblongata of a trout. *From The Janelia Archives.* BELOW RIGHT: Drawing by Ramon y Cajal of the retina and optical centers in the eye of a bluebottle fly. *From The Janelia Archives.*

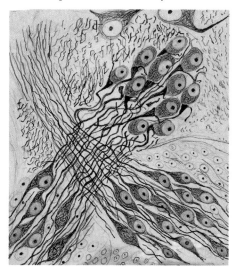

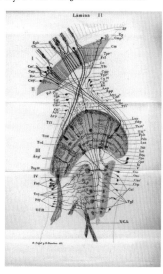

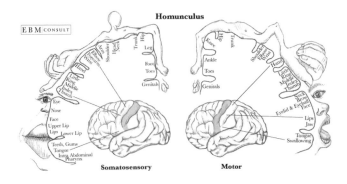

Homunculus

EBM CONSULT

Somatosensory   Motor

TOP: Homunculus of the the human brain. *From EBM Consultant.* CENTER: Alim Louis Benabid. French neurosurgeon who ushered in the implantable brain stimulator era and changed the treatment of Parkinson's disease. *From Wikimedia Commons.* BOTTOM: The inspiring Dowdy couple; deaf and blind but not really handicapped. *Photo by Lowell Handler.*

to procure the remaining cocaine doses. In every sense, Halsted was adrift at sea, and by the time they had reached the Florida coastline on the return trip, he was "plagued by nightmares, exhaustion, irritability, [and] outright suspicion of his fellow travelers . . ."[18] Truly a Greek tragedy, perhaps the only explanation for Halsted's survival is that the gods wanted Halsted to live, both to see the New York harbor, but also to rule in Baltimore someday.

Having failed deprivation treatment, it was clear to Halsted that more intensive medical intervention was required. Welch still believed in his friend, but insisted that he seek help at a sanitarium, at the time just becoming popular for treatment of drug addictions. Halsted checked himself into the Butler Sanitarium in Providence, Rhode Island, in 1886, and would spend seven months hospitalized for his cocaine addiction. A mainstay of treatment at the time was substitution of one drug for another, and Halsted was placed on a regimen of morphine—leading to a lifelong addiction to that as well. While there was an emphasis on healthy eating and outdoor activities, the introduction of morphine would lead to an awkward balancing act; "one drug heightened sensations and a feeling of omnipotence, the other a peaceful release from the world."[19]

Halsted was now in a precarious station in life. Instead of private wealth, his father's business was coming to ruination. There was nothing to fall back on; now six years removed from completion of his medical training, the previous great name he had made for himself was ruined. His lavish tastes firmly set in stone, Halsted desperately needed to revive his professional standing. In December 1886, just eight months from his disastrous oceanic experiment, Welch once again came to the rescue. Teetering, Halsted arrived by train in Baltimore and moved into the same boarding house in downtown Baltimore with Welch.

Johns Hopkins University was founded in downtown Baltimore in 1876 (later moving to a more suburban location in the early 20th century), and although two hundred years younger than Harvard and Yale, is considered the first research university in America. Following the lead of German educational leaders (particularly Wilhelm von Humboldt, founder of the Humboldt University of Berlin), President Gilman emphasized the role of research in education, both at the undergraduate level and among the graduate schools. This scientific discovery of new knowledge, as opposed to unimaginative recitation of outmoded facts, would greatly alter the American academy.

The metamorphological growth in understanding of human disease had changed stunningly in the preceding one hundred years. Morgagni first connected symptoms with anatomical conditions. Rokitansky and Virchow performed organ-based and cellular-oriented autopsies, thus furthering the understanding of morbidity. But it was the new science of bacteriology that was unlocking the comprehension of disease, and the new preeminence of laboratory science was poised to change healthcare in the new world, and Welch (under the watchful eye of Gilman) would establish on Laudenslager's Hill in Baltimore the pantheon of American medicine. It cannot be overemphasized that the first building was the *Pathology research building*. Instead of conceiving a medical campus with medical wards, surgical amphitheaters, and a lying-in clinic, the enterprise's cornerstone would be a building dedicated to the understanding of disease. Morgagni, Rokitansky, and Virchow would have been proud.

With the hospital not scheduled to open until 1889 (and the first medical school class not convening until 1893), Halsted's proposed role in life would be to work in the pathology laboratory with Welch. For the next several years Halsted focused on animal experimentation and basic science research, greatly advancing the science of surgery.

In Vienna, Billroth had blazed a trail in abdominal surgery, but his outcomes were still hit-and-miss. Halsted theorized that a better technique might improve the clinical results, but what was demanded was a scientific analysis of the technique and more important, a technique based upon microanatomy. Within a week of his arrival, Halsted and Franklin Mall began animal surgery and microscopic evaluation of the different layers of the intestines, discovering the heretofore unappreciated role of the seemingly mundane "submucosa."

The intestines have a three-layer configuration: an outer muscular layer, an absorptive interior lining layer, and a thin middle lining of connective tissue—the submucosa. Over the course of a few months, sixty-nine experiments were performed, evaluating the role of the submucosa and testing the strength of the new repair. In April 1887, Halsted presented a paper at the Harvard Medical School, advancing the understanding of how best to achieve robust tissue healing. This groundbreaking research (a cornerstone of general surgery to this day) made intestinal anastomosis (the sewing together of two ends) dependable, immediately impacting the survivability of abdominal surgery. Intestinal surgery would never be the same.

Curiously, Halsted couldn't be positive he was correct; he was not operating on humans. Welch had brought him to Baltimore to labor under his supervision, but as a surgical pariah, he was not to be trusted with real patients. The laboratory was his clinic, and dogs were his patients. Admirably, Halsted administered fastidious care to the dogs. In fact, he developed "a manner of dealing with animal experiments that soon became the national standard,"[20] his second major contribution to surgery in Baltimore in an amazingly brief time.

Welch had petitioned President Gilman's tolerance of Halsted's poor reputation; such was Welch's esteem of his New York companion. The dog surgery, gainful basic science research, positive reception at Harvard—by all appearances William Halsted was finally righting the ship, except he wasn't. Soon after his Boston presentation, Dr. Halsted checked himself into Butler Sanitarium again, this time for *nine months*. Over a twenty-two-month-long stint, "William Stewart" (his assumed name at Butler) spent sixteen months at the sanitarium, accomplishing nothing scientifically in the prime of his surgical career. No one taking bets at that time would have predicted he would ever resume surgical practice again, let alone assume the mantle of the most significant surgeon in American history.

Halsted clandestinely returned from his sophomore recuperation at Butler in December 1887, and quietly returned to work at the Pathology Building. The bachelor surgeon faithfully dined and socialized at the Maryland Club near his Baltimore dwelling, fastidiously researching by day with Welch and the young coterie of physicians. The previously energetic and gregarious surgeon had adopted a protective carapace, his old-fashioned pince-nez eyeglasses perched on his nose, shielding a dark secret of combined addiction to cocaine and morphine. Perhaps only Welch ever knew (for sure) that Halsted was an unrelenting, unrepentant, and partially incapacitated user.

In the classic German legend of the scholarly alchemist Faust, the malcontented intellectual makes a pact with the devil, exchanging his soul for unlimited knowledge and worldly pleasures. There is no doubt that Halsted's cocaine use started innocently, but he rapidly was bewitched by its powers. Prescribed morphine at the sanitarium, the Greek god of dreams, Morpheus, also entranced Halsted. Sadly, a pervasive personality change gripped Dr. Halsted to his grave. Serious, secretive, ashamed, and vulnerable, he retreated into a cocoon that few could break into. By our nature, surgeons are "control freaks," and the powerlessness must have tortured Halsted.

The routine at the "Pathology" had been comforting to William Stewart Halsted, and by early 1889, he had begun seeing patients and operating at local hospitals. He had arrived in Baltimore just two years before, shattered. Slowly, the former metropolitan bon vivant was reclaiming pride of profession, and a major decision now faced the administration at Johns Hopkins. Who would be hired as surgeon at the new hospital?

The Johns Hopkins Hospital opened in May 1889. Although the Pathology division had been functional under William Welch for several years, a full-time surgeon was only hired in the months before the hospital opened. Today, every hospital in America thrives or dies based upon surgical volume. In the 21st century, surgeons find themselves in very advantaged positions based upon the revenue generated by their work. In 1889, there were fewer than ten physicians in the United States whose practices were limited to surgery,[21] such was the nascent nature of the profession. It is not so surprising that President Gilman had already hired the scientist and pathologist Welch, and then in turn employed William Osler to be the founding internist. Osler, a Canadian by birth, moved from the University of Pennsylvania and would become the preeminent physician in the world, eventually settling at Oxford as the Regius Chair of Medicine.

Just months before the opening of the hospital there was still no surgeon on staff. The leaders at Hopkins determined that their fledgling institution would be one of the rare hospitals that would hire a fully engaged surgeon. The successor to John Lister in Glasgow, Sir William Macewen was a pioneering surgeon in his own right. An important innovator in brain and bone surgery, Macewen's prospects unraveled when his insistence on bringing his entire nursing staff was not acceded to. Now scrambling, the board of trustees at the hospital faced a difficult decision, hire the one surgeon known to them, warts and all, or scurry to find another European candidate.

After some deliberation, and more than a little trepidation, Halsted was invited to be the surgeon-in-chief to the dispensary and acting surgeon to the hospital in February of 1889. Three months later the hospital would open, and more than three years would transpire before the medical school commenced.

High on a hill above Baltimore, the domed redbrick buildings with intricate roof lines were rising in magisterial dominion. The dome of the administrative building, perched on the western edge of the campus, has become synonymous with Hopkins medicine, but Welch and Halsted were making scientific history at the nondescript hindmost of the university grounds.

At Bellevue in 1885, Halsted had overseen the construction of an operating room tent, but at Hopkins (for the first fifteen years), all surgical procedures were performed in the basement of Ward G, in a makeshift area lit by gaslight. All of the early ana- tomical and surgical theaters throughout Europe and America were at the top floors of academic buildings, harnessing the natural light streaming in from skylights and large building windows. Unglamorously, Halsted operated under improvised conditions in the women's ward near the Pathology building.

Halsted had an old German operating table from the Franco-Prussian War, equipped with a central trough that drained blood and the deluge of caustic pre- paratory solutions. On the table was stationed a stretcher that also served as the implement for patient transport. Instead of the classic black wool Prince Albert frock coat favored by surgeons, Halsted selected white duck cotton operating suits, with high collars and short sleeves, topping off the ensemble with a little cotton skull cap. Indeed, the garb was still primarily worn to protect the surgeon's finery beneath, but it was still a dramatic improvement over the blood-caked and detritus-covered coats.

There in the basement, dressed in white, incorporating what he had learned in Vienna, Berlin, and Würzburg, Halsted cobbled together a surgical practice from scratch. Having pioneered regional anesthesia and scientific animal experimentation, Halsted initiated a series of innovative changes that resonate in every hospital and academic institution to this day. In the opening years of the Johns Hopkins Hospital there were no trees, no medical students, and few surgical colleagues. "Halsted was a complex and isolated man, forbidding and nurturing; rigid, proper, and secretive; compulsive and negligent; stimulating and reclusive; addicted and abstemious; oblivious and solicitous; and always concerned with advancing the science of surgery."[22] If Halsted's story is familiar, it may be that you have seen *The Knick*, the television series whose central figure is Dr. John Thackery, a character who is very much based upon William Stewart Halsted. Forever burdened with his drug impulses, he pushed on, his revolution beginning almost immediately at Hopkins, starting with his care of a nurse's chafed and inflamed hands.

Lister had pioneered antiseptic surgery a quarter of a century prior to the naissance of Johns Hopkins Hospital. It had started with sponge dousing of carbolic acid to a surgical site, and then evolved to the use of an atomizer to spray the acid into the air. Next, the surgeons' hands were approached with special indignity, with application of strata of acids, oxidizing agents, and amalgams. Scrubbing, dipping, coating, and painting was consigning everyone's hands to becoming reddened and inflamed, if not purified. In particular, a well-to-do Southerner, nurse Caroline Hampton was troubled with dermatitis of her hands.

Welch had brought rubber work gloves back from Germany, using them for autopsy duty. They were too thick and clumsy for surgical work, so Halsted had another thought: could an American company help him develop a product better suited for the operating room? Halsted later recalled, "As she [Caroline] was an unusually efficient woman, I gave the matter my consideration and one day in New York requested the Goodyear Rubber Company to make as an experiment two pair of thin rubber gloves with gauntlets. On trial these proved to be so satisfactory that additional gloves were ordered."[23] It took a few years for everyone in the operating room to adopt Caroline's gloves, but the "single greatest advance in the history of sterile technique"[24] had occurred because of Halsted's concerns for his nurse, who, just a few months later, became his wife. Almost forty years old, Halsted broke the bonds of bachelorhood, with Welch (forever single) serving as best man.

Before his arrival in Baltimore, Halsted had performed a series of "firsts" that were simultaneously courageous and bizarre. He was the first surgeon to remove gallstones, performing the operation in the home of a privileged and elderly woman suffering a life-threatening gallstone attack. The patient happened to be his mother. Later, he performed the first blood transfusion, using the new implements of hypodermic needles and tubing. The patient was suffering from postpartum uterine bleeding, but her life was spared after the transfusion. The donor? Dr. Halsted himself. The recipient? Halsted's own sister. He also performed the first appendectomy, but tamely not on a relative.

The surgeon-in-chief Halsted quickly settled into a routine of bold surgical innovations with assistant surgeon J.M.T. Finney and his first resident Fred Brockway. Gerald Imber, in his book, *Genius on the Edge*, poignantly captures the first breast cancer operation performed at Johns Hopkins in June 1889, on a thirty-eight-year-old mother of ten who had furtively battled a growing, abscess-filled tumor for six months:

Halsted inscribed an extensive incision from the axilla, near the old abscess site, counterclockwise down and along the sternum, under the breast, encompassing the entire breast, and up the lateral aspect, meeting the original swipe and forming a giant, bloody teardrop.

Hardened by infection, the skin near the axilla was unusually difficult to reflect upward, and the lymph nodes under the arm couldn't be reached. The recent abscess had matted them down. Common sense dictated returning to the axilla on another day. Dissecting with a scalpel, Halsted mobilized the entire breast and much of the underlying pectoralis major muscle. He applied artery forceps to arteries and veins as they appeared, and secured the vessels with fine silk sutures to minimize blood loss and crushed tissue. He removed the anatomical specimen in its entirety and carefully examined it at the operating table. Having taken great care to avoid cutting into the tumor for fear of spreading the cancer, Halsted now rolled the mass between bare fingers, and cut through its substance, making careful mental notes of its consistency and appearance before sharing his thoughts with his assistant. He placed numerous suture tags on areas of interest before sending the specimen to Welch's pathology laboratory, where microscopic sections would be prepared for later examination.[25]

Halsted, like his mentors in Vienna and Halle, was attempting the first forays into the war against cancer. Pioneers like Virchow had identified the bizarre and warped cells that characterized tumors. These brigand cells, pathologic by nature, bunched into colonies and then whole masses, feeding off the host, multiplying, occupying, and eventually achieving a self-defeating coup d'état when the host succumbed. Richard von Volkmann, Billroth, and now Halsted, believed that cutting wider, deeper, and more aggressively was the answer to freeing the patient from the invader.

The ancient Greek word for cancer was *karkinos*, or *crab*. In 400 B.C.E., Hippocrates observed a tumor mass as a "clutch of swollen blood vessels around it, [reminding] Hippocrates of a crab dug in the sand with its legs spread in a circle."[26] Later writers would further embellish upon the idea of cancer as an alien invader, the crab outfitted with a tough carapace and bellicose pincers. With this mindset, the Halstedian concept of the surgeon's role as liberator, the chief function was to separate the patient from her disease. Unlike the Greeks, who had no microscopes and no conception of the cellular basis of disease, Halsted understood that cancers were comprised of abnormal cells. What he and other dauntless surgeons were missing was the pathologic process of *vascular metastases*, where the cancer cells spread through the bloodstream and not just through contiguous, centrifugal growth.

Only decades since the discovery of what cancer really was—the pathologic transformation and growth of cells—and decades before chemotherapy and radiation would become available, Halsted perceived his heroic task was to lacerate, harrow, and extirpate (and disfigure and dismember, if need be). In the few years between the launch of the hospital and the opening of the medical school in 1893, Halsted would enlarge the zone of excision, eventually removing the entire pectoralis, and occasionally ribs, the collarbone, and all the lymph nodes. He wanted to get at the root of the cancer, naming the operation *radical mastectomy*, from the Latin word for root, *radix*. We misconstrue the meaning of radical in this context when we assume it means "severe" or "profound." Clearly, radical mastectomy, in Halsted's hands, was severe and uncompromising, but he was conceptually digging deeper.

In the end, the radical mastectomy would lose its favored position among cancer surgeons. A few brave surgeons, like Geoffrey Keynes, George Crile Jr., and Bernard Fisher, bucked tradition, believing that simple, less aggressive surgery was just as effective, and certainly less morbid, than radical mastectomy. (This transformation is elegantly reported by Siddhartha Mukherjee in *The Emperor of All Maladies*.) While the technique itself is no longer performed, we can certainly understand why Halsted posited that it might work. Wide resection is still a mainstay in solid tissue oncology, where shockingly wide swathes of muscles, skin, and bone are removed. Halsted did dramatically improve cancer mortality, but a cure was (and is) still on the horizon.

If radical mastectomy did not have staying power, other initiatives of Halsted did. Halsted had been inspired by the German way of training surgeons, notably Langenbeck's and Billroth's ideas of robust, demanding inculcation. Osler agreed with Halsted, and the first formal residencies in American institutions were at Hopkins. By 1893 Halsted was granted the title "Professor," and he formalized his total immersion training system, where young men (required to be unmarried) would live at the hospital, and be available twenty-four hours a day, seven days a week. Old-time surgeons laugh at young American trainees who complain about "living at the hospital," when in fact they "only" work eighty hours per week, as limited by federal law. (My wife reported a sobering story to me while I was a resident, years before the law went into effect and I regularly worked more than one hundred hours per week. Driving by the Penn State Hospital, my young daughter cried out, "Mommy, that's where Daddy lives!" I actually shed a tear when I heard the story, pangs of guilt and remorse crushing me.)

Halsted's particular skill of identifying potential, imbuing confidence and crafting surgical skill potentiated his reach across the United States for decades, since so many of his disciples became surgical leaders themselves. Gone were the days of part-time hacks with no scientific training masquerading as surgeons. "Laboratories were integrated into the great hospitals, aseptic surgery was slowly accepted, and postgraduate training became available."[27] "Halsted's legacy was built on two equally potent, unimpeachably world-altering platforms. The first was the establishment of the school of scientific, safe, and anatomically correct surgery; the second, a working environment that shaped the education of generations of surgeons and propelled American surgery to its preeminent position in the world. The former brought about an undeniable surgical revolution, proven by consistently superior results . . ."[28]

As a young surgical resident, learning the delicate technical points of surgery, a professor of mine, displeased with my unrefined approach, said at one point, "Did you just hear that?"

Confused, I said, "No, I didn't hear anything."

My professor replied, "I think that was the ghost of Dr. Halsted rolling over in his grave, the way you just crushed that tissue in your forceps."

Through years of careful badgering, we all learned how to be elegant surgeons. "Aseptic technique, gentle handling of tissue, scrupulous hemostasis, and tension-free, crush-free, and anatomically proper surgery are the rules. And they are Halsted's rules. Although "Halsted" is not a household name, every individual in America who undergoes successful surgery owes William Stewart Halsted a nod and a deep debt of gratitude."[29]

As important as Halsted was in transforming "the shunned black sheep of the medical world into a specialty offering the promise of mightily alleviating the suffering of the human condition,"[30] helping reinvent a discipline and a surgical philosophy,[31] perhaps one of his greatest innovations was performing *elective surgery*.

From the beginning of time, prototypical surgeons were busy draining abscesses, applying salves, and engaging in bonesetting. These primordial healers, retaliating against the gods and evil spirits, were always reflexive, responding to bad fortune. The forebears of surgery were the physicians to the gladiators and military doctors. With advancement, surgeons turned from hopeless emergency cases to conditions where their intervention might actually be beneficial. Patients submitted at a point of in extremis prior to Billroth, but cancer patients in Vienna, ill and racked with cancer, allowed him to remove parts of their bowels. Confidence in surgery was growing.

In the same month that Halsted executed his first Hopkins mastectomy, he performed, arguably, the most important elective operation ever attempted.

Children are born with inguinal hernias at a rate of 5 percent, but the incidence rises to almost 15 percent among adults, with men eight times more susceptible. "Turn your head and cough," is the stuff of legends, but prior to successful surgery, hernias led to millions of deaths around the world. Prior to safe operations, hernia belts ("trusses") and postural exercises were the only remedies offered, even if ineffective.

A hernia is a loop of bowel spouting into a weakened portion of the abdominal wall. This can occur at the umbilicus or at a surgical incision, but most commonly in the groin. An *inguinal hernia* occurs when "a ring in the lower abdomen through which the spermatic cord exits is unnaturally expanded to allow the insinuation of bowel, which follows the path of the spermatic cord through the external inguinal ring and into the scrotum."[32] Stated simply, a weakness in the lower abdominal wall allows your "guts" to billow through. Every medical student remembers the photo of a man with a hernia so massive it spilled dozens of pounds of intestines into his scrotum, necessitating ambulation with a wheelbarrow in front of him. I shudder just thinking about it.

The loop of bowel that pokes through the abdominal wall can become strangulated, hastening death, but typically, the hernia patient presents with non-life-threatening complaints of pain and unsightly swelling. With no effective treatment, and a tacit "you're not dying," physicians were accustomed to brushing away patients. That all changed on June 13, 1889.

Halsted was not the first to attempt hernia repair, just the first to report (in English) a dramatically effective technique. Of course, it was based on scientific analysis of the anatomical structures. He had performed numerous fastidious dissections, comprehending that deep repair of the tough fascia and muscles to the stout inguinal ligament, and reinforcing the internal abdominal ring, were the keys to a successful repair. Making an incision in the groin area, Halsted would discerningly tease the tissue layers apart, protect the spermatic cord, excise the hernia sack, and suture the appropriate layers together. The Halsted repair was born.

Halsted presented his technique in late 1889 at the Johns Hopkins Hospital Medical Society, and published his report in January 1890. In retrospect, Edoardo Bassini, an Italian surgeon in Padua, had independently published a similar procedure (in Italian) a couple months before, leading some academics to give dual credit to Bassini and Halsted. Rapidly, patients came streaming from all over the country to have Halsted repair their hernias.

Further refinement by Halsted, and his star residents, drove the success rate even higher. One of Halsted's great legacies was Henry Cushing, another Yale graduate, who would refine the hernia repair of Halsted to include the use of cocaine anesthesia. (Cushing would become the father of neurosurgery, and his face can be seen on the logo of the American Association of Neurological Surgeons.) Today's technique of hernia repair is technologically different than the Halsted/Bassini approach, but the significance of Halsted's intervention for a serious condition, under elective scheduling, shocked the trajectory of hospital care. Just a few decades before, hospitals were only death houses, and no one (in their right mind) would consult with a surgeon when relatively well.

By the 1890s, the Professor, with dozens of residents in the wings, was overseeing a vast surgical empire. For only a few cognoscenti, Halsted's travails with cocaine and morphine addiction were known. He would disappear every summer for months, leaving assistant surgeons and surgical residents in charge of the most sophisticated hospital in the world. His absence was ideal training ground for a life of surgery, at times the loneliest profession.

In the 20th century, there were a small number of hospitals in America that had become surgical meccas, including Johns Hopkins and the Mayo Clinic. Medicine and surgery, under the leadership of German thinkers and refined under American devotees, was no longer dependent upon divination and sorcery. Medical education reform, by way of the Flexner report of 1910, had revolutionized the way medical schools operated, and had resulted in the closure of half of American medical schools by 1920. Flimflam no longer tolerated, only schools with true academic missions would be accredited, facilitating the transfer of the mantle of leadership of surgery from Europe to America. This handover was complete with the cataclysms of the world wars.

Halsted passed in 1922 at the Johns Hopkins Hospital. Days shy of his 70th birthday, he died childless, but with numerous professional heirs and innumerable philosophical debtors. Patients had reason to believe in medical sciences by the Roaring Twenties, even before the advent of antibiotics. And a great deal of that confidence is owed to Professor Halsted. "For the few who knew of his ability to navigate uncharted waters [drug use] while the siren song rang in his ears, his journey was nothing short of heroic. If a single person can be considered the father of modern surgery, the only contender is William Stewart Halsted."[33]

ELEVEN

Vitallium

"My plan was to return home from Sweden to Chicago in 1967, and to spend the rest of my life conducting biomechanical spine research and performing spine surgery. But when I spent two weeks with Dr. Charnley in Wrightington, England, watching the first 'modern' hip replacement surgeries, it hit me like a thunderbolt. I would never have dreamed it was all possible. I knew I had seen the future, and in a moment I changed the course of my life, and returned to Chicago to spend the rest of my life performing hip and knee replacements."

—Jorge Galante, MD

"Dr. Neer became disenchanted with the end results of patients with fractures of the proximal humerus treated with resection of the humeral head. He mentioned this to Dr. Darrach who said, 'Smiley, why don't you do something about it?'"

—Charles Rockwood, MD

The mosquito nets that hang down from the ceiling in the open-air hospital ward are like spinnaker sails over an armada of hospital beds. In Rwanda, like most African countries, the hospital wards have limited walls and windows, which facilitate the free flow of air across the dozens of patients whose cots are lined up in neat rows. Tuberculosis is thwarted with the exchange of air, but open windows permit mosquito entry. Each cot has a mosquito net anchored overhead to a rafter; the netting itself is opaque and dense enough to block the tiny mosquitos that transmit the parasites that causes malaria. The nocturnal routine is always the same, where every inhabitant unfurls the twisted-up netting to cover all four corners of their sleeping area. In the morning, the process is reversed, and the white nets are wound up with a dozen twists and moored to a corner of a bed.

I arrived late last night from another outpost hospital in the western provinces of Rwanda with our team of American doctors and nurses and Rwandan surgical residents. I have traveled to Central Africa to mentor the young physicians, always heeding the advice of a veteran surgeon before my first trip overseas years ago: when you prepare to teach in a third-world country, throw your modern textbooks out the window. They are outdated in two respects: the proposed treatments are based on current technology and newfangled gadgets (which are not available in developing countries) and the diseases are distinctive (namely, TB is still a major problem around the globe, much less so in America). A much better approach is to find a fifty-year-old textbook, where TB of the bone is addressed and simple tools are used to solve everyday problems.

Waking early, I join the local medical staff in the outdoor atrium for morning rounds. Men, women, and children are all housed in the main ward, which is surrounded by massive trees, heavy vines, and exotic-sounding birds. My team and I (joined by a Rwandan orthopedic resident, Paul) enter the main ward, where I witness most of the patients twirling up their mosquito nets. Some patients are listless, and I'm told that there has been a gastrointestinal illness that has been rampant through the region in the last few weeks. Along one edge of the building are the malaria sufferers, obvious to the trained eye, owing to their utter catatonic state and complete lifelessness. In medical school, I had learned about the flu-like symptoms, including headache, fever, joint pain, and even convulsions, but seeing malaria patients up close, suffering so severely, has given me a profound respect for the little plasmodium that causes malaria and a new fear of mosquitos.

Paul and I walk along one aisle of patients, flanked by little cots in neat rows. Most Westerners want a private hospital room; here there are sixty cots in this high-ceilinged, open-air ward. As we come to the end of the row I spot a young man, probably twelve years old, who is lying on his back, propped up on his elbows, with a few blankets under his back and head. The nurse here at Kibogora Hospital tells me that Joseph has been here, in that cot, for two years. His TB is so severe it has spread to his right hip, knee, and tibia (his shin bone), with the result that his pelvis and groin pain won't allow him to lie flat or sit in a chair. He has reached a truce with his infection, but is imperceptibly ceding ground as his leg pain becomes more crippling. Joseph has very dark skin, an anxious smile with beautiful teeth, and piercing, despondent eyes. I know he was told about my coming here, and in his adolescent mind, I wonder what his hopes are. I pull back the white sheet covering his legs, and with my coterie of nurses and aides crowding in, quickly ascertain which leg is infected. His left leg is twice the size of his right, and he holds it perfectly still. His right leg is smaller, at least in part, because he's not been out of bed once in two years. But his left leg is swollen, with tight skin from his waist to his toes. The nurses warn me not to move his leg, and I'm rapidly losing hope that his leg is treatable. Through the interpreter I tell him that I will look at his X-rays and will return to discuss the plan. Quietly he replies, "murakoze," which is "thank you" in Kinyarwandan.

Here in this remote hospital on the shores of Lake Kivu, there is small X-ray unit and a certified X-ray tech. The equipment and the film quality are like stepping back in time, and as I snap the X-rays onto the viewing box, my heart sinks. Joseph's femur and tibia are three times their normal size, wavy and deformed, moth-eaten and crooked. From his hip joint to his ankle, his bones are being eaten alive by the TB bacteria. I wonder, what would I do right now in America? Would long-term antibiotics and multiple operations save his leg? Or would we end up replacing his hip and knee after multiple operations? Would an orthopedic oncologist be able to save this leg and get Joseph out of bed and off his back?

After contemplating Joseph's plight, his two years in bed, the fact that both parents were murdered in the 1994 genocide, understanding that he is an orphan in every sense of the word, I think the only reasonable treatment is amputation. If Joseph is ever going to have a chance of getting out of bed and leaving this hospital alive it will be as an amputee. How do you tell a child (through an interpreter, no less) that he must lose his leg in an attempt to save his life? It's a short conversation surrounded by patients, staff, and concerned family members from the surrounding inpatients. Joseph only says, "Okay."

Cutting off Joseph's leg is among the hardest things I have ever done. Performing an amputation at the hip joint is challenging technically. There are enormous blood vessels that must be treated expertly so that Joseph doesn't bleed to death on the operating table, and the diseased leg is so swollen and edematous that dissection is difficult. I have no electrocautery so every tiny vessel is tied off by hand. I carefully deal with flaps of muscle so the wound will close well, and after ninety minutes of tedious exploration, the moment of final amputation occurs. With the hip joint disarticulated and the limb finally free, I hand off the leg to a nurse. I loathe this moment. I have amputated arms and legs over the years, and it always feels like abject failure, even though I know it's the right thing. In war, in business, in parenting, we "take our losses," but amputation emotionally always feels like comprehensive defeat.

The surgery is accomplished under spinal anesthesia, so that Joseph is conscious and awake during the operation. He seems to be uttering the same phrase again and again, and I ask one of the nurses what it is. Solemnly she responds, "He is saying, 'Please take me home with you.'" Blindsided, I'm a jumble of emotions, saddened by Joseph's plight and filled with hatred for tuberculosis.

Dr. Charles Neer performed his first shoulder implant arthroplasty on January 26, 1953, on a fifty-four-year-old housewife who three years earlier had suffered a severe fracture of her left shoulder. The patient had initially been treated with simple physical therapy by another orthopedic surgeon, leaving her with almost no motion of her shoulder. Encumbered with unrelenting pain and poor function, Mrs. "T.M." submitted herself to Dr. Neer's care, becoming the first person in the world to receive the "Neer shoulder implant." Her outcome was excellent—she later told Charlie Neer that "the old pain is now gone"—with dramatically improved shoulder motion and function.

Dr. Neer's first shoulder publication, "Fracture of the neck of the humerus with dislocation of the head fragment," which had reported on the poor results of treating severe shoulder fractures with removal of the humeral head, was published in the March 1953 issue of the *American Journal of Surgery*. Submitted and accepted for publication in 1952, it merely showed a picture of the shoulder implant, while allowing for the fact that it had never been used. Therefore, by the time the publication had been received by librarians and surgeons around the world, patient "T.M." had already become the pioneering guinea pig. While it is accurate to call her the *index case* of Dr. Neer's series of shoulder arthroplasties (from Greek, "to form or mold a new joint"), she was not the first human to have a shoulder implant inserted.

Tuberculosis is still a worldwide infectious disease conundrum. Caused by the bacteria *Mycobacterium tuberculosis*, TB is spread from person to person through the air when infected patients cough, spit, or sneeze. Unlike a *staph* infection of the skin, which is characterized by a reddened, often pus-filled abscess, TB does not gain entrance to the body via the skin, but through the lungs. Today, one-third of the world's population has latent TB, which means that the bacteria have established a foothold in the pulmonary tissues, but have not yet caused illness (nor are the patients able to transmit the disease). Because of the typically slow onset of TB, patients may linger for years with a mild cough, night sweats, and progressive weight loss. If bacteria were capable of strategic planning, this would be their plan for establishing a beach-head to all of humanity: make the disease tolerable at the onset while targeting the lungs, where infestation and irritation will trigger coughing and hacking, leading to the conveyance of aerosolized droplets of bacterial colonies in the new crowded cities of the world in the post-industrialized era. Touching someone with a *staph* infection can get you infected, but stand one foot away and you're safe. Conversely, touching someone with TB *cannot* spread the disease; share a room or workplace with a TB patient and you can die.

Mycobacterium tuberculosis is a bacterium that has been afflicting mankind for at least ten thousand years, and is still an important worldwide pathogen (killing over a million people per year). TB is simultaneously one of the most deadly, curable, and preventable diseases in the world,[1] and the World Health Organization estimates that over two billion people are infected with the organism. Most Americans and Europeans today are shielded from the realities of the global epidemic, but until recently, no one was sheltered from the scourge of *consumption* (as the pulmonary form of TB infection was called). Like a dental cavity that eventually degenerates into an abscessed tooth, chronic pulmonary TB can transmute into infectious tubercles in the bones of the spine or extremities, or even brain tissue.

When a mycobacterial infection spreads to the bones, a progressive deformity, with years-in-the-making, slow-motion churning and gurgling of bone (like bubbling tar) develops, and the warped limbs and joints gradually become more painful and dysfunctional. In 1720, Benjamin Marten, while living in London, theorized that consumption was caused by an infection of the lungs, and that the provocative agent to be blamed was an *animalcule*, small enough that it couldn't be seen with the microscopes of the day. It was Robert Koch who finally identified the bacterium that causes TB in 1882, which ended up earning him the Nobel Prize in Medicine. It surprisingly wasn't until the 1880s that there was a consensus on the germ theory among learned men; prior to that time simple issues of sanitation and hygiene were mocked. With no medicines to control patients' TB infections (streptomycin wasn't discovered until 1943), it was left to surgeons to intervene bravely, if not cavalierly, in the lives of advanced TB sufferers.

Themistocles Gluck (1853–1942) started practicing medicine in 1882, just when the European medical community was awakening to the linkage of germs (bacteria, parasites, and viruses), infections, and illness. Gluck had trained in Berlin under legendary surgeon Bernhard von Langenbeck and pioneering pathologist Rudolf Virchow. Like thinking about genetics without knowing about DNA, it's difficult to conceive of practicing medicine without believing in microbes. The newly minted Dr. Gluck was launching his career in the months that Koch elucidated the bacterium that causes TB; while no pharmacologic treatment would be identified for over half a century, Gluck became obsessed with the orthopedic treatment of those afflicted with bony TB. Appointed head of surgery at Berlin's Emperor and Empress Friedrich Paediatric Hospital, Gluck initially practiced in an era where there were no X-rays. Flying blindly, he and his colleagues had to imagine what they would find once they incised the skin and dissected deeper.

Gluck's early animal experimentation focused on organ resection and transplantation. Vastly ahead of his time, experimenting on animals, he made the observation that the loss of one kidney would result in the other kidney redoubling its function, heralding the body's ability to adapt to loss. If Prometheus, one of the Greek mythological Titans, could regenerate his liver nightly after having it pecked at during the day by an eagle, and if Mary Shelley's Victor Frankenstein (published, amazingly, in 1818) could create a creature ex nihilo, what was a German surgeon capable of now that anesthesia was becoming practical? As a veteran war surgeon, Gluck and his colleagues had seen the success of treating battlefield wounds, and if traumatic skin lacerations could reliably heal after sewing the edges together with silk and catgut sutures, could deeper structures (muscles, tendons, organs, even bones?) also heal after being connected or stabilized? To answer these questions, Gluck continued his animal experimentation, finding that separated tissues could be brought together and heal, in a process he called "autoplasticity."[2] During the Serbian-Bulgarian war (1885), Gluck was able to treat a soldier's femur fracture with two steel plates and screws, amazed at the rapid recovery and early motion afforded by his intervention. The next logical philosophical leap would be the introduction of other foreign materials in the body as *substitutes* for bones, taking autoplasticity to previously unimagined (and potentially catastrophic) echelons.

Science fiction writers can, on occasion, resemble prophets. Jules Verne said, "Anything one man can imagine, other men can make real," and it's possible that it was Dr. Frankenstein who animated Dr. Gluck's inkling about tissue regeneration. In the late

1880s, Gluck's animal studies concentrated on tissue replacements, and his attempts to develop what he termed "guide rails" for regrowth of diseased or damaged tissues turned his hand to aluminum, wood, glass, nickel-plated steel, and ivory. We now know that the primary mineral content of ivory, hydroxyapatite, is identical to the dentin of teeth and the main non-collagenous substrate of bone, but in Gluck's time, the visual similarities and abundant supply (the Ivory Coast was then a German colony) of ivory made it an obvious selection for investigation. Ivory became his material of choice for bone substitution, and in short order, in 1890, he began implanting carved and machined pieces of ivory into diseased joints, starting with a knee replacement in a seventeen-year-old girl, followed by a wrist replacement three weeks later. He reported performing fourteen arthroplasties that year (including knee, hip, wrist, and elbow), with all patients suffering from TB receiving an artificial joint.[3]

The initial results of Gluck's ivory joint arthroplasties were remarkable. Patients were pain-free and movement had been restored. Gaining confidence, Gluck was eager to present his short-term successes to the Berlin Medical Society, but trouble was brewing. Surgeons in Germany already regarded his use of silk sutures to repair tendon injuries as scandalous, and now, this madman was proposing a presentation on his series of joint replacements. All of his patients had previously endured long-standing tuberculous joint infections (which typically do not generate pus), but following surgery, all of his patients were doomed to infections of another kind: draining-puss and red-hot joints which could lead to systemic bacteremia (bacteria in the blood), sepsis, and even death. In response to Gluck's desire to present his findings at the Society, the president of the German surgical association wrote him, saying, "As the leader of German surgery I cannot allow that you discredit German science in front of a platform of international surgical specialists. My pupils and I will fight you with all means." Gluck was forced to abandon not only his presentation, but the work on arthroplasty in its entirety. Medical societies function well when avant-garde ideas are treated with skepticism; it is the long-lens of history that allows for careful analysis of revolutionary treatments, and Gluck's innovation of replacing a joint with a hand-fashioned ivory apparatus was truly *too* ahead of its time, predating antibiotics, modern metallurgy, and implant sterilization. While Themistocles Gluck never tried replacing a joint again, he pursued other advances in general surgery, enjoying a long career. Later in life, the man who rightly is regarded as an "unrecognized genius" and the "first arthroplasty surgeon," observed, "We can certainly make the observation in medicine, as often also in other scientific disciplines, that certain facts have been known as such for a long time before their value is truly recognized."[4]

No one is sure if Gluck ever implanted one of his ivory implants in a shoulder, but there is no doubt that Jules-Emile Péan implanted the first metal shoulder replacement in a human in Paris on March 11, 1893. Dr. Péan was a celebrated surgeon in Paris, renowned for his dexterity, inspirational teaching, and dramatic flair. The French had elevated medicine from a lowly professional status in the early 19th century to a respected scientific enterprise, but had relinquished the role of ascendancy in medical thought to the Germans in the last half of the century. As recalled in the *Journal of Shoulder and Elbow Surgery*, Péan made a lasting contribution to the world of shoulder surgery when he implanted a custom-made platinum shoulder implant in a thirty-seven-year-old baker who was dying from a severe TB infection:

> The prosthesis was designed and constructed with some speed (after Péan had earlier resected part of the infected humerus) by Dr. J. Porter Michaels, a Parisian dentist. The shaft component was made of a platinum cylinder with two ridges and several holes for attachment of the periosteum and muscles. There were screw holes at the distal end for attachment to the bone stump. The head consisted of a ball of rubber previously hardened by boiling in paraffin for twenty-four hours. The rubber ball contained two equatorial grooves arranged at right angles. Each groove contained a metal loop, one attached to the glenoid and the other to the proximal end of the shaft component. It was believed that this prosthesis provided ample strength combined with freedom of movement.[5]

In what initially appeared to be an unqualified victory, the first signs of trouble appeared at one year, when the patient developed redness and draining about the elbow. An operation was required to alleviate the symptoms, and after three additional surgical debridements, it seemed as though the infection was under control. A modern surgeon, upon presentation of this case, would conclude that unlike Gluck's experience, the Parisian baker had not experienced a postoperative skin-borne infection (such as *staphylococcus*), which would have led to draining pus from the joint within days of the first operation, but instead, had recurrence of the TB infection around the implant. With no antibiotics to control the local recurrence, the implant was fated for failure.

Two years after implantation, a draining sinus of pus had developed around the arm. No amount of antibiotics (even today) could reverse a chronically draining prosthetic joint infection—only implant removal could mollify the affliction. Before the surgeon Péan removed the platinum and rubber implant, he did something else momentous: *he took an X-ray*. Röntgen had discovered the illuminating power of electromagnetic rays while working in Wurzburg, Germany, in 1895, and in the world's first arthroplasty X-ray, Péan took an image showing a "long and resistant osseous shell around the prosthesis."[6] The implant that was removed in Paris 120 years ago, instead of being buried in some rubbish pile, is in Washington, DC, at the Smithsonian Institution, ready for viewing for anyone that wants to see the progenitor to all the millions of implants that find their way into human bodies every month in our world.

Like heeding a "Don't Tread On Me" flag, surgeons were properly menaced by the thought of chronic infection in any joint wherein they contemplated implant surgery. Decades passed after Péan removed his platinum and rubber shoulder implant; the only implants being inserted were metal plates and screws in the occasional fracture patient, although scientists had never established which metal alloy was best indicated for human application. However, in the early decades of the 20th century, surgeons were awakening to the ravages of joint arthritis, particularly of the hip and knee. Although the incidence of TB was decreasing with improved living conditions and the emphasis on fresh air, X-rays were allowing doctors to see arthritis with their own eyes as never before.

Every joint in the human body shares several characteristics: at least two connecting bones, a joint capsule which functions as a membrane that encompasses the lubricating fluid, stabilizing ligaments that hold the joint together, and most miraculously, articular cartilage. The cartilage simultaneously provides a cushion and a smooth, gliding surface that enables painless motion—providing that the cartilage is healthy. Articular cartilage is the slipperiest material in the world, although with age it begins to lose its astounding properties. When a physician tells someone they have arthritis of a joint, the implication is that the cartilage is diseased, either through "wear and tear" degeneration, or alternatively, through an auto-immune process (like rheumatoid arthritis) wherein the body's immune system attacks and destroys the articular cartilage. In either case, once the cartilage is worn or destroyed, arthritis (literally, "inflammation of a joint") results in pain, stiffness, progressive loss of motion, formation of bony outcroppings on the bone ends, loose cartilaginous bodies floating in the joint, and, oftentimes, deformation of the joint. Prior to 1895, physicians could only examine a patient through direct observation, but "X-ray vision" empowered practitioners to visualize arthritis, and start thinking about not just treating TB infections, but to start thinking about *surgically treating arthritis*.

◆

Marius Smith-Petersen (1886–1953) spent his entire professional career in Boston, operating at the Massachusetts General Hospital until his death at age sixty-seven. Dr. Smith-Petersen was a native of Norway who attended high school and college in Wisconsin, then completed his medical schooling and orthopedic residency training at Harvard. Renowned for his singleness of purpose, personal magnetism, and dramatic flair while operating,[7] his greatest contributions regarded the hip. Even while still a resident in orthopedics, he developed a novel surgical approach to the hip, which has been utilized around the world for decades. Frustrated with the poor results of the rudimentary treatment of hip fractures of the femoral neck he had been taught in residency, Dr. Smith-Petersen innovated the tri-flanged hip nail that vastly improved the lives of millions of patients, and continues, with slight modifications, to be the treatment of choice today. It was this device that Dr. Charles Neer reviewed in his first paper.

Dr. Smith-Petersen's greatest discovery was an implant that rendered itself obsolete, and sparked an insurrection against one of mankind's supreme foes, arthritis.

Once physicians could see hip arthritis on an X-ray, the question became: what is of greater consequence, the bone spurs that tentacle out like coral from the femoral head and acetabulum (hip socket), or the bone-on-bone articulation that results from the loss of the articular cartilage coating? It's not surprising that surgeons concluded that the surrounding bone spurs might be the culprit, as the critical biomechanical studies had not yet been performed that would reveal how astonishingly slippery healthy cartilage is. Like lifting the hood of a car, applying automotive rubber cleaner on belts and tubes, and expecting the engine to run better, the bony projections around the hip joint were attacked in hopes of improving the performance of the hip. In the 1920s, surgeons began making large skin incisions around the hip, deeply exploring the muscles around the hip socket, and using hammers, chisels, and large pliers-like biters to excavate and gouge away the carapace that entombed an arthritic hip joint. While this improved some patients' hip range of motion and addressed some of their pain, surgeons like Dr. Smith-Petersen were unconvinced. Advancing the technique further, a small group of surgeons modified the procedure to include a makeshift biological covering of the femoral head. Harvesting a swatch of the broad tendon on the side of the thigh (the *fascia lata*), a surgeon could festoon an improvised gliding surface where the articular cartilage had been lost. Surgeons were hoping (in vain) the body would accept the tendinous tissue and refabricate a new gliding surface; in short order, those who cared to scrutinize their results realized that cartilage was more special than they had previously thought.

While Smith-Petersen was not ready to completely abandon the newfangled arthroplasty of the twenties, of sculpting the bones around an arthritic joint, a chance observation kindled a thought about the body's ability to form new tissue. In 1923, the Harvard surgeon had seen a young man who complained of a painful mass in his back. Smith-Petersen took the patient to the operative theater and found a piece of glass that was encased in the patient's own scar tissue. Recalling later, he said, ". . . a piece of glass was removed; it had been there for a year. It was surrounded by a minimal amount of scar tissue, lined by a glistening synovial sac, containing a few drops of clear yellow fluid. This benign reaction to an inert foreign body gave rise to the thought there was a repair which might be applied to arthroplasty. This first thought gradually developed and the idea of the "mould" was conceived. A mould of some inert material, interposed between the newly shaped surfaces of the head of the femur and the acetabulum, would guide nature's repair so that the defects would be eliminated. Upon completion of repair the mould would be removed, leaving smooth, congruous surfaces mechanically suited for function."[8] Smith-Petersen therefore imagined an operation where the rounded end of a glass cylinder would be positioned over a newly shaped femoral head; the body's reaction would be to grow a new fibrous tissue, ideally like the hyaline cartilage of the femoral head, that would provide new gliding surfaces. The glass mold was intended to be an incubator of sorts, with the intention of coming back for a second operation, between fifteen and twenty-five months after surgery, for simple removal of the mold.

Through trial and error, over a ten-year period, the cup was made of glass, viscaloid (a form of celluloid), Bakelite, and Pyrex, and were tested with variable success, starting in 1923. Breakage of the glass, reaction to the viscaloid, and infections were occurring. Pyrex was much more stable and unbreakable, and patients were actually growing a cartilage-like structure on the head of the femur. Fifteen years after the first cup, or mold arthroplasty, it was clear that glass and polymers were not strong or durable enough for the enterprise.

Realizing that the first attempts at mold arthroplasty were not up for the task, Smith-Petersen was discussing his failures with his Boston dentist, Dr. John Cooke, who suggested Vitallium, a newly reported metal alloy that was gaining favor among dentists. Austenal Laboratories had been founded by Dr. Reiner Erdle and Dr. Charles Prange with the express purpose of creating dental castings from new metal alloys. Since the bronze age, man has tinkered with combinations of elemental metals (such as copper, tin, iron, and zinc), but with the discovery of rarer elements, such as chromium, cobalt, and molybdenum, "stainless" or "rustless" steel became a possibility.

We are used to seeing the steel girders of an industrial factory or bridge rust, a process of "oxidation" where the ambient oxygen in the air combines (in the presence of water) with the iron in steel, forming *iron oxide* through a series of chemical "redox" reactions. Surprisingly, these same reactions can occur *in the human body*, where oxygen and water are obviously present. A regular steel implant would catastrophically fail in a person; it would be the work of metallurgists in the 20th century to identify which metals were practicable for human use. Trial and error with the available elements would determine which alloys were suitable, based on their workability, stiffness, and affordability, but it would be a surgeon from San Antonio who would unlock the secrets of alloys for Dr. Smith-Petersen, Dr. Neer, and every surgeon who followed them until today.

Fracture care was dramatically improved with the advent of radiography; surgeons could actually see, and then classify, fractures of every bone of the body. Eventually, trauma doctors learned which broken bones healed with plaster of Paris casts and which fractures required operative intervention. Metal screws and plates (long, thin slabs of metal with screw holes) were initially made of pure metals (e.g., aluminum, silver, gold, and tin), but these proved to be too soft and weak to do the job of buttressing bone while it healed. Iron was much stronger, but the chemical reaction in the surrounding tissues made it impractical. Hoping to minimize corrosion, the next logical step was to electrochemically coat the plate with chromium ("galvanized steel"), but disastrously, where the screw threads engaged the plate, the chromium plating was scuffed away, exposing the steel to corrosion.

The only acceptable orthopedic implant would therefore be a through-and-through alloy, but no biologic testing had ever been performed. The Virginia-raised Dr. Charles Venable and his two San Antonio orthopedic surgeon partners dedicated years testing screws in the forearm bones of dogs, evaluating them with X-rays and microscopic analysis of the bones and surrounding tissues. Experimenting on twenty-four dogs, a dizzying array of metal screws were placed close to each in the forearm bones of dogs—realizing that each metal has its own "electromotive force," or EMF, a mathematical expression that describes its electric potential in relation to other metals. Screws comprised of metals with widely divergent EMF's placed close to each other create an electric potential, which creates galvanic action between the two screws. Of all the metals and alloys tested, there was one that was best tolerated by the bone and adjacent tissues: Vitallium.[9] Unlike other stainless steel alloys, Vitallium had no iron, and was exclusively comprised of cobalt (65 percent), chromium (30 percent), molybdenum (5 percent), with trace manganese and silicon. Vitallium demonstrated the least electrolytic reaction and greatest healing.

In their concluding paragraph, Venable and his coauthors stated, ". . . we suggest an alloy for bone work that is similar to Vitallium . . ." (provided that its chromium did not end up being toxic in the human body). The die, so to speak, was cast. Austenal Laboratories had been advocating the use of Vitallium to dentists for some years, but Venable's publication was the first in a surgical journal to mention the alloy. Dr. Smith-Petersen (with personal assurance of Dr. Venable, who told him that it was "entirely suitable"), immediately set about making Vitallium molds, and after some trial and error, was able to implant the first mold in June 1938. Just ten months later, his most significant paper was published[10]—he already knew the principle of mold arthroplasty was a major innovation—but the breakthrough touch was the Vitallium material. Now that there was a resilient and durable implant, he would ponder an intriguing eventuality in the main body of his paper: "The question now arises whether the original mold [sic] principle may not have to be sacrificed; if Vitallium proves to be inert, even when interposed between two moving surfaces, there may be no reason for the second stage, that is, the removal of the mold [sic]."[11] Smith-Petersen was guessing that the new "resurfacing" operation might be enough, but only time would tell. As science fiction writer Jules Verne said, "Science, my lad, is made up of mistakes, but they are mistakes which it is useful to make, because they lead little by little to the truth."

The conceptual leap from a mobile cap covering the femoral head to a substituting metal head, in retrospect, is not surprising. Surgeons were used to dealing with fractured and crumbling femoral heads; the mechanized emotional response to those patients was a fatalistic acceptance of a life in a wicker wheelchair. The enterprising surgeons practicing orthopedics circa World War II could envision the use of a metal head to completely *replace* a necrotic and collapsing proximal femur, and the first to do it was Dr. Harold Bohlman, of Baltimore and the Johns Hopkins Hospital. He had read Venable's and Smith-Petersen's publications with great interest, and within months of the Vitallium mold arthroplasty paper, Bohlman was able to configure a Vitallium alloy ball mounted on a metal stem, resembling a Tootsie Pop.

Within a year, surgeons around the world were scrambling to design femoral head replacement prostheses similar to Bohlman's, oftentimes from metal but occasionally from newfangled polymers such as acrylic. The initial enthusiasm would be tempered within a few years, when surgeons started to see loosening, subsidence, and failure of the pegged metal balls. Previously crippled patients, whose hip pain and disability had been (temporarily) addressed with implantation of a Vitallium head, were becoming disabled once again. Toward the end of World War II, just a few years from the introduction of improved alloys, the other astonishing development that emboldened surgeons was the discovery, and production refinement, of penicillin. Although steam sterilization (developed in the 1880s) of surgical instruments and implants had mitigated the risk of infection, antibiotics affected a change in every corner of the medical world, further awakening an interest in more profound answers to hip maladies. If infections with molds and small, pegged femoral heads were relatively negligible, pioneering surgeons considered whether mega-prostheses might perform better in the long-term.

The irony in science and technology is that many of the world's greatest innovations occurred in lonely, backwater locales by individuals who had an inkling, an inspiration about how to solve a problem. For every Mozart from Salzburg, there is a Bob Dylan from Hibbing, Minnesota. And when it comes to surgery, groundbreaking divination emanates from unlikely characters from places like Oklahoma City and Columbia, South Carolina.

The American Academy of Orthopedic Surgeons (AAOS) is the largest, and most important orthopedic organization in the world. The AAOS has an annual meeting, and for the first fifteen years it was usually held in Chicago, its national headquarters. But in 1950, thousands of surgeons attended the meeting at the Waldorf Astoria Hotel in New York City, and in attendance was a surgeon from Oklahoma City, with a sixty-five-year-old female patient in tow. Dr. Earl McBride was there to present an exhibit about his new hip implant that was unlike anything that had ever been conceived. Oddly, at his side was the first patient to undergo replacement with a Vitallium head that was perched atop a long metal stem that was positioned down the hollow femoral canal. Rarely done today, physicians in the 1800s would parade patients before their colleagues to demonstrate rare diseases or unusual outcomes. It's safe to surmise that Dr. McBride brought his patient all the way to Manhattan from Oklahoma City because he wanted surgeons to see for themselves her remarkable outcome.

Within two years, his publication in the *Journal of Bone and Joint Surgery*[12] detailed the use of his "door-knob" prosthesis, a joint replacement where the implant was impacted down the canal of the femur. This was the first time since 1890, when Dr. Gluck had stuffed ivory implants into long bone canals, except of course that McBride was using Vitallium. He wrote:

> Even though she had essentially a full range of motion, no pain, and was able to walk on the streets [of New York] without support, there was a great variety of assertions such as: "It just won't work; Too much metal; The acetabulum won't stand up; There will be pressure necrosis; It will work loose."[13]

In the 1952 publication, all twenty-two patients (including the original woman who accompanied him to New York) were doing well with no infections and no disloca-tions. The naysayers were wrong. The threaded stem of McBride's implant would *not* stand the test of time, but the impact of his innovation is best appreciated when you consider the rise of stemmed implants for the hip and shoulder that exploded in the months after the February 1950 meeting. It wasn't until the 1970s that the United States Food and Drug Administration (FDA) would have a Bureau of Medical Devices to regulate such things. Until then, there was a free-for-all with all medical devices, with the result being that any surgeon or engineer could imagine, design, craft, and implant a device with essentially no barriers. Not always good for patients, the restriction-free environment in the United States (and the rest of the world), combined with the technological explosion and the advancement of antimicrobial drugs, set the stage for the rise of implants.

In preparation for his planned presentation at the International Surgical Conference in Berlin in 1890, Themistocles Gluck collaborated with medical instrument manufac-turers and the royal furniture craftsmen to fit a human skeleton with ivory shoulder, elbow, wrist, knee, and ankle replacements. One can imagine the hours of labor required to assemble the human bones together with the handcrafted ivory replacement joints, but after being rejected by his German surgical counterparts (fearing his work would disgrace German science), Gluck wasn't allowed to present his exhibit, which came to be known as the "Skeleton of Paris."[14] It was displayed around Europe for decades, but was lost to the Soviets after World War II. Conceptually, Frankenstein's monster had gone from a fantastical horror creation to a skeleton that was an osseous and ivory amalgam.

Within a few decades of the war, a television show premiered in 1973 in which a US astronaut named Steve Austin is severely injured in an experimental aircraft crash. Barely alive, he becomes the *Six Million Dollar Man*. Portraying the next promethean creator, the mysterious government operative Oscar Goldman intones, "Gentlemen, we can rebuild him. We have the technology. We can make him better than he was. Better . . . stronger . . . faster."

TWELVE

Oversight and Entitlement

"I think we just delivered the South to the Republican Party for a long time to come."

—President Lyndon B. Johnson upon
signing the Civil Rights Act of 1964

One of the great daily challenges every medical student faces is hunting for a quiet corner of the world to hunker down and study like a madman. My solution as a first-year student had been to bypass the library and make a quest for an abandoned hallway in one of the older buildings on the historic University of Kansas medical campus in Kansas City. The oldest medical school west of the Mississippi River, KU has been perched atop a hill overlooking the railyards and Kansas River below, and nestled up against State Line Road for over a century. The redbrick buildings are a mishmash of clinical, research, administrative, and inpatient wards bearing the names of the East Coast pioneering physicians who forsook prestige in Boston, New York, and Philadelphia, and made their way to the cattle town at the confluence of the Missouri and Kansas Rivers.

My favorite haunt has become the Eaton Building, whose upper stories lie dormant, and the speckled, worn marble floors hint at a past life of a clinical ward or hospital wing. Although there is no sign that says, "Do Not Enter," I'm not entirely sure I'm welcome in this abandoned building, but it's quiet and I have cobbled together a decent desk and chair, and after a few weeks it's starting to feel like it's my space.

Every night I come to Eaton, and am happy that my spot remains my little sacred study hollow. I like the smell of this oversized room, a faint essence of iodine (which always reminds me of my veterinarian father) mixed with bygone cleaning supplies and old-fashioned floor wax. Not that anyone has cleaned this room, or entire unit, for years, but the isolation is exactly what I want. There are over fifty buildings on this campus, and I can't believe my good fortune in finding a no-fuss, no-drama, study bastion.

Looking out the window of this scholarly domicile, through crooked blinds with broken strings, I see the massive Bell Memorial Hospital of the University of Kansas. It's actually the fourth hospital to bear the name, but this 1979 edifice is massive and modern, constructed of white concrete slabs and large window panes with exposed vent tubes and interior stairwells that recall the Lloyd's of London building. Its exterior is lit by flood lamps, and the contradiction between its modernity and my little brick building, shadowy and quiet, stirs something monkish within me, invigorating contemplations of the inner workings of the human body.

Lost in thought, memorizing the origins and insertions of the muscles around the shoulder joint, I faintly hear a scratching and clanging sound emanating from the hallway. As happens when you're isolated and alone, an odd sound supercharges my senses, and I feel like a sonar technician on a submarine. The reverberations of a sandy-scraping and metallic knell are unfamiliar and disquieting; now I'm fully attuned to the shuffle-shuttle-clang *coming my way. The hallway outside my room is dark, and turning in my chair to face the doorway, an inky silhouette dissolves into view, accompanied by the syncopated motif.*

Adjusting my eyes, I am gazing at an aged, friendly African American woman, bent-over and crooked, bearing the scars of years of labor and arthritis. Her black leather "old lady" shoes are worn and unevenly eroded owing to her angled and warped knees and ankles. Her brown dress is a little ragged, and her black overcoat is draped over her sagging shoulders, an odd clothing choice for August, but typical for an urban woman getting dressed for an important meeting or church. In her leathered hand is a rusty length of rebar, the textured metal rod used in construction sites. This explains the clanging sound I heard emanating from the hallway, but the heavy metal rod seems like an odd choice for a cane, with no handle and its significant weight.

Out of breath and disoriented, the woman's relief matches mine, hers from having found a living soul in this vacated building, and mine from her not being a phantom. She finally asks me, "Do you know where Room 312 is?"

Standing up, I make my way to my vagabond friend, and discover that her name is Mrs. Robinson. Her frilled black hat rests askew on her head, with gray hair springing out from underneath its brim. Her dark eyes, yellowed sclera, and fatigued visage cannot conceal her sociable character, and catching her breath, Mrs. Robinson tells me her grandson was in a car accident. Her family had told her that Vernon was being housed in Room 312, and with that mention, Mrs. Robinson glances up at the number over my doorway. Realizing that Mrs. Robinson is lost, and has somehow found herself in this unoccupied structure, I point to the illuminated modern hospital viewable from the windows of our room. "That's the University Hospital, and that's where your grandson's room is."

"But Vernon is colored, so I knew he would be here."

Now I'm curious, wondering if there is actually some reason Mrs. Robinson is here. "But why would he be here?"

"Because this is the Negro Ward, where I used to work as a nurse, and this is where all the colored doctors and nurses cared for the negro patients."

I am speechless. Kansas was founded as a free state in the midst of the Civil War, just across the border from its contentious neighbor, the slave state of Missouri. To the surprise of many, Civil War battles were fought within miles of the KU medical campus, but Kansans prevailed, never allowing slavery. Sadly, segregation thrived, even in Kansas (the landmark Supreme Court decision Brown v. Board of Education *was situated in Topeka, Kansas), but I never considered that the University of Kansas had a separate hospital for some of its citizens.*

I take Mrs. Robinson over to the main hospital, hearing stories about life before and after segregation and the Civil Rights movement. I am pleased to be the one to inform her that all patients, without respect for race or religion, are housed together in the main hospital. After dropping her off at the elevator bank (still clinging to her clanking cane), I head to the cafeteria for more coffee. The hospital cafeteria is built on the western side of the modern hospital, with a large wall of windows jutting out from the main structure. Looking outside, across a small grassy garden, I am staring at my study haven, the Eaton Building. I never noticed before, but there is a walled-off former main entrance on the east side of the old brick building. I look closer, and chiseled above the entryway (that leads nowhere) is the identification, EATON BUILDING. This building used to be on the backside of the medical campus, and that set of doors used to be the main entrance. Peering through the dark summer evening, Midwestern fireflies providing traces of light, I squint a little more at the Eaton Building sign, and realize that not long ago, carved into stone were the words "Negro Ward."

Following the standoff at the University of Alabama between Governor George Wallace and the federal government regarding the admission of two African American students, President John F. Kennedy on June 11, 1963, gave a heartfelt plea that would become known as the "Civil Rights Address." Initially standing "in the schoolhouse door," Governor Wallace relented, but that night, President Kennedy delivered a thirteen-minute speech about equal access to public institutions for black Americans, saying, "I am therefore, asking the Congress to enact legislation giving all Americans the right to be served in facilities which are open to the public—hotels, restaurants, theaters, retail stores, and similar establishments. This seems to me to be an elementary right. Its denial is an arbitrary indignity that no American in 1963 should have to endure, but many do."

Five months later, President Kennedy was assassinated, and while there are boundless disastrous consequences associated with his slaying, there are at least two surprising accomplishments that likely would not have occurred without his murder. Before Kennedy's original term had ended, and within seven months of his slaying, the Civil Rights Act of 1964 was signed into law by President Lyndon B. Johnson. One year later, Medicare and Medicaid were simultaneously established by the signing of the Social Security Amendments of 1965. The two most significant acts, part of what Johnson called the "Great Society," share an unanticipated linkage, enjoying a symbiotic relationship not fully appreciated, even today. In President Kennedy's Civil Rights Address, when cataloguing the various institutions (hotels, restaurants, theaters, and retail stores) that barred equal access to Americans of African descent, he never mentioned hospitals. In 1963, it was too much to ask for, and not realistic, but within a few short years, most of the country's over five hundred "Negro Wards" would be shuttered, rendered impractical following the formation of Medicare.

Medicare is the single largest payer for health care in the United States; and it just celebrated its 50th birthday. If Medicare were human, it just reached eligibility for AARP. Once a senior citizen begins receiving Social Security benefits, she is automatically enrolled in Medicare Part A. One is "entitled" to receive Medicare benefits as a United States citizen; one most opt to forego *all* of their Social Security benefits to *not* be enrolled in Medicare. Hence, the term, "entitlement" spending when considering Medicare benefits. This type of "compulsory" health insurance began in late 19th century Germany, where the philosophy of "social insurance" was introduced by the leadership team of Chancellor Otto von Bismarck.[1] The German Sickness Insurance Act was passed into law in 1883 (the year after Robert Koch discovered the bacteria that causes tuberculosis), establishing a compulsory health care system financed by employer and employee contributions, in which the wealthy contributed more than the poor.[2]

Many European countries adopted similar forms of compulsory health coverage, including Great Britain in 1911. The British National Health Service was later formed in 1948, ensuring health care coverage for all Brits (emulated in Canada in 1968). The social insurance program initially constructed in Britain in 1911 was no doubt inspirational to American intellectuals, including the leader of the Progressive (Bull Moose) Party, Theodore Roosevelt.

Roosevelt's politics had morphed from a strict, conservative "blue blood" Republican agenda to progressive, social equality initiative orientation, and national health insurance was a party platform for the Progressive Party in 1912. Both Roosevelt (Progressive) and Taft (Republican) lost the 1912 election, with Democrat Woodrow Wilson securing the presidency from 1913 to 1921. While national health insurance had been an issue of some interest in 1912, the momentum for its passage was stalled by World War I.[3] Decades would pass until a form of national health insurance would become a political possibility.

Justin Ford Kimball, a native Texan, had graduated from Baylor University and law school at the University of Michigan, and started his career as a high school teacher and principal in small Texas and Louisiana towns. Successful at every turn, he became a school superintendent before returning to Waco to practice law. Critically, Kimball then worked as counsel for the receiver in the case of a bankrupt chain of insurance companies, which exposed him to actuarial sciences. In a field where analytical skills are paramount, combining mathematical and statistical methods to analyze and manage risk, Kimball was a natural.

A gifted administrator, "crusty and colorful . . . a worldly erudite man who claimed kinship to half the population of the state and was well connected to the Dallas upper crust,"[4] Kimball became the superintendent of schools in Dallas from 1914 to 1924.

Halfway through his tenure, a worldwide plague of biblical proportions rattled an already fatigued nation.

The horrific influenza pandemic of 1918, one of history's greatest natural disasters, had spread around the world in the final year of World War I, resulting in the deaths of more than fifty million people worldwide. More Americans died from the flu (675,000), than perished overseas fighting in the Great War. With no flu vaccine and no medicines to treat the lethal pneumonia that accompanied the disease, Americans felt particularly vulnerable to the contagion. While less than one thousand people died in the Dallas area, sickness and lost workdays were critical issues that confronted Superintendent Kimball.

Kimball created a "sick benefit fund" for the city's teachers to protect their livelihoods during the influenza pandemic, "where a membership contribution of $1 a month entitled those who fell sick to compensation of $5 a day, which offset lost earnings after the first week of illness."[5] His experience in evaluating and mitigating risk using statistics had prepared Kimball to formulate a program for his teachers. It also forced him to carefully tabulate numbers of subscribers and their health-related statistics.

In 1929, Kimball became the vice president of Baylor University's Dallas medical campus, overseeing education programs at the colleges of medicine, dentistry, and nursing, with an additional responsibility of shoring up the shaky finances of the university hospital. Hospitals were just transforming from almshouses and places of death and dying to sanctuaries of recovery and wellness, but at a price. Baylor University Medical Center was in deep financial trouble—even before the collapse of Wall Street on October 29, 1929—and Kimball was the kind of pragmatic, multitalented Texan for the job.

As school superintendent, Kimball had full knowledge of the numbers of teachers "insured" through their sick benefit fund, and now as administrator of the Baylor Hospital, he gained access to the hospital's finances, its costs and revenues, and, in particular, how much money Baylor was bleeding when caring for Dallas teachers. One of Kimball's young protégés, Bryce Twitty, conjectured, "Why we couldn't do for sick people [teachers] what lumber camps and railroads had done for their employees . . . [referring to] company doctors"[6] who tended to local workers, benefitting both the companies and the well-being of the labor force.

In the early fall of 1929, Kimball approached his old friends at the Dallas schools administration, proposing a type of *hospital prepayment program*, where teachers could make a similar monthly payment to budget against future hospital bills. There were no national actuarial data to guide Kimball; life insurance companies had always shied away from health care, and no one had worked out the statistics between healthcare demand and costs. How much to charge teachers? No one knew, but Kimball did have the thorough records from his days as superintendent. "Those records . . . were the only actuarial material I could find anywhere in the U.S. I had designed the forms myself to extract this information, having been an insurance lawyer. [During the fall, advertisements were circulated among teachers] that if 75 percent of the teaching group would sign up and send in 50 cents each month beginning with their November sick-benefit dues, Baylor Hospital would accept the amount as prepayment for hospital care when needed."[7]

Serendipity for both the teachers and the hospital dictated that the stock market crash and the launching of the program happened within hours of each other. Not surprisingly, teachers subscribed in droves, and by December, more than 75 percent of all teachers in Dallas subscribed, and on December 20, 1929, the plan went live, coinciding with Christmas vacation. The success of "The Plan" was immediate, and employees of Dallas's Republic National Bank and the *Times Herald* soon joined; in fact, 408 employee groups with 23,000 members subscribed over the next five years. Kimball had saved Baylor University's hospital, rescuing it from insolvency and, like a pied piper, ushered a steady stream of patients to its doors. As medical costs were beginning to explode, patients were happy as well: an accident or serious illness no longer meant financial ruination.

With America teetering on the brink of the Great Depression, plans began to spring up around the country. Initially centered around single hospitals, joint hospital programs started in larger cities, and small "prepayment" premiums were well received by Americans who were beleaguered by war, pandemic, and financial collapse. More sophisticated insurance products would be developed over the next several decades, like indemnity (specific cash benefit, or "cap"), service benefit (care for a number of days for specific conditions), and major medical insurance (supplemental, "catastrophic"), but in the 1930s, the plans served as the only *health insurance* that most Americans had ever heard of.

In Minnesota, the local organization was called the Blue Plan, and its administrator E. A. van Steenwyk sought a new symbol for his company. After some deliberation, a blue cross was chosen, and of course, in time, the cross came to symbolize, nationwide, the hospital prepayment programs that would eventually become a centralized national organization. Justin Ford Kimball, who was neither an actuarial scientist nor trained hospital administrator, took the reins of a hospital as a fifty-seven-year-old neophyte, and started a revolution in health insurance that within a few years became Blue Cross, now a multibillion-dollar industry that dominates the federal and commercial healthcare landscape, and paved the way for Americans to afford major elective (implant-based) operations that would have been unthinkable in the Roaring Twenties, when Kimball was just a school superintendent.

Within a decade of the founding of Blue Cross, the American Medical Association (AMA) and its physicians decided that a similar prepayment program should be considered for physician office visits. There had been longstanding antagonism between the very powerful American Hospital Association and the AMA, particularly on the issue of health insurance, where hospitals were almost immediately receptive to the idea of national health insurance (even government-controlled single-payer insurance), and the AMA was fighting tooth and nail to reject government-administered health insurance.

Advances in medical and surgical sciences were "shifting the locus of care from the home or doctor's office to the hospital,"[8] and as "house calls" were vanishing, there was a growing public demand for medical service plans for physician services. In time, physicians warmed to the (initially) not-for-profit Blue Cross idea of healthcare prepayment, and *Blue Shield* was born. Years later, Blue Cross and Blue Shield nationalized and conglomerated into one large corporation.

The foundation of Blue Cross and Blue Shield had been built upon relationships that hospitals and physicians had with employee groups, like teachers' unions, factory and steel mill workers, and police associations. Separately, a surprising collaboration between employers and their workers during the administration of President Theodore Roosevelt had resulted in sweeping reformation in almost every state in workers' compensation insurance. Concessions were made by both sides; employers were motivated to establish a system of workplace injury insurance to minimize their legal exposure to employee lawsuits, and employees finally enjoyed workplace protection from punitive work hours, unsafe working conditions, and lack of healthcare for injuries sustained on the job. The combined effect of work-derived health insurance and workers' compensation insurance meant that many workers enjoyed health care services that had been absent just one generation before. In the early 1940s, a set of wartime wage and benefit laws and regulations stipulated that employers provide healthcare as a "fringe" benefit, furthering the link between a job and health insurance.

Unfortunately, because the reforms had almost exclusively centered on workers, Americans who were unemployed or elderly were still out in the cold in the late 1950s, and as medical care got more expensive, hospitalization was becoming ever more threatening to one's financial health. Listerism, the technique of cleansing the skin and surgeon's hands with carbolic acid, forever changed the notion and reality of which conditions could be surgically broached, and the half-century from the 1880s to the 1930s witnessed a titanic shift in the vulnerability of mankind. However, serious infections usually hastened death, and it was only the practical introduction of sulfa drugs and penicillin in the 1940s that broadened the safety zone for surgical interventions that could be hazarded in hospitals. The combination of aseptic surgery and antibiotic treatment proved simultaneously heroic and expensive. It's quite cheap to let someone die—it's very costly to save someone's life.

Wartime American employment, production, and innovation continued at a dizzying pace. "The stunning growth of the nonprofit Blue Plans was not lost on the commercial insurance industry, especially those companies that were already selling life and casualty coverage to employee groups. . . . Between 1940 and 1946, the number of group and individual hospitalization policies held by commercial companies rose from 3.7 to 14.3 million."[9] Similarly, group surgical indemnity coverage grew from 2.3 to 10.6 million policies. There was very little American hospital construction prior to the end of the year, but the 1946 Hill-Burton Hospital Survey and Construction Act launched a nationwide hospital boom. "Between 1946 and 1960, the number of voluntary and state and local government hospitals had increased by 1,182 . . . federal spending under the program began at $75 million a year in 1948 and rose to $186 million by 1961."[10] One critical stipulation of the Hill-Burton Act was that hospitals (which had received grants) were required to provide free care for twenty years to persons unable to pay for medical services. For those initial recipient hospitals that received funds in 1946, it is no small wonder that Medicare's activation in 1966 came at just the right time.

The explosion and expansion of gleaming hospitals and the swelling numbers of insured lives couldn't conceal the plight of the those "left out by reason of age or economic status . . . the indigent and the unemployed."[11] The first group to be seriously addressed would be the elderly. The Great Depression had ravaged the financial stability of millions of Americans, and the monumental Social Security Act of 1935 established a permanent national old-age pension system through employer and employee contributions (still reported in box 4 of your W-2 form). It was natural that President Harry Truman's head of the Federal Security Administration (later renamed the Department of Health, Education, and Welfare, and still later split up into multiple federal departments) would scheme with his colleagues at the Social Security Administration that the same program that provided old-age benefits could be configured to provide healthcare coverage benefits. FSA administrator Oscar Ewing stated, "The proposed benefits would give [the aged], through their own contributory insurance system, badly needed and valuable hospitalization insurance . . . that would reduce federal, state and local expenditures . . . and reduce deficits of hospitals."[12] Just like Justin Ford Kimball's Dallas teachers setting aside a little money with each paycheck as a prepayment of hospital expenses, Ewing was proposing a system where Americans would prepay money to be used in the event of sickness as an elder. Uttered in 1951, it would take a decade and a half to codify into law.

The most strident opposition to nationalized healthcare for the elderly had arisen from the AMA, calling Truman's initial proposals "un-American" and "socialized medicine," fearing that congressional oversight of medicine would lead to poor doctor pay. During the forties and fifties, "government solidified the private health-care system through corporate tax breaks that subsidized companies offering insurance to their workers. More workers were brought into the private system through this indirect and hidden form of government assistance, creating even greater resistance to the idea of the federal government directly providing insurance."[13] While the AMA and its physicians had little power in the 1800s, the modernization of medical education, and the purging of pretender medical schools in the aftermath of the

(Carnegie Foundation–funded) Flexner Report of 1910, resulted in the monopolization of power by physicians, as Pulitzer Prize–winning medical historian Paul Starr has advanced. [14] Flushing snake oil down the drain after the Pure Food and Drug Act of 1906 and shuttering counterfeit medical schools after the Flexner chronicle endowed doctors with ever-increasing prestige and negotiating power.

The first Congressional bill to address healthcare for the elderly was the Kerr-Mills Act (Social Security Amendments of 1960). Although the dream of compulsory national health coverage for all Americans had died decades earlier, even limited coverage for the elderly had been a slog. Oscar Ewing's chief advisers Wilbur Cohen and Isidore Falk, therefore developed an "incremental approach" to accomplish their (as of 2019, still unrealized) goal of universal coverage. "The idea [of incrementalism] was to bring about the passage of a modest program of insurance for a small number of people, and then gradually to expand that program until it covered the entire population." [15] The Kerr-Mills Act had tremendous bipartisan support, as opposed to Sen. John F. Kennedy's alternate compulsory health insurance proposal, to be financed with an increase in Social Security taxes. While there was an increasing number of retirees who had health insurance (31 percent in 1952, 44 percent in 1956, and 53 percent in 1959), [16] the Kerr-Mills Act depended upon states agreeing to participate and the efficiency of federal administration, two of the factors that limited its effectiveness.

Incrementalism is a fruitful tactic if a restricted or pilot program thrives; alternatively, incrementalism can also be a winning strategy if the maiden program doesn't succeed, since proponents can argue that the limitation itself had hamstrung their pet project. The Kerr-Mills Act presented universal coverage agitators both options: they could point to spotty coverage for seniors across the country while showing that many were still left out in the cold. Within months of its signing by President Dwight Eisenhower, the act was challenged by newly inaugurated President John F. Kennedy in his 1961 State of the Union address when he called for a federal Social Security–linked program to provide hospital insurance to the fourteen million Americans over age sixty-five.

The King-Anderson bill was introduced soon after Kennedy's speech, with the proposal of compulsory coverage for hospital and nursing home care for seniors. Dubbed "Medicare," the King-Anderson bill enjoyed support among unions and liberals, but was opposed by the AMA, business groups, and conservatives. The co-sponsor of the Kerr-Mills Act was Wilbur Mills, a Democrat from Arkansas who had risen to the powerful position as the chair of the House Ways and Means Committee during the Kennedy administration. At the time, there were still many "conservative Democrats" in the House and Senate, and Chairman Mills was one of them. From the beginning, the King-Anderson bill faced an uphill battle, starting with the slim margin of victory that Kennedy had secured in his defeat of Richard Nixon in 1960. Representative Mills worked for years to modify the legislation and secure the necessary votes to get it out of committee, but roadblocks by the AMA and other lobbies stalled the bill (saying nothing of Mills's own recalcitrance).

Less than three years into his term, President Kennedy was assassinated on November 23, 1963. "Two days before Kennedy's death, *Washington Post* columnists Rowland Evans and Robert Novak wrote, 'As long as Mills keeps opposing health care financed through the Social Security system, President Kennedy's plan is doomed in the Ways and Means Committee.'"[17] Democrats enjoyed landslide victories in the House and Senate, and President Johnson vowed to make civil rights and Medicare a priority as part of his "Great Society"; he knew that the emotional time was ripe during the "honeymoon" first session of his presidency.

The passage of the Civil Rights Act (1964) occurred in the "twilight of a New Deal dispensation that stretched back thirty years,"[18] but which could not have happened during the Truman or Eisenhower (or even FDR) administrations. The Democratic Party had dominated national politics for decades, having won seven out of the last nine presidential elections, while averaging "a whopping 424 Electoral College votes compared to Republicans' 101."[19] The question arises, then, why the struggle to pass proposals by Truman or Kennedy? The answer is the ferociously cohesive Southern Democrats, who "protected segregation, fought unions, and subverted most social reform, [combined with the] seniority rules and the South's pattern of reelecting its members."[20]

The Democratic Party had two main coalitions during the New Deal era: Northern liberals, who were fond of crafting socially innovative proposals, like "Social Security expansion, national insurance, robust labor protections, child welfare programs, and so on . . ."[21] and Southern conservatives, who functionally ruled Congress. Political scientist Ira Katznelson has described it as "a coalition of Swedish welfare state and South African apartheid, dominated by the latter."[22] When the Civil Rights Act did pass, it only did so after a filibuster of ninety days was broken by a cloture vote, where twenty-seven of the thirty-three Republican senators joined forty-five Democrats to break the Southern resistance.

Under a similar arrangement, many political pundits believe that the Civil Rights Act facilitated, even enabled, Medicare's passage. Republicans sensed that Medicare was a fait accompli, and sponsored an alternative bill nicknamed "Bettercare," a voluntary insurance program that would cover doctors' fees, financed, in part, by general tax revenues. For its part, the AMA recommended a different plan, entitled "Eldercare," which would have functioned as an expansion of the Kerr-Mills program, covering doctor visits, nursing home care, and prescription drugs. In essence, Eldercare was the forerunner to Medicaid.

The competing proposals of Medicare, Bettercare, and Eldercare were simultaneously contradictory and complementary. Hospital coverage, physician office insurance, and expanded indigent care were the three pressing needs that had never been considered by Congress, not even by those in favor of universal hospital comprehensive insurance for seniors. It would be impossible for all three proposals to pass at once, adding a huge obligation to the people and their government. Actually, that's exactly what happened.

Chairman Mills pulled one of the greatest legislative coups in history, when on March 3, 1965, he proposed combining the main aspects of all three bills. During a meeting of the House Ways and Means Committee, Mills turned to Johnson's representative, Wilbur Cohen, and asked why they "could not put together a plan that included the Administration's Medicare hospital plan with a broader voluntary plan covering physicians and other services?"[23] Cohen later recalled, "The federal government was moving into a major area of medical care with practically no review of alternatives, options, trade-offs, or costs."[24] After a few months of deliberations, both the House and the Senate passed the bill, known colloquially as "Mills three-layered cake," but formally as amendment Titles XVIII and XIX of the Social Security Act. Title XVIII was comprised of two parts, Parts A and B, outlining hospital and supplementary medical coverage (like physician's office visits), respectively. Medicaid was established through passage of Title XIX, but has never been referred to as "Part C;" that would come thirty years later with the passage of the Balanced Budget Act of 1997, formalizing managed capitated-fee health plans, initially called "Medicare+Choice," and later called "Medicare Advantage."

Understanding the history of the "baking of the cake" explains many of the confounding details of Medicare and Medicaid. For instance, why is hospital coverage defined as "Part A?" Because hospital coverage was defined under "Part A" of the Medicare Act (that had originally been the King-Anderson bill). Why is physician coverage reimbursed under "Part B?" Because doctor visits are processed through "Part B" of the Act (officially Amendment Part B of Title XVIII of the Social Security Act). Why is Part A funded through Social Security taxes? Because the bill was passed, from the beginning, as an add-on to Social Security, and explains why Part A costs are reimbursed from the Social Security bucket. Conversely, Part B payments come from general tax revenues, as was originally proposed.

After years of wrangling, a full half-century since Theodore Roosevelt had proposed universal coverage, Medicare was signed into law by President Lyndon Johnson on July 30, 1965, while sitting in the Truman Library in Independence, Missouri. Sitting next to him was the first-ever recipient of Medicare, Harry S Truman, who received his official Medicare card that day.

Civil rights activism had fundamentally reshaped the way America thought about health care for the poor, unemployed, and seniors. "Medicare, the result of a landslide election propelled by the passage of the Civil Rights Act and the civil rights movement that shadowed its implementation, was a gift of that movement."[25] Within a decade, "[h]ospitals became the most racially and economically integrated private institutions in the nation . . . all but four or five of the once more than 500 black hospitals had either been closed or converted to other purposes."[26, 27] What is more difficult to comprehend: that there were still segregated hospitals in America in the 1960s (like Eaton Ward at my alma mater), or that the Medicare Act helped close those hospitals?

It would take a full year for Medicare to "go live," and in that year, one would guess that the behemoth US federal bureaucracy, like a titanic ocean vessel, would have steered and tilted its way toward a new horizon, exercising control over the thousands of physicians and hospitals that served the tens of millions of newly insured lives. Shockingly, there was no dominion, as Wilbur Cohen, one of the chief architects of Medicare, lamented later: "The sponsors of Medicare, myself included, had to concede in 1965 that there would be no real control over hospitals and physicians. I was required to promise . . . that the Federal Agency . . . would exercise no control."[28]

When Medicare passed, legislators had codified the then-ruling payment policies of the Blues, in which the not-for-profit Blue Cross hospital insurance plans during the 1930s functioned as a "stable conduit of money to the [hospital industry]."[29] Critically, the state-based Blues plans typically *reimbursed* hospitals for costs incurred while treating patients—no matter the cost—so that there was little incentive to constrain costs at a time when healthcare was entering into an explosive growth phase. Uwe Reinhardt of Princeton University has argued that this orientation around "reimbursement" and not "payment," where "hospitals would have to manage their line-item costs against external constraints,"[30] fostered an inherently inflationary system. The state-based Blue Shield plans (for physician reimbursement) paid physicians at his or her "usual, customary, and reasonable (dubbed 'UCR')" fee, again exhibiting almost no cost control.

The inflationary arrangement did not stop at hospital and physician reimbursement; "Medicare was required to reimburse each individual hospital retrospectively a pro rata share for all the money that the individual facility reported to have spent on capital investments in structures and medical equipment . . . and a pro rata share for whatever its operating costs might be."[31] With guaranteed reimbursements securing a rate of return, it is little wonder that investor-owned hospital growth took off. Medicare adopted the Blue Shield UCR-style physician reimbursement, paying physicians according to "customary, prevailing, and reasonable (called 'CPR')" fees, with only slightly more rigid restrictions.

"In effect, then, in return for acquiescing in the passage of Medicare into law in 1965, healthcare providers extracted the key to the U.S. Treasury from Congress," argues Reinhardt.[32] With reimbursement—and not payment—the watchword, annual Medicare outlays immediately, and annually like a metronome, vastly surpassed predicted aggregates. Ironically, it would be stalwart Republican presidents (Nixon, Ford, Reagan, and George H. W. Bush) who "sought to bring the hospital and physician sectors to their knees, in ways that Democrats would never dare to do."[33] In the late 1970s, the Carter Administration agreed to the "Voluntary Effort" of the hospital industry to control costs, but this naïve promise failed to make an impact.

Twenty years of "reimbursement" came to an end during the Reagan Administration, when Medicare rules were changed to a more business-oriented methodology. "The very idea of *retrospective full-cost reimbursement*, an approach that would look strange to anyone accustomed to normal business principles, particularly vexed the administration."[34] Researchers and policymakers, therefore, arranged medical conditions into slightly over five hundred "diagnosis-related groups," or "DRGs," permitting the remuneration of hospitals based upon a preset fixed sum per case, allowing a "fair profit." This was truly revolutionary, and has been copied by countries around the world, and even private insurers in the United States.

Reimbursing hospitals by a DRG case–based accounting system was the opening salvo by the federal government in ending the decades-old approach of hospitals and physicians charging (and receiving) limitless sums of money for guideline-free "customary" care. With economists and statisticians gaining power, Medicare funded a major study on the "relative costs" of providing various physician services, aiming to identify the time, skill, and risk involved in treating a large number of distinct medical vignettes.[35] This led to the "resource-based relative value scale" (RBRVS), paving the way for 1989 legislation that formalized a physician fee schedule based on RBRVS (with "geographic adjustors for variations in labor costs, malpractice premiums, and the [cost of office space]").[36] Forever gone would be the old notion of fat cat hospitals and doctors tugging on Uncle Sam's purse strings, whimsically charging usual or customary fees for services.

Medicare-initiated innovations like DRGs and RBRVS have been adopted by the private insurance sector, and further, when hospitals and physicians negotiate with not-for-profit and for-profit carriers, their rates are based upon a particular Medicare fee schedule year (a physician might say, "our new Cigna contract is 135% of 2015 Medicare"). Although Medicare accounts for "about 20% of total national health spending, at $572.5 billion of a total of $2.793 trillion,"[37] it is the power broker in payment reform for both public and private healthcare spending. More recently, Congress has attempted to corral physician costs on a *global* scale, setting a target for overall Medicare spending linked to the growth of the Gross Domestic Product (GDP). Utilizing this calculated "Volume Performance Standards" (VPS), the program stipulates a reduction in physician fees in the present year if the previous year witnessed a budget excess. While this Draconian arrangement, the Sustainable Growth Rate (SGR) system, should be easy (and powerful) to implement, it is one of the last frontiers where physicians have exercised legislative influence, and the "doc fix" has rarely been activated. Disregarding the SGR has almost become an annual congressional rite of passage in DC, and some have criticized the near-perpetual state of abeyance of the SGR as a contributing factor in the "unsustainable" increase in healthcare spending.

Medicare accounts for one-seventh of total federal spending: about $588 billion of the $3.9 trillion budget funded Medicare in 2016.[38] In its fifty years, the percentage of the federal budget spent on Medicare has steadily increased, topping over 15 percent in 2016 (and 3.2 percent of GDP), and projected to surpass 16 percent of the budget and 3.6 percent of GDP by 2024. Between 2010 and 2050, the population over sixty-five will double, from about forty million to eighty-four million people, and a large tranche of that group will be seniors over eighty, who are typically very expensive to care for. Uwe Reinhardt posits, "The current debate on US fiscal policy clings to the notion that the fraction of overall government spending as a percentage of GDP must be kept at or below a given percentage—regardless of the funds needed . . . *This* is the idea that is unsustainable."[39]

Skeptics can rightly criticize the poor value Americans receive for their healthcare dollars when compared to most Western countries (so admits this surgeon-author). We pay too much for drugs, implants, and procedures, but in this new era of cost-consciousness and outcomes tracking, Americans will see an improvement in "getting what we pay for." However, there is simply no other place in the world where most economists, actuarial scientists, policy-makers, and physicians themselves, would want to be cared for when suffering from a heart attack, cancer, or trauma, but improved cost-control initiatives will need to be nurtured.

Understanding the genesis of the FDA and Medicare is essential to understanding the "perfect storm" in the explosion of implants. Improved materials sciences, the discovery of antibiotics, the supervision of implants by the FDA, the government-facilitated launch of thousands of new hospitals following World War II, the invention of health insurance and the formation of Medicare all coalesced within a few decades. Patients needed health insurance to pay for the new expensive operations; hospitals, physicians, and implant manufacturers needed a reliable flow of insured patients. In 1965, who could have guessed, in their wildest dreams, what was about to happen? Of course, Medicare costs have always exceeded budgeted predictions, but Wilbur Mills and his colleagues cannot be blamed for not reading the tea leaves when the three-layered cake was made. Revolutions are tricky things to predict.

THIRTEEN

Device Clearance

I am still acclimating to the Greenwich Mean Time zone, having arrived in London only two days ago. After a short hackney cab ride from my flat to the Euston Station, it was a quick two-hour high-speed train ride on one of the busiest rail routes in Europe, connecting London to Manchester Piccadilly station. A connection in Manchester placed me on a spoke of the TransPennine Express to the town of Wigan, where I hailed a taxi to travel through the Lancashire countryside to the parish of Wrightington. The view out my taxi window was of gently rolling English farmland, with hedgerows of compact bushes and gnarled trees separating newly tilled, corrugated fields.

Wrightington's pastoral setting does not evoke thoughts of world-class medicine, but on the outskirts of town lies a hospital that has drawn me here. My taxi whorls around the roundabout and coming into view is the Wrightington hospital complex. The original estate, Wrightington Hall (rebuilt in 1748), is composed of cream-colored stone, its stately manner and imposing structure in keeping with its regional importance in former days. The remaining buildings on the grounds are one or two stories tall, and unlike the estate, are composed of red brick. My taxi comes to a stop and my driver, a slim, older gentleman from Pakistan, tells me, "You know they say this is a really important hospital in history, the place where joint replacement surgery was invented."

Getting out my wallet to pay my driver, I tell him, "That's right. And it's why I'm here, to see where it all happened, and to pay my respects to the memory of Sir John Charnley."

A very old stone wall flanks a macadam parking lot, and on the other side lies a one-story 1960s building, the famous Centre for Hip Surgery. Hundreds of surgeons who have visited this mecca have stood in that doorway and posed for a picture with Charnley or the other surgeons who have helped make this place so famous.

Going through the doorway, I meet with Mr. Bodo Purbach, a German-born orthopedic surgeon and the greatest devotee to Charnley, an enthusiastic disciple who is eager to show me around at the end of a busy workday. With a young registrar (surgical resident) in tow, we visit with some patients who received their hip replacements earlier in the week. Glancing *around, I'm struck by the simple décor and no-frills nature of the hospital buildings. Originally proposed by Florence Nightingale as a means of promoting air flow, these long buildings are organized in parallel rows; the furniture, paint, lighting, and double-occupancy of patient rooms, appear as though they were last updated thirty years ago.*

Mr. Purbach looked at me and said in his slight German accent, "And now the fun part!"

Pulling out a cluster of keys on a large key ring, we walk to the end of one of the buildings down a dark hallway. Now accompanied by Mr. Lennard Funk, a shoulder surgeon, Mr. Purbach unlocks the large door and flicks on the lights. As the fluorescent lights hum on, I'm in a museum of sorts, with display cabinets of hip implants, dusty old curios of collected medical device boxes, and representative samples of the history of hip implants going back almost a century.

This is a bonanza for anyone interested in the history of medicine and surgery, and for me; I feel like Howard Carter at the moment of Tutankhamun's crypt unveiling. Over against one wall is the customized polyethylene wear testing machine that I've seen in journal publications over the years in a stylized cartoon, except now I'm seeing the real thing with my own eyes. Then, to my left is a fifty-year-old box of Charnley's acrylic cement. I continue to walk through the room, peering down at odds and ends in class cabinets, the metal hip stems that Charnley designed over the years. And then I see the polyethylene and Teflon cups.

Perhaps Charnley's greatest contribution to mankind was the use of polyethylene as a bearing surface for total joint replacement. Its discovery was an accident of fate, a salesman for a German polymer company showing up with a briefcase of gear samples, guessing that Charnley might be interested in polymer parts for his testing machines. But his lab assistant John Craven (and in time, Mr. Charnley), thought polyethylene was the breakthrough material they had been searching for. To test this new polymer, Craven developed a multi-station wear testing device . . . and there it sits on the countertop in front of me. Pure world-changing history sits on this countertop with a thin coat of dust.

In the adjacent cabinet, I see the culprits that almost destroyed all of Sir John Charnley's work. The Teflon cups that were retrieved from human subjects lie on a shelf, with massive erosion and uneven wear. If polyethylene was the answer to prayers, Teflon was the evil plague that shook Charnley's confidence to the core. By 1962 it was becoming clear that Teflon was a long-term failure in every case, and while the metal alloy hip stem was performing well, the hip socket Teflon cup was essentially "melting away" with wear from the metal ball. I lean over, my face close to the white, waxy-appearing Teflon, and I realize I am face-to-face with the actual implant I have seen in a photo in a Charnley publication. This implant was the final straw in a string of failures confronting Charnley. He never performed animal studies, and in 1962, there was no regulatory body in England that oversaw the approval of medical devices. Charnley would conceive an idea, fabricate the implants (either by himself or in concert with his corporate partners) and proceed to surgery. There were no tracking mechanisms and no outcomes measures to analyze patients.

Charnley would operate another twenty years after his polyethylene discovery, but the fact that he kept the failed Teflon cups gives us another clue about the man. He was a true pioneer, able to meet with, "Triumph and Disaster and treat those two impostors just the same," in the words of Kipling, and the curios of Teflon, instead of being objects of shame, are an important chapter in the development of joint arthroplasty. The greatest innovators have always diligently sought legitimacy, embracing the truth of their outcomes, scraping through the patina of early promising results and unearthing reality.

Theodore Roosevelt was born in New York City in 1858, famously struggling with childhood asthma attacks before "making his body" and embracing strenuous exercise as an adolescent. After graduating Harvard College in 1880, and returning to the city to attend Columbia Law School, the exuberant Roosevelt was elected a New York assemblyman as a baby-faced twenty-four-year-old. A free-market capitalist, Roosevelt ascribed to the conventional conservative theories of the day, favoring a laissez-faire doctrine of low wages, low taxes, and few social services. But in 1882, Roosevelt had a revelation, instigated by union activist Samuel Gompers.

A bill was introduced in the New York legislature in 1882, that would have banned tenement-house cigar manufacture, on the grounds that it put an unsustainable burden on cigar workers.[1] At the time, the cigar companies in New York City required workers to "take their work home with them," storing their tools and, worse, housing wet bundles of tobacco leaves in their already overcrowded tenement apartments. Gompers challenged the skeptical Roosevelt to see conditions for himself, and liking "the idea of testing arguments against evidence,"[2] he toured the New York tenements with Gompers.

What Roosevelt discovered astonished and revolted him. Not only were conditions appalling for workers, but the living conditions for families with children unconscionable. Roosevelt could not abide a laissez-faire attitude that was as rotten as the pungent tobacco stacked in immigrant workers' living quarters. The young Harvard graduate returned twice to the tenements by himself, and in the end said, "Instead of opposing the bill I ardently championed it."[3] A man who loved the strenuous life, and could never countenance weakness, had learned that "self-sufficiency and competitive spirit, important as ideals, are negated when life is unfair."[4]

Upton Sinclair, while not a core member of the muckrakers, published a book in the spring of 1906, that epitomized the "literature of exposure," *The Jungle*, about the wretched world of Chicago meatpackers.[5]

To write *The Jungle*, Sinclair had spent seven weeks in the slaughterhouses and meatpacking plants of Chicago, donning grubby clothes and carrying a lunch bucket to mix in with the immigrant workers. During the daytime, he visited the squalid, lethally dangerous workplaces, documenting the indifference of management to the workers' hardships and the lack of government oversight. In the evenings, he knocked on the workers' doors, his pencil ready to record their accounts.

Sinclair wanted to steer attention to the plight of the exploited Lithuanian immigrants in Chicago's "Packingtown," and his novel was a melodramatic yarn of desperate working-class life. The passages in the book that caught the popular fancy, however—though they spanned just a few pages—told a more particular story about the meat that Americans were consuming. Sinclair told of rats scampering across heaps of rotting flesh, leaving droppings; of tubercular meat packaged and sold at market; of acid corroding workers' flesh; and, most shockingly, of men tumbling into cooking-room vats and ignored "till all but the bones of them had gone out to the world as Durham's Pure Leaf Lard!" Those gut-churning images would outlive any ideological message about the workers that Sinclair intended. "I aimed at the public's heart," he later wrote, "and by accident I hit it in the stomach."[6]

After a two-year stint as governor of New York, Roosevelt briefly served as US vice president, until McKinley's assassination six months into his second term. Thus did Theodore Roosevelt become the youngest president in US history, just three years after his foray into soldiering, "the great day of my life" in the Battle of San Juan Hill. The forty-three-year-old now turned his progressive attention toward the unchecked industrial growth of the preceding decades. After securing his own term in 1905, President Roosevelt told Congress, "Traffic in foodstuffs which have been debased or adulterated so as to injure health or to deceive purchasers should be forbidden." No doubt Roosevelt remembered the awful provisions of the US Army, but remembering his own incredulity over the austerity of the tenements or over battlefront foodstuffs, he knew he needed a breakthrough to achieve his legislative proposal. Whether as police commissioner, governor of New York, or president, Roosevelt tempered his ardor for domination and conquest with a sensibility for fairness and a zeal for stewardship. While the National Park Service is the most obvious and visible example of Roosevelt's enlightened, balanced approach to enterprise and restraint, perhaps the most impactful and daily (even hourly?) transformation of the American existence was the creation of the Pure Food and Drug Act of 1906, and later the FDA, our first citizen-protection agency.

There had been a war between doctors and the "patent medicine" salesmen for a century before the 1906 bill. "The demand for 'secret formula' medicines began in the Colonies with medicines from England. But when, during the Revolution, English patent medicines could not be imported, American businessmen filled English bottles with almost anything that seemed a suitable imitation . . . it was all a packaged fantasy, and the package was the vital part."[7] With the creation of the new science of chemistry came the Promethean ability to create additives and preservatives, with the power to dilute, or confuse. Rotten milk, decaying meat, and decomposing vegetables could all be disguised with the newly discovered chemicals.

After jousting with Congress and battling with the meatpacking, livestock, and "pharmacologic" industries, the 1906 Pure Food and Drug Act was signed into law by President Roosevelt on June 30, 1906. Not only was inspection of meat the law of the land, but regulation of the drug trade had metamorphosed with the stroke of a pen. The new law created the nation's first regulatory agency, with *safety* of food and drugs being paramount, but also emphasizing the battle against *fraud*. The new bill outlawed misrepresentation, demanding accurate labeling and prohibiting adulteration of ingredients. It would be wrong *only* to credit the 1906 Pure Food and Drug Act with reformation of the meatpacking industry and the diminution of "snake oil" patent medicines. More importantly, it was a change in policy that there was a role for government to protect its citizens from unchecked commerce. However, in a more meaningful sense, the solution came about "not only because greed and abuse forced action, but also because there was a new hope created by the advances in science."[8] It was precisely because there were, for the first time, real medicines that could heal patients that fake potions were worth destroying. Paraphrasing Irving Fisher, Phillip Hilts concludes, "The world was gradually awakening to the fact of its own improvability."[9]

The 1938 Food, Drug and Cosmetic Act was signed into law by President Franklin D. Roosevelt, largely in response to the 1937 S.E. Massengill company's distribution of its lethally tainted sulfanilamide. The crisis had highlighted the flaws (and weaknesses) of the 1906 Pure Food and Drug Act. America was at a crossroads, just emerging from the Great Depression and awakening to the idea that specific drugs could be crafted to treat particular diseases. Chemistry was real, and the molecular basis of disease, springing from the German idea of the cellular basis of disease, meant that cellular dysfunction must be understood (and treated) based upon the molecules in, and administered to, our bodies. As the glow of enlightenment flickered on, how could swindling patent medicine "cure-alls" possibly be trusted? The pendulum was swinging, and companies, to survive, would need research departments, scientists, and laboratories. In the 1920s, there were a few thousand scientists at drug companies, but by the 1940s, there were almost sixty thousand, even before the postwar innovation explosion.[10]

The 1938 Food, Drug and Cosmetic Act was a "landmark in civil governance, not just for the United States, as it turned out, but for democratic governments around the world. In the years to come, each nation of the developed world would adopt its central principles. It was the first law that required analysis of drugs before they went to market. And it put into law the notion that the scientific approach—not the commercial, not the anecdotal, not the approach based on authoritative opinion—would be the standard for modern society."[11] That can be said for pharmaceuticals, but not for implants, for there was no thought of implants in 1938. It would take decades for the law of the land to meaningfully modernize the "medical industrial complex."

The FDA was moved from the Department of Agriculture to the Federal Security Agency, later named the Department of Health Education and Welfare (HEW). It now resides within the Department of Health and Human Services. It has continued to oversee drug manufacturing (including testing, factory inspections, labeling, marketing, packaging, and long-term safety analysis); food safety oversight; vaccine, blood, and serum supervision; and it ensures the safety of cosmetics and products that emit radiation. The FDA has power of administration over all these areas, whether they be human or veterinarian. Additionally, the FDA advances public health by regulating the manufacturing, marketing, and distribution of tobacco products, and plays a significant role in counterterrorism capability (by ensuring security of the food supply and fostering the development of therapeutic medical products.) But it was only in the 1970s that the FDA acquired the power to regulate medical devices.

The definition of a medical device prior to the 1940s never included the concept of something "implantable." Hundreds of years ago, devices tended to be fraudulent or magical, completely lacking in scientific merit: divining rods, nose straighteners, patent metallic tractors (essentially magic wands), heated rubber applicators, and unjustifiable knee braces. Because most of these types of appliances worked (if they did) through the placebo effect, there was little harm in not condemning them, other than loss of money. With the discovery of X-rays by Röntgen in 1895 and radioactivity by Marie and Pierre Curie in 1898, it did not take long for "curiosity devices" that used radioactivity to hit American and European markets. Many of the early radioactivity researchers succumbed to illnesses brought on by encounters with the elements, and in time, safety regulators at the FDA realized support was needed to minimize radiation exposure. The Bureau of Radiological Health was transferred from the Public Health Service to the FDA in 1971, benefitting from a more similarly aligned intellectual framework.

The great challenge that the FDA faced in the postwar technological boom regarding devices was similar to its challenge in regulating drugs prior to the 1938 Act: there was no oversight supremacy regarding pre-market testing, review, or approval. The enforcement resources of the FDA were spent, retroactively, policing gadgets and gizmos that often were viewed as harmless shoe-store X-ray machines, but were becoming ever more dangerous. In 1962, President Kennedy proposed changes to the way medical devices entered the market,[12] but attention shifted to the debate over Medicare and drug scares (including the Thalidomide nightmare), and device regulation was set on the back burner, ostensibly awaiting a crisis.

President Richard M. Nixon endorsed medical device legislation early in his first term, organizing the Cooper Committee—chaired by Theodore Cooper, MD, director of the National Heart and Lung Institute, former academic heart surgeon, and later Upjohn executive—to formulate a legislative proposal for overseeing the medical device industry. In 1970, the ten-member committee (all government officials) published the official Cooper Report, emphasizing the need to treat devices differently than drugs. Thus began a classification of all marketed devices, organizing them according to specialty, and categorizing them based on the level of scrutiny required to achieve approval. By 1973, the FDA had established fourteen various classification panels of experts, and by 1975, published a general notice advising device manufacturers about the classification procedures.

While the FDA continued its preparatory work, several crises arose that helped bring the public's attention to the need for federal regulatory oversight. The Dalkon Shield was a metal contraceptive intrauterine device that had an ominous appearance of a swimming parasite with a long flimsy tail. The "tail" portion of the device was a porous thread, later determined to be the cause of infections. Inserted into millions of women starting in 1971, it led to thousands of hospitalizations, pregnancy complications, pelvic inflammatory disease (PID) cases, infertility, and even death. By 1975, it was obvious that a catastrophe was brewing, and over 150,000 women filed lawsuits against the manufacturer, A.H. Robins Company, the largest tort liability case since asbestos.[13] The Dalkon Shield calamity was the type of crisis that forced Americans to ask, "How can we let this happen?" Instead of young children dying, as in the Sulfanilamide elixir tragedy, the victims were young women of childbearing capability, now permanently scarred from PID and infertile. The outrage over the lack of regulatory surveillance of the Dalkon Shield, and similar concerns over faulty pacemakers, demanded a change from Congress.

After several years of administrative and legislative wrangling, the Medical Device Amendments of 1976 were signed into law by President Gerald R. Ford—the most significant legislation ever passed regarding medical devices. The Bureau of Medical Devices was formed, overseeing the implementation of the 1976 Amendments. The law meant that any *new* implantable device had to undergo a "premarket approval" (PMA); no longer could a tinkerer fabricate an implant in his workshop garage and implant it the next day, as John Charnley had been doing with total hip implants for twenty years in Wrightington, England. As with drugs, the new law stipulated that an application be filed by the device manufacturer. For devices already on the market, a "grandfathering-in" was outlined, but for new devices, a new PMA application was now necessary.

In 1982, the medical device and radiological bureaus were merged into one organization, the Center for Device and Radiological Health (CDRH), a division within the FDA. The CDRH regulates all medical devices and radiological testing machines, and any device that emits electromagnetic rays, such as microwave ovens and mobile phones. Perhaps its most important role is sorting through applications for medical implants, such as pacemakers, deep-brain stimulators, orthopedic joint replacements, cochlear implants, and cardiac valves.

Similar to the dual pressures faced by the drug approval process, manufacturers and patients are often frustrated by the FDA's toilsome exactitude and slow pace, but in the rare instance of a drug or device malfunction or calamity, the FDA is criticized for lax standards or haphazard clearance. Among the worst FDA failures ever was the inappropriate clearance given to the makers of Thalidomide, which horrendously caused phocomelia in newborn infants, a condition in which the intrauterine development of the babies' arms and legs is short-circuited, leading to grotesquely shorted limbs with the abnormal hands and feet attached just inches from the shoulders or hips. In the worst cases all four limbs were affected, and given the fact that the defects occurred in so many children, with lifetime dire consequences, the FDA's "never again" response was understandable.

There have been similar failures among approved implants, and in these instances, manufacturers, hospitals, physicians, and the FDA have blood on their hands. One of the most prominent examples of systemic failure is still ongoing, with injured patients all around us. Over 40,000 Americans and almost 100,000 patients worldwide underwent total hip replacement with the DePuy Articular Surface Replacement (ASR) hip implants, which was finally withdrawn from the world market on August 26, 2010. Settlements from almost ten thousand plaintiffs will likely require DePuy (a Johnson & Johnson company) to shell out almost $4 billion. How did a tragically flawed total hip arrive (and stay) on the market for seven years?

It is arguable that hip replacement is the most successful operation in the world, considering the dramatic decrease in pain and improvement in function, combined with a low incidence of infection and need for re-operation. Some have even called it the "operation of the century."[14] In some ways, however, the spectacular success of hip replacement in the elderly led to a temptation on the part of surgeons and patients to push the boundaries of mechanical durability, attempting replacements in ever younger and more active patients. From the 1960s to the 1990s the gold standard in joint replacement was a stemmed implant down the femoral canal (secured with bone cement), an acetabular (hip socket) component impacted into bone, and a polymer (polyethylene) liner in the metal cup. This proved to be long-lasting in the elderly, but the polymer lining, like any mechanical bushing, was prone to wear and eventual failure. Mechanical engineers continued to ponder "alternative bearing surfaces," wondering if there was another combination of metals and plastics that could allow for arthroplasty in younger, more active recipients.

The breakthrough idea, as often happens in medicine, was a revisitation of an old concept. The first successful hip arthritis operations were performed by Marius Smith-Petersen, chairman at Massachusetts General Hospital, who had performed "cup arthroplasty" of the femoral head by capping the femoral head with glass and metal cups. The mixed results of cup arthroplasty were greatly surpassed by Charnley's *total* hips, but the problem of hip dislocations and polyethylene wear had inspired surgeons and manufacturers to (again) consider replacing the head with a large metal ball. Derek McMinn, an Irishman trained in London and practicing in Birmingham, England, since 1988, thought about reincarnating the idea of hip replacements *without* a polymer lining, as originally proposed by McKee and Farrar in 1960.[15] McMinn began using (in 1991) a large smooth metal head that rotated within a polished metal hip cup, hoping that careful placement of the components would result in a thin layer of fluid interposing itself in a self-lubricating fashion in between the metal ball and metal cup.

Originally calling his hip the "McMinn Hip" in 1991, the intrepid surgeon began in 1997 to use the newly designed "Birmingham Hip Resurfacing" (BHR) implants. The early results were excellent, and the allure of bone preservation (less bone is removed in BHR), greater hip stability (because of a larger metal head), and lack of polyethylene wear made the BHR a tempting alternative to regular hip replacement among young, active patients. The timing of the BHR couldn't have been better, owing to the failures of the 3M Capital hip, which was removed from European markets in the same year.

Derek McMinn published the early results of his Birmingham hip in the British *JBJS* in 2004. At an average of 3.3 years from surgery, only one of the original 440 patients had revision surgery, an astounding 99.8 percent having avoided failure of their implant. By the year 2000, Mr. McMinn was implanting over two hundred BHRs per year, and his secret was out. The electricity in the air regarding a hip that could survive and not dislocate energized engineers and surgeons in America and around the world, and they scrambled to design a "metal-on-metal" (MoM) hip replacement for themselves.

DePuy Orthopedics, based in tiny Warsaw, Indiana, is the oldest implant manufacturing company in the world. DePuy was founded in 1895, and was originally a splint manufacturing company before becoming an orthopedic giant. Like most orthopedic companies, DePuy determined that designing a MoM hip was imperative. While the FDA did not approve the Birmingham hip until 2006, it seemed inevitable that every company would be distributing poly-free hips in short order.

The initial step in manufacturing any implant is the assemblage of a team of engineers, designers, market experts, and consultant surgeons. A design team gathers on a regular basis over the course of a few years, but initially must analyze market needs, existing products, and the "freedom to operate" regarding existing patent laws in the United States and internationally. The "intellectual property" (IP) that a company owns (once a patent has been granted) establishes a barrier of imitation that is *supposed* to be obeyed by manufacturers in internationally law-abiding countries. If a competing company has strong IP surrounding an implant, the new design team is forced to innovate new styles, shapes, techniques, and technology, so as not to infringe upon a competitor's legal rights. Sometimes the alterations in design offer an improvement in performance (while creating a "work-around" in the barely legal, imitation game), but occasionally the modification creates serious crises.

While implant companies typically employ "in-house" engineers and market experts, the surgeons on a design team are independent physicians who are consulted for their expertise in the field. Typically, design surgeons are university-bound, academic surgeons, practicing at the pinnacle of their profession, training young surgeons, and instructing at meetings (often about implants they have designed). Understandably, companies engage "thought leaders" who not only use a great deal of the product themselves, but can influence other surgeons as well to become high-volume users. Almost without exception, surgeons—like physicians who are involved in drug trials, and scientists who hypothesize and experiment in scholastic and commercial settings—practice ethically and without compromise. But the payoff for surgeons who are fortunate enough to be a member of a successful design team is so huge that one's ethical standards can be severely tested.

DePuy assembled a team of surgeons from Australia, Ireland, England, Germany, and the United States, initially meeting at the turn of the millennium. Derek McMinn's original MoM hip had debuted in 1991, and the successor Birmingham Hip was implanted as early as 1997, and the incentive to get the next major MoM hip onto the market as rapidly, and safely, as possible was immense. DePuy business leaders (and their surgeon consultants) knew that a successful hip implant launch could generate billions of dollars of revenue, and no doubt were buoyed with optimism with McMinn's early orthopedic meeting presentations.

DePuy submitted the ASR to the FDA for approval, and in a unique twist, two versions of the ASR hip were submitted, both of which used the same hip cup (acetabular component), but differed on the femoral side in having either a long stem that would be impacted down the femoral canal like a railroad spike ("ASR XL"), while the other "stemless" option would essentially be a cap on top of the femoral head with a small central peg designed to go down the femoral neck. It is surprising that the FDA never approved the latter, stemless, option that most resembled the Birmingham Hip, instead only authorizing the ASR XL in 2003. Both versions gained approval in Europe and Australia that same year,[16] and soon the marketing campaign was ramped up, with glitzy brochures and technique guides being distributed to surgeons around the world.

Surgeons in Australia and throughout Europe began implant ASR hips in 2004, and in America in December 2005,[17] but it only took a few years (outside America) to know that something bad was happening with ASR patients. The FDA, and its CDRH, were oblivious early on about early implant failure within our borders, because *there is no national implant registry in the United States.*

◆

The first registry on joint prostheses was created almost fifty years ago at the Mayo Clinic, the same year that one of its physicians became one of the first surgeons in

America to implant Charnley's total hip.[18] In 1975, Göran Bauer, professor of Orthopedics at Lund University, Sweden, conceived a plan of establishing a *national* total
knee registry (the Swedish Hip Arthroplasty Register followed in 1979), requiring
all orthopedic surgeons in the country to collect baseline information and faithfully
track all patients. Bauer had trained in Sweden but spent seven years in New York at
the Hospital for Special Surgery where he served as director of research. He returned
to Sweden in 1969, just as joint replacement was becoming an accepted procedure,
and in decentralized southern Sweden, he sensed the need for standardization and
outcomes assessment. What started as a multicentered endoprosthetic study among
small Swedish hospitals turned into the world's first national joint registry.[19]

Because the regular ASR was resurfacing only, the FDA considered it a new
technique, and instead of considering it as a 510k device, it forced DePuy to perform
a clinical study that involves much greater scrutiny ("investigational device exemption," or "IDE"), thus greatly slowing the rollout of the ASR in America. While the
ASR patients in the IDE study were being carefully followed by DePuy, the ASR XL
recipients were not. It will be surprising to readers that there is no formal mechanism
for implant manufacturers to track the success of their products in America. It is more
likely for Toyota to know about faulty exhaust pipes in a Prius than DePuy to understand how a new hip implant is performing in the United States.

In a joint registry, at first glance, it would appear that surgeons are analyzing
patients, but instead, it is the *survival of the implants* that is studied. This is different
than disease-based registries (like diabetes or heart disease) that follow the vital status
of the patients. At the time of surgery, the patient is anonymously entered into a database, along with information about the device, the surgical technique used, and the
name of the doctor. A national joint registry, therefore, provides valuable information
about the longevity of an implant when used in real-world conditions by all surgeons.
Because of this, there is no "cherry-picking" of cases that are excluded from a surgeon's
publication on an implant he may have designed. Surgeons are notoriously unreliable
at self-reporting poor outcomes; performance-oriented professionals are reticent to
admit failure (even when an innocent participant) or reveal an error in judgment.
National joint registries, therefore, are a critical tool for analyzing the short, medium,
and long-term outcomes of a joint replacement.

As of this writing, there are only eleven national joint registries in the world:
Sweden, Finland, Norway, Denmark, New Zealand, Australia, Canada, Romania,
England/Wales, and the Netherlands. The Nordic Arthroplasty Register Association
(NARA) is a collaboration among the knee and hip registries of Sweden, Norway, and
Denmark. There is a movement, promoted by the International Society of Arthroplasty
Registries and the FDA's International Consortium of Orthopaedic Registries (ICOR)
to create uniformity of data collection, enhance sharing of data, and to create a universal
bar code for implants. Despite Mr. Charnley's primacy in inventing joint replacement,

England only created a national joint registry in 2002. Amazingly, the United States still has no national registry. John Callaghan, MD, a renowned orthopedic surgeon from Iowa, has stated, "We are No. 1 both as a provider and user of implants. We should be the leader in the follow-up of them."[20]

Surgeons in Australia began implanting the ASR hip in 2004; by 2007 the Australian National Joint Replacement Registry (ANJRR) reported that the DePuy implant had a higher than expected revision rate. Governing bodies have concluded that hip prostheses are expected to fail less than 1 percent per year, meaning that five years from implantation, 95 percent of patients should still have a well-functioning hip in place. However, the Australian data showed a 5.16 percent revision rate at two years,[21] a startling pace for an implant that was supposed to be the solution for younger patients who needed a particularly robust hip. Professor Steven Graves, director of the ANJRR, notified DePuy of the sobering news, calling it an "unambiguous end point—nobody can argue about it."[22] But as Deborah Cohen points out,[23] that's exactly what DePuy did for the next several years, despite mounting evidence that the ASR was a ticking time bomb.

What does it mean for an implant to fail? When a pacemaker fails (electronic misfire, battery failure, or wire breakage, for example), the results can be immediately lethal. If an implant has not been properly sterilized or the packaging has been compromised, a patient can succumb to an infection. But in the case of the ASR hip, the failure of the implant meant that patients were exposed to the particles of metal scrapings that were generated from the flawed design or implantation of the components. The microscopic molecules incited the membranes around the implanted metal components to inflame, creating swollen, irritated cyst pockets that eventually led to destruction of the muscles, tendons, ligaments, and bones around the hip. Patients who trusted their surgeons to take away their arthritic hip pain were sometimes damned to experience even worse pain than they started with. A failed hip, therefore, necessitates even more complex hip surgery, demanding the surgeon take out the flawed implants, and then implant revision components that (hopefully) can find firm foundation in the remaining bone. Every patient who has undergone more than one operation on a joint has further compromise in the dynamic function of the muscles around the joint, with atrophy, scar indentation, loss of mobility, and less significant pain relief. In the end, a failed implant is more like getting life-threating food poisoning at your favorite restaurant and less like a tasteless meal at a boring bistro.

Barry Meier is a healthcare and business reporter for the *New York Times*, and has written over fifty articles on the failure of the ASR hip and other medical catastrophes over the last decade. His first article on joint registries appeared in 2008,[24] and by 2009 he reported on a proposed bill in the US House of Representatives that would have created a national hip and knee registry.[25] In 2010, Meier's "Concerns over metal on metal hip implants"[26] sounded the alarm over the impending calamity over metal-on-metal hips, even at a time that DePuy was defending the ASR. Meier quoted leading surgeons from Rush University and the Mayo Clinic, who voiced concerns about the catastrophic failures they were seeing in patients who had MoM hip replacements. By 2010, over one-third of hip replacements in the United States were MoM, and the prospect of tens of thousands of artificial hips failing was alarming, indeed.

In Australia and New Zealand, approval for the ASR was withdrawn in 2009; their joint registries had both identified the unacceptably high need for revision surgery. Barry Meier did not specifically reference the market rejection in his 2010 article, but he identified the worrisome trend in MoM hip failures. He quoted DePuy's defense, "as with other materials, metal-on-metal wear debris may cause soft tissue reaction in the area of a hip implant in a small percentage of cases."[27] The problem with DePuy's defense in 2010 is that the national joint registries in Europe and Australasia had objective statistics backing their conclusions, and further, once court trials began, the email conversations among DePuy engineers and business leaders and designing surgeons would be discoverable—and damnable—regarding the indefensible ongoing sale of an imperfect product.

The high revision rate of the ASR was predictable to one surgeon in particular: Derek McMinn, the designer of the Birmingham Hip. In 2005 in Helsinki, Mr. McMinn criticized the ASR (in a video posted on his website[28]) and presciently predicted the downfall of the ASR in the same year it was celebrating its unveiling in American markets. In the video, McMinn laments that "design changes have sex appeal but history records several major blunders," and proceeds, with surgical precision, to disembowel DePuy for its many missteps: an acetabular component (hip cup) that is too thin and deformable, an undersized cup that will concentrate too much force directly on the edge of the cup (which he predicts will lead to early failure), a design that is too conforming (which decreases the margin of error), and the manufacturing process of heat treatment of metals that has been shown to increase wear. While often acknowledging the impulses behind their individual design innovations, McMinn predicts that, combined, the changes will doom the ASR implant system to failure, particularly if the surgeon deviates at all from perfect placement of the devices. In the end, with evangelical zeal, Derek McMinn implores, "a major redesign is urgently required . . . only this time back a winner, and copy the Birmingham Hip precisely."

In 2009, the National Joint Registry of England and Wales issued a report of high revision rates following ASR hip replacement: 7.5 percent at three years. In April 2010, the Medicines and Healthcare products Regulatory Agency (MHRA, the United Kingdom's FDA) issued a Medical Device Alert, noting "a small number of patients implanted with [MoM] hips develop progressive soft tissue reactions to the wear debris associated with MoM articulations." As now has become evident, surgeons were communicating their concerns with DePuy, and in one famous case, Mr. Antoni Nargol, a surgeon from the northeastern England town of Tees, was told that other surgeons were not seeing the same failures he was. Nargol testified that DePuy "told me there were no other problems."[29]

Until there were.

Facing an onslaught of registry data that showed high early implant failures, and perhaps more important, an increase in negative scientific presentations at closed-door orthopedic meetings that resulted in plummeting sales of ASR hips, DePuy "decided in 2009 that it would be discontinuing the ASR System as a result of declining demand."[30] But is wasn't until August 26, 2010, that DePuy recalled the ASR, *a full year after internally deciding that the project was a bust*, significant because thousands more patients would receive an implant known to have flaws. The economics of such a decision are monstrous. To recall a device, the company is voluntarily waving a white flag and admitting that the device is flawed and dangerous. The FDA classifies Class I recalls "a situation where there is a reasonable chance that a product will cause serious health problems or death." A Class I recall automatically stops manufacturing of the device, and is supposed to (physically) recall back to the company all of the worldwide inventory of the devices. Viewed pessimistically, it is an admission that the company has been supplying a dangerous implant, and is the first strategic move in the legal war that will ensue. DePuy had been defensive even months before the 2010 recall, but the time had come to stop the bleeding, and ready themselves for the storms of litigation.

Almost 100,000 patients had an ASR hip implanted between 2003 and 2010, and while many are satisfied and will receive a lifetime of benefit from their operation, it is likely that tens of thousands of patients will need to have their implant removed and revised. The total settlement costs for DePuy could surpass $4 billion, and this will depend upon the results of the bellwether cases in the Multidistrict Litigation (MDL) hearings. MDLs are different from class action lawsuits, where a representative group of plaintiffs file a complaint alleging that they have been harmed in a similar fashion. Examples of class action lawsuits include runners who received $8 million from Vibram relating to false claims regarding their FiveFingers shoes, and investors who have sued corporations for misstating earnings. Class action cases rarely go to trial, and consumers often get jilted, with plaintiff attorneys often getting huge payoffs to drop the case.

MDL trials are organized by a federal panel who transfers a small handful of cases to one federal district court. The federal judge responsible for the MDL court identifies the bellwether cases to be held throughout the country, and appoints plaintiff and defendant attorneys to serve on various committees. Plaintiff attorneys who specialize in MDL product liability cases (e.g., involving implant or drug manufacturers), or securities fraud cases covet MDL committee positions. Once on such a committee, an attorney is only too proud to broadcast his committee credentials, hoping to attract more even more plaintiffs. The bellwether cases, taken as a whole, determine the settlement amount that will be offered to plaintiffs. Injured patients who qualify for a settlement may choose to receive the specified payout from the manufacturer, but patients sometimes choose to sue the manufacturer individually if they conclude their case is particularly compelling. While the results of some early bellwether cases may result in jury decisions for the plaintiffs in the hundreds of millions, most patients in MDL settlement decisions receive only hundreds of thousands of dollars. The real windfall happens to the plaintiff attorneys, especially the fortunate ones who are on the prized MDL committees and who have the reputation for "getting patients the money they deserve."

Judge David Katz, of the Northern District of Ohio, was chosen as the presiding judge over the MDL involving the ASR hip. Judge Katz selected the bellwether cases in 2010, and the first case was scheduled to go to trial in San Francisco in 2013, but shortly before trial, DePuy settled the case for an undisclosed amount. In the first ASR case to actually go to trial, a Los Angeles jury awarded a Montana man $8.3 million (for medical expenses and for pain and suffering). DePuy won a case in Chicago shortly after the Los Angeles defeat, but by the end of 2013, three years after the worldwide recall, DePuy agreed to settle approximately eight thousand qualified lawsuits in the United States for $2.5 billion after a meeting with Judge Katz and the court-appointed committee of attorneys. DePuy also agreed to set aside almost $500 million for added payments for severely affected patients, and to pay claims from private insurers and Medicare for costs associated with medical treatments and additional operations. All told, DePuy and its parent company agreed to set aside $4,000,000,000 to settle the ASR debacle. For a company whose total orthopedic business that same year was $9.5 billion, a settlement that equaled almost half of all orthopedic revenues, was shockingly costly.[31]

Patients who hear of a $4 billion settlement must surely conclude that their ship has come in; but after lawyers' fees, patients will only receive about $160,000 to compensate for their pain and suffering.[32] Lawyers in America will have received about one-third of the settlement, or $800 million, with the chief negotiating attorneys and committee members getting a bonus of $160 million dollars. For patients who unwittingly found themselves ensnared in one of the biggest medical disasters in American history, the frustration over a paltry payout is understandable. Imagine enduring several failed hip replacement operations and getting a check for $160,000, while some high-volume law firms got over $20 million while representing hundreds of clients, many of whom required little additional legal work. Perhaps a more salient frustration among injured patients is the sense of betrayal by the system—the implant manufacturers, the FDA, the courts, the hospitals, and even, surgeons—who seem *not* to be paying attention.

"The *medical implant system* is part of a larger healthcare system and consists of diverse relationships between regulatory and other governance bodies, commercial organizations, practitioner stakeholders such as surgeons, and patients," writes

Matthias Wienroth and his coauthors in their analysis of the ASR failure.[33] In our "technological society,"[34] society itself operates as a laboratory, where the experimental space has extended to include the market and consumers.[35] Testing of implants should not stop in the lab; quite to the contrary, it is crucial that continuation of testing and surveillance is mandatory. Even under the best of circumstances, patients are, in effect, part of a large, uncontrolled experiment,[36] and when the regulatory framework in a society is flimsy, sluggish, or nonexistent, is it any wonder that patients can be harmed by the thousands?

To this surgeon, it is unconscionable that there is no national joint registry in the United States of America. One can be a free market capitalist and simultaneously recognize that every stakeholder in American medicine must do better. Eleven countries in the world track every single total knee and total hip replacement performed, and it was these countries that alerted American patients and manufacturers that something terrible was occurring with the ASR hip. Momentum is slowly building for an American joint registry, but until it becomes mandatory, I can guarantee readers that regulatory rigor mortis will continue. The law of unintended consequences has meant that the Health Information Portability and Accountability Act of 1996 (HIPAA) actually complicates the tracking of outcomes while protecting patients' privacy, but serious work is being done to establish a national and international joint registry. The FDA's International Consortium of Orthopedic Registries (ICOR) was launched in 2010 to coordinate registries in fourteen nations, however ironic it was to lead a consortium when the founding nation does not, itself, have a registry.

Of course, an American joint registry should not stop with tracking knee and hip replacements. For the millions of patients who undergo implantation of pacemakers, heart valves, cochlear implants, shunts, catheters, stimulators, anchors, vertebral cages, organs, and electronic implants, a bona fide surveillance system is *vital*, in both senses of the word. The implant revolution will lead to possibilities we simply cannot fathom, and the recipients of implants (be they alloy, polymer, or tissue, and regardless if they are large chunks of metal or injectable nanobots), must have the faith that the medical implant system is tracking the performance of every implant, obsessed with the *end result idea*.

FOURTEEN

Medical Industrial Complex
and Medical Devices

In the decades following the Civil War, the western expansion of railroads connected forests, farmlands, oil production facilities, and ports across the continent,
enriching industrialists like J. P. Morgan, Cornelius Vanderbilt, and Jay Gould,
and spurring the growth of interior American cities. While oil production was
relatively modest prior to 1901, the discovery of major oil deposits in the Midwest
transformed the petroleum industry from a light and lubricant enterprise into a
fuel production business. Houston, Texas, had grown from a small western outpost
to a major port with the development of rail networks and deep-water shipping
channels to the sea.

At the beginning of the 20th century, and less than a century after its founding,
Houston had become an important transportation center, and its banks, "fat on the
profits of oil, cotton, and lumber enterprises, were solvent and able to assist business
ventures with much needed loans."[1] Two pairs of brothers decided to move their cotton
trading business, Anderson, Clayton and Company, from Oklahoma City, hoping
to strike it rich in the burgeoning southeast Texas town. The company, founded by
Monroe and Frank Anderson with Frank's brother-in-law Will Clayton (and later,
Ben Clayton), had enjoyed a measure of success since it began in 1904.

Only five years old when his father died, Monroe Anderson had lived in Jackson,
Tennessee, his entire life. With essentially no college education, Monroe gained
commonsense banking knowledge from his years as a bank cashier in Jackson, and
after a few years of remotely managing the books of Anderson, Clayton, decided to
move to Houston in 1907. While the firm would not relocate their headquarters there
until 1916, Monroe became a fixture in the booming port city for three decades. He
modestly lived in small hotel rooms his entire life, never marrying and fastidiously
saving his money.

With the Houston Ship Channel's completion in 1914, the port of Houston became one of America's busiest hubs, and the balance of power in cotton trading shifted to the United States after World War I. European firms, handicapped by weakened markets, perilous shipping, and lack of warehousing, ceded power to cotton trading centers like Houston, and firms like Anderson, Clayton flourished. The partnership not only weathered the war, but was positioned for meteoric growth in the roaring 1920s. In the midst of unbridled success, Frank Anderson succumbed to appendicitis in 1924 (at age fifty-six), a reminder of the fragile existence every human clung to in the era before antibiotics, highlighting the importance of medicine in Monroe's mind.

In 1936, Monroe Anderson, his health declining, established a foundation while his privately held firm was worth roughly $100 million. The hotel-bound Anderson struggled with kidney and heart disease, even though he had never smoked or drank, and was piously frugal.[2] In 1938, he suffered a stroke, and only then moved into a house south of town near the Rice University campus where he would live out the remaining months of his life until passing away in 1939, aged sixty-six. Having industriously built the largest cotton trading firm in the world, and scrupulously saving and investing, his $20 million estate was positioned to make a difference. The savvy men who were named the executors of his foundation knew that small gifts spread across many well-deserving charities would be meaningful, but perhaps not impactful on a grand scale. What they accomplished over the next few decades is nothing short of astonishing.

The reader may wonder why Monroe's name is not more familiar until they consider his full name: Monroe Dunaway Anderson, from which came the foundation commonly known as M.D. Anderson.

In 1941, the Internal Revenue Service recognized the legality of the M.D. Anderson Foundation (MDAF), and in the same year, the Texas legislature passed a bill authorizing the establishment of a state-sponsored cancer hospital. The first significant distribution of the MDAF was a $500,000 matching gift for the establishment of one of the first dedicated cancer hospitals in the world, with the stipulation that the hospital be built in Houston. For more than a decade the hospital operated from the Oaks, the former residence of James Baker, a prominent Houston attorney (and grandfather to James Baker III, the chief of staff and treasury secretary to President Reagan).

In 1925, the Hermann Hospital was built four miles south of downtown Houston on a "tract of soggy land,"[3] isolated and remote. The city of Houston owned 134 acres of wooded land next to the Hermann Hospital that had never been developed. In 1943, the MDAF fashioned a proposal of $500,000 for a cancer hospital, twenty acres adjacent to the Hermann Hospital, $1 million for new building construction for the Baylor Medical and Dental schools to relocate from Dallas to Houston, and $100,000 per annum for ten years for medical research to Baylor. The proposal was accepted, and construction of the initial buildings began soon after the end of World War II. By 1948–49, a flurry of construction was occurring "out in the country," with hospitals, clinic buildings, and a medical school rising from the woodlands south of town.

It was a bellwether year in 1954 for the little plot of land. Several major buildings, including the Baylor Medical School and the M.D. Anderson Cancer Center, opened. While only fifteen years old, the MDAF had already distributed $14 million, and somehow had grown to assets of $24 million through shrewd investments and management. The original $19 million had doubled in managed assets and distributions, and the concentration of gifts for the establishment of a complex of medical institutions was achieving a kind of miracle in the decade of possibilities.

However impressive the first years of construction were, they are positively transcended by what stands there now. Now called the Texas Medical Center (TMC), what started as a 1950s building boom became the largest medical complex in the world, and is the greatest visual exhibit on planet earth of what has changed in the world of medicine since the implant revolution. Skyscraper hospitals, clinic buildings, and thoroughfares comprise a mini-metropolis that dwarfs most downtowns in America. It is simply a wonder that medicine has changed so dramatically in a span of decades.

Today, the TMC occupies 1,345 acres and hosts 280 buildings and 50 million square feet of office space.[4] There are three medical schools (Baylor, University of Texas-Houston, and University of Texas Medical Branch) and over nine thousand beds in its many hospitals, which include Memorial Hermann, Methodist, St. Luke's, Texas Children's (the world's largest children's hospital), the University of Texas M.D. Anderson Cancer Center (the world's largest cancer hospital), the Texas Heart Hospital, the Methodist DeBakey Heart Hospital, the DeBakey Veterans Hospital, Shriner's Hospital, and many more. There are over 100,000 employees and 10 million patient encounters per year, with 750,000 emergency visits per year, and 25,000 births per year. The TMC is the eighth-largest business district in the world, with a Gross Domestic Product of $25 billion (ahead of many countries).

The two square miles that the TMC occupies is a technological and logistical spectacle, particularly when one considers the seed money originated from a bachelor who lived unostentatiously in downtown Houston hotels. (The MDAF has donated more than $81 million to Texas Medical Center institutions[5] and more than twice that outside the TMC.) From a broader context, the labyrinth of buildings has one description: A *Medical Industrial Complex*. In Houston, and in other prominent places like Boston, New York, and Rochester, Minnesota, the obvious visual representations of the industrialization of medicine is breathtaking. But every American city and hamlet is dotted with clinic buildings, labs, hospitals, rehab centers, therapy offices, surgery centers, and business offices.

Another, more powerful way of contemplating the reach of the medical industrial complex is to consider the number of patients who have been treated, and the most salient question is how many Westerners have received, in their body, an implant? In other words, instead of thinking about hospitals and medical complex buildings, how many humans have pieces of the medical industry in their own bodies?

First: what constitutes an implant? In this book, implants are either *temporary* or *permanent*. For centuries, catgut and silk sutures were the mainstay in primitive surgery, and both of these sutures rapidly degrade in the human body. Weeks after placement in the body there is no evidence of their presence due to rapid degradation and destruction by the immune system. Obviously, these types of suture are temporary implants, and are clearly not the main thrust of this work. *Permanent implants* are the focus of this book, and every medical specialty utilizes implants that are meant to stay in the body till death. The simplest form of a long-lasting implant is "non-absorbable" suture, typically made from synthetic polymers (e.g., polyester or polyethylene). Almost anyone who has had an operation beyond a superficial skin biopsy, removal of a small cyst, or lancing of a boil likely has permanent suture in their body. Prior to the 1950s, the "tipping point" of the implant revolution, synthetic, non-degradable suture was unthinkable, and today it is almost ubiquitous.

Second: what are implants made of? Implants are made of plastic, metal, organic, biologic, and electronic material.

Organic implants are the nonliving tissues that are transplanted into a patient's body, most commonly in orthopedic operations for musculoskeletal deficiencies. A common example is a patient who receives purified, processed, and sterilized human bone graft into their own bone defect, such as a non-healing fracture. Another example is a young athlete receiving a cadaver (deceased human donor) hamstring into their knee for ACL reconstruction. These grafts do not have to be tissue-typed (donor matched) because they are not alive and do not need to be biologically functioning. The final example is a *xenograft*, an implant from another species; pig heart valves are commonly used in patients with heart valve disease but who are not good candidates for intrinsic valve repair or mechanical valve replacement.

Biological implants are living tissues and organs, specially procured and processed and transported for immediate implantation. An organ transplant, like a kidney, heart, liver, lung, pancreas, and intestine is a live, functioning organ that is surgically implanted, with reconnection of the blood vessels and soft-tissue connections that allows the organ to function. Nondegradable sutures and metal clips are used to secure the organ in place, but the centerpiece of the transplant is, of course, the organ. The organ comes from a donor, related or unrelated, brain dead or very much alive, who has been "matched" to the recipient, thus lowering the risk of organ rejection or even worse, the risk of graft-versus-host disease (where the immune cells within the transplanted organ rise up in the recipient's body and wage full-scale war on the host tissues).

The epitome of a biological implant is another human being. Although a temporary (nine-month) implant, the notion that we humans can combine biological materials (semen and ovum) in laboratory settings, and store the fertilized byproduct in sub-freezing conditions in a large metal cylinder, and months, or years, or even decades, later, implant the microscopic embryo in a woman and grow another human is dizzyingly overwhelming. It sounds like Greek melodrama, with the gods bringing forth life from a frozen cauldron, but has become so commonplace that the wonder of it has evaporated. In vitro fertilization (IVF) is no less wondrous than a proposal of a colony of humans on Mars, even though it happens over 60,000 times a year in the United States, and 350,000 times a year throughout the world. In recent years, we have passed one million live births in the United States from IVF, and almost 2 percent of all American births are from IVF.[6]

It can be easily argued that a "test tube" baby is a temporary, nine-month implant, but surprising new findings show that all women may harbor the cells of their babies through the process of *fetomaternal transfer*, wherein fetal cells cross the uterine-placental barrier and form "microchimeric" nests of fetal cells within the mother.[7] Previously thought to be impossible, the transfer of the cells into the mother may have positive, protective effects, but also may represent a threat to the host. Whether naturally conceived or developed via IVF, fetuses are capable of "living forever" within their mother, even if they are not properly "matched" (which normally results in cellular rejection). In short, the process of fetomaternal transfer may be an ancient phenomenon we are only recently appreciating, but the new art of IVF may, for the first time, allow completely foreign cells to inhabit a woman's body (with no shared genes between fetus and mother if the egg was donated from another woman, as is common among older women undergoing IVF). The ultimate expression of a biological implant (an embryo) may take decades before we truly understand its biological consequences, but until then, the miracle of creating life with glass pipettes, liquid growth medium, and frozen storage chambers will remain a beatific phenomenon.

Plastic (or polymer) implants are made of organic polymers—most commonly from petrochemicals (i.e., from petroleum)—that are synthesized from chemical reactions to create molecules that are long chains of repeating building blocks, like a chain of paperclips. The assembled chain of simple structural units can be extremely long—thousands of units—and the chains themselves can be made to link together sideways, forming a block of polymers. Nature does have its own polymers, like horn, hair, or cellulose,[8] but the ability to synthesize plastics (through the process of polymerization) in the years following World War II has led to revolutions in packaging, shipping, manufacturing, clothing, and medicine.

The long-chain polymers that form snakelike backbones are generally chemically *inert*, nonreactive and non-biodegradable. The very thing that makes plastics hard to recycle (conventional petroleum-based plastics never really go away, even when they break down into microscopic disintegrating particles) is what makes them ideal implant material, namely, the ability to remain stable and incognito to biological systems that peruse the biological landscape for intruders. Our immune systems simply have not evolved to recognize polymers as foreign, and as long as a polymer implant is sterile and has structural integrity, it will be tolerated, accommodated, and even welcomed by the body.

Polymers can be made hard as a rock or into a pliable plastic. Chemical engineers can configure almost any shape and size implant, and special-order particular performance under varying stress, heat, and friction conditions. Alchemists spent centuries trying to turn chunks of earth into gold and precious metals; while we know this to be foolhardy due to the conservation of matter principle, scientists today are able to produce wondrous materials out of petrochemical sludge, nothing short of an alchemical sleight of hand.

Metals are commonly used in the body for orthopedic, cardiovascular, and electronic applications. The main ingredients in metal implants are cobalt, chromium, molybdenum, titanium, nickel, carbon, platinum, gold, and others. Mining companies search the planet for unique mineral deposits, with particularly rich concentrations in the *continental shields*, areas where the world's ancient crust still exists at the surface, like interior Africa, Western Australia, and Canada. These mines are worked for ore, yielding trace amounts of desired minerals for processing, purification, alloy-manufacturing, and eventual implant production.

Minerals are "solid, naturally occurring, inorganic substances with a crystalline structure, which possess a uniform chemical makeup regardless of their point of origin."[9] In a sense, it is not possible to "make a metal"—the minerals are found, mined, and processed. For instance, the greatest deposits of chromite in the world are in South Africa, India, Kazakhstan, and Turkey,[10] while more than half of all cobalt stores are in the Democratic Republic of Congo.[11] Through smelting (application of heat and chemical reducing agents to extract metal), pure metals are isolated and bar stock of the raw material is created. Implant manufacturers use this raw supply of metal to make implants—an international smorgasbord of materials readied for surgical implantation. A person with a total hip replacement likely has metals from Africa, Asia, and North America, and polymers from North America.

The shoulder prosthesis used by Péan in Paris in 1893, was made of platinum and rubber, but half a century passed until joint replacement with a large metal alloy stem became a reasonable enterprise. In 1937, just as antibiotics were about to become clinically useful, pioneering orthopedic surgeons tested various alloy screws in animals, finally settling upon an admixture of metals that is similar to what we use today. Processing and sterilization greatly improved, and orthopedic implants earned a proper place in the armamentarium of disease treatment, but were limited to joint replacement and plating of fractures. In the infancy of the implant revolution, the only sensible application of minerals was with bulky hunks of metal, but that would change by decade's end with the insertion of wires into the body.

Today, millions of patients around the world have cardiac devices implanted in their bodies. This was unthinkable just a few decades ago, when patients suffering from heart attacks or arrhythmias were simply allowed to die. Almost every reader can recall a story of a forebear who was struck down by a heart attack or stroke, and such accounts of angina and apoplexy are so common that we forget how remarkable it is that we expect (demand?) to recover fully after a cardiovascular event. The most common use of metal in the heart is in coronary artery stents and the use of wires in pacemaker applications. It is commonly stated that over 500,000 coronary stent operations are performed every year around the world,[12] (I will show that this is dramatically underestimated) and implantation of pacemakers is expected to exceed two million units by 2021.[13] Stent utilization trailed pacemaker placement by more than thirty years—in essence, it was easier to poke wires into the muscle of the heart than to thread an expandable mesh cage into a minuscule coronary artery.

Virtually every last medical intervention involves some type of modern material and newfangled technique. Don't believe me? Take, for example, a simple visit to an urgent care center after scraping your leg on a rusty nail. You'll be greeted by a triage nurse who will enter your information into an electronic medical records database (invented and refined in the last few years), an IV (with modern, pliable polymers) will be started to run antibiotics (developed in the last half century), delivered from a bag constructed of specialized, nonreactive plastics. An injection of tetanus booster from a syringe will be given. The syringe itself is dizzyingly complex, composed of petrochemically-derived polymers from around the world, sterilized and specially packaged, and shipped to this little "doc-in-a-box" urgent care center. The anti-tetanus medicine itself requires vast sophistication, complex machinery, and laboratory specialization, with layers of regulatory oversight. Your wound, to be cleansed properly, will require a mini-operation of sorts, with specialized kits opened to scrub your skin and drape the injured area with surrounding disposable blue paper drapes. These drape kits likely were assembled in Puerto Rico for American distribution, as were the fluids for the IV (which explains the mini-calamity to American healthcare following Hurricane Maria, saying nothing of the tragedy to the island). The sterile gloves donned by the ER doc are a marvel of technology, composed of latex-free polymers, sterilized and packaged in a cutting-edge factory in Chicago. The nylon sutures are made by Johnson & Johnson in Cornelia, Georgia, and are mated to a miniature suture needle made of space-age metal alloys, and sterilely packaged. The tools used for passing the needle through the skin are enclosed in a purpose-assembled little "suture kit," also composed of unique plastics and economical metals. Even the little Band-Aid that will cover your IV site after it is withdrawn has its own little clever story of entrepreneurial chutzpah and empire-making by the Johnson brothers (yes, it launched the J&J domain). A careful blob of Neosporin ointment will be slathered on your cut, and the history of its development, manufacturing, packaging, and distribution would require its own chapter. An hour later you'll go home, free of anxiety that you'll die of tetanus or lose your leg to a freak infection like our grandparents might've worried. While nothing will be implanted following your small snafu, the medical industrial complex has risen to the occasion.

A much greater marvel is to suffer a heart attack and be resuscitated. We all take for granted that an ambulance will materialize at our doorstep in a jiffy after a 911 call and ferry us to a hospital cardiac angiography suite. On the way to the hospital, the EMT in the ambulance will have transmitted the Medtronic EKG rhythm strip that, by itself, gives profound information about the type of heart attack someone is suffering. Once in the "cath lab," a cardiologist and his team (also miraculously present and poised for lifesaving) will rush to get the massive IV started in the groin. Within minutes a thin, flexible wire will be snaked up the aorta to the heart, entering the coronary arteries, where balloon angioplasty will be performed to provide dilation of the artery.

In all of mankind's existence, an acutely blocked coronary artery has been a certain path to the grave, but since the 1980s, cardiologists have been able to stand next to a human body in a cath lab, and manipulating the controls at the end of a wire, unclutter a small artery with an inflatable, tiny balloon and prop it open with a springy-cage metal alloy stent that expands upon command. All the details of polymer IVs, sterile drapes, fluoroscopy (live-motion X-rays), specialized packaging, and anesthesia are similar for all such interventions, and become so routine and dependable that we forget what a sensation each one is. But it is typically the implant itself that garners the greatest attention. While there are data now that question the wisdom of stenting arteries in patients who are *not* experiencing an acute heart attack and have little or no chest pain but who have an abnormal EKG—it is clear that there is, and will be, a huge worldwide market for stent placement under properly indicated situations.

The vast array of materials available to scientists, including polymers, metals, and space-age electronics (saying nothing of the innovations in organic and biologic implants), combined with advanced manufacturing and packaging processes, has altered the way that companies approach acute and chronic diseases. Among implant manufacturers, similar to the pharma industry, market analysis drives research and consideration of device innovation. If the numbers justify an investment, a working group of scientists and physician consultants will be formed, hoping that in a few years' time, after thousands of hours of development, a technical breakthrough with market potential and clinical impact will have been achieved. Usually, such a project leads to failure, as it is *extremely* difficult to invent a new device that: 1) works, 2) is safe, 3) is novel, 4) is capable of being reliably manufactured, 5) would yield a profit, and 6) doesn't infringe upon someone else's patent.

If a sort of miracle has occurred, a medical device invention will change lives *and* achieve financial success. As has been argued in this book, America is embarrassingly behind the times in tracking implants in device registries, but we must face the facts, and, in the process, achieve greater patient safety and efficacy, and in turn, drive greater device innovation and better health outcomes (with or without devices). It's inarguable—as has been highlighted in Jeanne Lenzer's book, *The Danger Within Us*—we don't have an accurate accounting of device implantation in our world.

To understand the full scale of the medical industrial complex and the implant revolution, an investigator has to turn his gaze *not* just at the edifices (like the Texas Medical Center) and the financial numbers, but to the recipients, to those walking among us who carry the implants in their bodies. Shockingly, no one has ever attempted to calculate the numbers of devices that are implanted on an annual basis, and certainly no one has estimated the percentage of Americans or Europeans who have an implant or device in their body.

Until now.

FIFTEEN

Surgery of the Heart

In 1896, Stephen Paget, a renowned British surgeon claimed, "Surgery of the heart has probably reached the limits set by nature to all surgery; no new method, and no new discovery, can overcome the natural difficulties that attend a wound of the heart."[1]

The story of the first implantable cardiac device is typical of the advancement of science and medicine, with early disappointment, courageous explorations upon sacrificially willing patients, catastrophic failure, renewed investigations, and eventual triumph by a small group of investigators with almost pathological determination.

Electronics were, for decades, only utilized in cardiac medicine, starting in the 1950s. Today, implantable electronic devices enjoy widespread use in general surgery, urology, otolaryngology, neurosurgery, orthopedics, and even gynecology. The story of their implantation is a synthesis of improved surgical techniques, advanced anesthesia, antibiotics, upgraded metallurgy, and modern electronics, and particularly, the development of the transistor.

Even as surgical treatment of gastrointestinal and musculoskeletal issues was improving in the 1930s, no one dared operate on the heart. Risk of brain damage, sudden death, and failure of surgery was so grave that surgery was simply untenable. In 1938, Robert Gross at the Boston Children's Hospital performed ligation of the ductus arteriosus—the small artery that connects the pulmonary artery and the aorta in utero, helping the fetus bypass the nonfunctioning lungs; it should spontaneously close in the days following birth, but its ongoing patency is debilitating to an infant.

In 1944, Alfred Blalock, at Johns Hopkins, performed palliative treatment for a child suffering from the cardiac condition known as the "tetralogy of Fallot," a cardiac defect where the pulmonary valve is too constricted, leading to a hypertrophied right ventricle, a hole in the heart between the right and left ventricles and "overriding of the aorta," in which the aorta empties both the right and left ventricles, instead of just the left ventricle. Simply stated, the constellation of physical defects in the heart of tetralogy of Fallot patients makes it impossible to adequately oxygenate the blood, resulting in a "blue baby," a child whose oxygenation is so compromised they acquire a bluish hue to their skin. To be curative, incising the muscular wall of the heart—and looking inside the heart—was necessary, but no surgeon in the world could conceive of a method of "opening" the heart without killing the patient. Tetralogy of Fallot was a death sentence, but Blalock's work-around palliative surgery did improve patients' lives by connecting large vessels on the *outside* of the heart.

The world's first "open-heart surgery" was on September 2, 1952, at the University of Minnesota when F. John Lewis operated on a five-year-old girl using *total-body hypothermia* and *inflow stasis*. This was accomplished by placing the child in a horse-watering tank full of ice water in the operating room, cooling her to 82°F and, after surgically opening the chest, clamping the blood vessels entering the heart (inflow stasis). A quick operation to close a pathological hole between heart chambers was performed, and after warming, the child was resuscitated and survived the operation. More than fifty children with an abnormal passage between heart chambers were treated in this manner, but concerns over the ability to properly normalize the heart rhythm while rewarming led the Minnesota surgeons to consider another way of unlocking the heart.

Heartened by Lewis's progress, Dr. John Gibbon at the Jefferson Medical College in Philadelphia corrected the same cardiac defect in an eighteen-year-old in 1953, while using an artificial device to oxygenate the blood. The *screen oxygenator*, later named the Mayo-Gibbon heart-lung machine, was large, complex, and expensive, but did achieve success in the first application. The machine was the size of a hotdog vendor's cart, connected to the patient through a series of plastic tubes, whirring the blood to and from the patient with DeBakey's roller pump (stay tuned). Not only did the patient survive the world's first open-heart operation using cardiopulmonary bypass, she lived another forty-seven years before dying at age sixty-five.[2] Sadly, Gibbon's next three patients all died in the operating room (or shortly thereafter). It had taken nineteen years of laboratory research, with countless animal operations and endless hours of investigation to develop the machine, but by 1954 he decided to suspend all open-heart operations for at least a year while attempting to improve outcomes. It must have been a crushing defeat for Dr. Gibbon; in fact, he never performed another open-heart operation again.[3]

The surgeons in Minneapolis were working on their own cardiopulmonary machine (with little progress) and also innovated with biological solutions. In early 1954, during a brainstorming session among clinicians, one of the young surgeons reflected on his pregnant wife's ability to support her fetus with blood flow to the womb, and the idea of "cross-circulation" was born. The team investigated the ability of a dog to function as an external, biological, cardiovascular machine to keep another dog alive during surgery. After dozens of sham dog operations, Dr. Walton Lillehei at the University of Minnesota performed an operation on a one-year-old boy who was connected to his father, who served as a "biologic oxygenator." One can only imagine the anguish of a mother watching her husband and child wheeled away on gurneys to the operating room for the cross-circulation operation, or the gallantry of a father placing his own life at risk while sustaining his infant son. The "general discouragement about open-heart surgery changed drastically"[4] that day, even though the young patient succumbed to pneumonia eleven days later. Undeterred, Lillehei and his team performed forty-five operations over the next year using cross-circulation, each time the pediatric patient being sustained and saved through a parent (who risked their own life), with two thirds of the patients surviving surgery and being discharged from the hospital. Major cardiac defects were treated, including atrial and ventricular defects ("holes in the heart"), and even tetralogy of Fallot. A year earlier, these conditions were unqualified death sentences. After a year's experience, in 1955 (a decade after Alfred Blalock's original blue baby operation), Dr. Lillehei presented his data on the first tetralogy of Fallot patients, with Blalock in the audience in Philadelphia at the American Surgical Association meeting.

In the ten-patient series, six had excellent outcomes, but four died, all within hours of surgery. Such a failure rate would not be tolerated today, but in 1955 this was nothing short of a triumph. Dr. Blalock commented after Walton Lillehei read his paper, "I suspect it's a mistake for an old conservative surgeon to discuss this paper. I must say I never thought I would see the day when this type of operative procedure could be performed. I want to commend Dr. Lillehei and Dr. Varco and their associates for their imagination, their courage, and their industry."[5] Occasionally, in science, a groundbreaking presentation in a darkened conference center is given, and the magnitude of the moment is comprehended by every colleague in the room. Like aerospace engineers embracing and waving flags after a successful rocket launch, the surgeons in that room in Philadelphia were no doubt wiping away tears of joy, and were energized with a renewed optimism. Thirty years later at the 1985 meeting of the Society of Thoracic Surgeons, Dr. Lillehei presented the long-term results of those forty-five patients. Remarkably, seventeen of his twenty-seven patients who suffered from ventricle septal defects were still alive, a stunning finding in light of the impossibility of their survival without surgery. Dr. Denton Cooley, one of the great pioneering surgeons in the history of the art, spoke from the podium afterward, saying, "Dr. Lillehei provided the can-opener for the largest picnic thoracic surgeons will ever know."[6] While cross-circulation cardiopulmonary support was only briefly used, it ushered in the ability to perform open-heart surgery, and by the end of the 1950s, surgeons around the world were inspired to consider the impossible . . . maybe *nothing* was impossible.

Walton Lillehei and his team demonstrated that serious cardiac defects were treatable, reigniting a quest to perfect the mechanical cardiopulmonary machine. The

Gibbon machine, modified at the Mayo Clinic, was considered too costly for practical use, and Dr. Lillehei turned to his young crew of lab assistants to develop a better machine. The chief architect was Richard DeWall, a recent medical school graduate who envisioned a life of research and laboratory medicine. Starting with a clean slate, DeWall became the "MacGyver of medicine"[7] by assembling a "Rube Goldberg" contraption of twisted hoses, pumps, needles, and oxygen tanks. Instead of fragile glass tubes, DeWall used polyvinyl hoses, which had the dual advantage of being cheaper and surprisingly less reactive with blood than glass was. His shoestring budget actually became an advantage, energizing an openness to the polymer revolution; the polyvinyl tubing for pumping mayonnaise had come from a nearby factory.

DeWall's bubble oxygenator was inexpensive and reproducible, and more importantly . . . worked. During the proving grounds of 1954, while using cross-circulation, Lillehei had shown that open-heart surgery was possible, but by the middle of 1955, the bubble oxygenator became the chief method of sustaining life during the groundbreaking operations. Today, in every major hospital in the Western world, cardiopulmonary bypass machines are the vehicles that keep humans alive while surgeons operate on the heart. The full history of the development of the bypass machine is beyond the scope of this book, but not unlike the development of any device, occurred across continents, was pioneered by self-financed tinkerers, and was finally achieved after many, many failures. For a time, the only open-heart operations being performed were at the University of Minnesota and ninety miles away, at the Mayo Clinic. As successful as the operations themselves were, there was still one critical issue: postoperative complete heart block.

Scientists have been fascinated for centuries with the concept of the electrical nature of the human body and nowhere more so than the heart. The electrical impulse from the Sinoatrial node, also known as the natural "pacemaker of the heart," communicates to the Atrioventricular node, which drives the contraction of the ventricles. Your brain does not tell your heart to contract—it has its own metronome, a built-in electrical timer, rhythmically firing across the muscles of the heart. Put your hand on your chest, and the slight flutter you feel is your heart contracting and squeezing the blood to your entire body. Slide your hand up to your neck, and feel the pulse by your windpipe—that rhythmic beating is the echo of your heart valves slamming shut, in regularity, in response to the syncopated muscular contractions of the heart. In complete heart block, wherein the electrification of the heart has gone haywire, the patient suffers severe bradycardia (low heart rate), hypotension (low blood pressure), and extremely compromised cardiac function. In essence, the muscular pump that is the heart is uncoordinated and dysfunctional, and unless properly electrified, unable to sustain life. The Minnesota surgical team had been researching ways of dealing with cardiac pacing emergencies, and were able to save a child's life using a simple laboratory electrical testing device, the Grass Stimulator.

In Minneapolis in January 1957, following open-heart surgery to repair a ventricular septal defect, a child was crippled with *complete heart block*. A physiologist at the University of Minnesota had recommended that the heart team stimulate contraction of the child's heart with their lab machine, the Grass Stimulator, which produces a small-voltage electrical charge. In physiology labs and introductory classes around the globe (to this day), scientists use the stimulator to send a small pulse of electricity through the wires into a test subject. To make the leg of a frog jump, a physiologist pokes tiny wires into the leg muscles, connects them to a Grass Stimulator, and adjusts the voltage and timing to make the muscles contract. After preliminary tests on dogs, Lillehei's team was hopeful that the stimulator could work on a child suffering complete heart block.

When an open-heart child suffered complete heart block in 1957, Dr. Lillehei and his team inserted an insulated wire into the heart muscle of the patient, connected it to the stimulator, and realized—in triumph—that he could control the beating of the heart. By turning the dial, Lillehei was able to increase the beating of the heart, a contrivance that William Harvey would have savored. While this represented real achievement, it is astonishing to consider the real-world logistics of applying the life-supporting electrical current. The Grass Stimulator was the size of a microwave oven, requiring an AC outlet and extension cord. In fact, to venture from the operating room to the recovery room, a *one-hundred-foot extension cord* was required to keep the equipment plugged in and the patient alive. Imagine the cardiac team, venturing from the open-heart room with a small child on a postoperative gurney, the anesthesiologist monitoring the breathing of the unconscious patient (still intubated), and the surgeons feeding out orange extension cord down the hallway to keep the heart pacing at a life-sustaining rate.

The AC-powered cardiac stimulator was a lifeline for those patients who had suffered complete heart block, but a disaster occurred on October 31, 1957, when a municipal power outage in Minneapolis led to loss of electrical power to the patient ward rooms, resulting in the death of a young patient. The loss of the patient must have been crushing to Lillehei, and frustrated over being bound to a wall socket, he asked a local electrical engineer and electronics consultant to investigate the possibility of miniaturizing the stimulator and creating a battery-powered unit. Recognizing that most complete heart block patients returned to their own sinus rhythm within a couple weeks, Lillehei hoped that some type of innovation might untether patients from a wall socket and serve as a bridge to normal cardiac function. In one of medical history's great moments, Lillehei turned to the young engineer, Earl Bakken.

Earl Bakken and his brother-in-law, Palmer Hermundslie, had founded a company in 1949 to maintain and repair electronic equipment in the Minneapolis area hospitals, but in their first month had only $8 to show for their efforts (the servicing of a centrifuge).[8] In the burgeoning field of electronics and transistors, these handymen figured that someone would need to fix all the gadgets in the hospitals around town.

Earl was a Minneapolis native, and after graduating high school in 1941, enlisted in the Army Signal Corps and became a radar instructor and maintenance technician. His lifelong interest in home electronics logically led him to a military posting in the field, and even in his nineties, Earl says that he finds "a deep, almost inexpressible joy in the sight, sound, feel, and even smell of old radios, machines, and electrical equipment. There is a magic about those devices that a person can appreciate only when he knows them inside and out, and when he loves them not only for what they do, but how they do it."[9] This sentiment recalls Steve Jobs and every entrepreneurial tinkerer, and proves the point that almost all innovators are "garage guys" who can't stop turning over an idea in their mind and fabricating the solution with their own hands.

Upon returning to Minnesota at the conclusion of the war, Earl Bakken attended the University of Minnesota and earned undergraduate and master's degrees in electrical engineering. All who knew Earl were not surprised with his career choice, being a child prodigy with gadgets and models. As a young boy, Earl saw the movie, *Frankenstein*, and was captivated with the life-giving power of electricity. He later recalled, "I was simply awestruck by the fact that electricity, properly applied, could do a great deal more than light up a room or ring a doorbell. I realized that electricity defines life. When electricity flows, we're alive. When it doesn't, we're dead."[10]

While Earl Bakken was completing his coursework at the university, he would often walk across the street to the academic hospital, forming relationships with the scientists and technicians whose jobs were increasingly dependent upon electronic equipment. This led to the formation of Earl's company in 1949, but years of low-paying contract work and crude business development left his company in precarious health. The breakthrough opportunity came in 1957 (after the power outage), and when Dr. Lillehei charged Earl Bakken with developing a solution, he immediately began tinkering with an idea to provide battery-powered, rhythmic, electrical pacing.

Recalling an issue of *Popular Electronics* one year earlier, Earl Bakken drew inspiration from an article explaining how to make an electronic, transistorized metronome. To a lifetime fan of electrical gizmos and contraptions, Earl's simple challenge was finding a circuit that he could construct. In the April 1956 magazine article, the simplified circuit diagram with two transistors was presented, and Bakken cleverly innovated the amusement into a lifesaving device.

A circuit is an electronic grid, composed of wires, resistors, capacitors, and transistors. But it is the transistor part of a circuit that has revolutionized all of electronics, communications, and medicine. The invention of the transistor is the "central artifact of the electronic age."[11] The unreliable vacuum tube of earlier electronics was a power hog and generated too much heat; what was needed was an electronic device that could amplify an electronic signal that was smaller and more energy efficient. Bell Laboratories was the industrial research arm of the American Telephone & Telegraph Company, and as Alexander Graham Bell's telephone patents were facing expiration, an intense research effort to develop improved transcontinental communications led to the evolution of Bell Labs into the foremost scientific development organization in the world. Numerous Nobel Prizes have been awarded to Bell Labs researchers, and numerous revolutionary technologies were developed there, including the laser, solar cell, communications satellites, and the transistor.[12] The group that created the transistor was the contentious, eventual Nobel Prize–winning threesome of William Shockley, Walter Brattain, and John Bardeen, although none of them monetarily benefitted from their invention. Shockley departed for Palo Alto, California, where he founded Shockley Semiconductor, which employed the eventual founders of Fairchild Semiconductor and Intel Corporation. Silicon Valley germinated from companies innovating telegraph and radio technology, and was further fueled by the semiconductor and computer companies that were founded in the 1950s. Transistors and integrated circuits permitted miniaturization, decreased energy consumption, and enhanced computing power, all of which powered the space race, made possible the personal computer, and gave firm foundation for the modernization of medicine and the implant revolution.

Earl Bakken designed a two-transistor circuit and enclosed it in a crude aluminum box that was only four-inches-square and an inch-and-a-half-thick—about the size of a small stack of coasters or a deck of cards. Instead of multiple controls like the Grass Stimulator, there was only an on/off toggle switch and pulse rate and current output rheostats. On top of the unit were the exposed terminals to connect the wires to the patient, and inside was housed a powerful 9.4-volt mercury battery. The wires that emanated from the device were designed to pass through the skin and into the heart, so that when they were not needed they could simply be withdrawn at the patient's bedside.

Four weeks of experimentation yielded a device that was fit for experimentation at the university's animal lab, and a single day of trialing in dogs raised expectations that refinement could lead to a device that could be implanted in a human. In his autobiography, Bakken recalls returning to the hospital the very next day to work on another project, and "I happened to walk past a recovery room and spotted one of Lillehei's patients. I must have done a double-take when I glanced through the door. The little girl was wearing the prototype I had delivered only the day before! I was stunned. I quickly tracked down Lillehei and asked him what was going on. In his typical calm, measured, no-nonsense fashion, he explained that he'd been told by the lab the pacemaker worked, and he didn't want to waste another minute without it. He said he wouldn't allow a child to die because we hadn't used the best technology available."[13]

I am gobsmacked when I consider that an American surgeon was able to implant a device in 1957 without any FDA device clearance, but none existed at that time. The 1950s was the Wild West of device development, with no laws and no sheriff. While it was personally risky for Lillehei and Bakken to implant a "MacGyver" implant, it wasn't against the law. Today you would quite literally go to jail for such an offense. But in 1957, medical devices were about heroic gallantry and optimism, and the ambulance-chasing enterprise of personal injury law was yet to be born.

The world's first battery-powered, wearable cardiac pacemaker had come about from a confluence of new transistor and polymer technologies, and the evolution of batteries and new coating materials, and the prepared mind of Earl Bakken. What was the name of Earl's struggling little medical electronic service company? Medtronic. It has grown into the world's largest medical device company, with annual revenues approaching $30 billion, employing over 80,000 people, and boasting a market capitalization of about $100 billion.[14] Pacemakers were soon made implantable, with the main application for elderly patients suffering from age-related cardiac arrhythmias. Almost 500,000 pacemakers are implanted every year, and it's a virtual guarantee that you, Dear Reader, know someone with a tiny pacemaker implanted under their skin, resting upon their pectoralis muscle, covertly flickering away. They are so efficient, it doesn't seem revolutionary to an individual, but it does feel like a miracle.

Aged ninety-four and living in a grand house on the Kona coast of Hawaii, one of the great pioneers of the electronic implantable medical device industry is at peace. Earl Bakken has invited me to his exceptional home on the big island of Hawaii. For the man who understands how electricity can sustain life, it is not that surprising that he is completely "off the grid," owning the largest personal photovoltaic farm in the world. Overlooking Kiholo Bay, Mr. Bakken is self-powered and even makes his own fresh water with desalination machines. He favors "high touch and high technology," and is worried that mankind is losing the sense of the mystical and becoming too obsessed with newfangled technology.

Although Earl Bakken hasn't granted an interview in years, he was receptive to a visit by this surgeon once he learned the purpose of my project. (Earl Bakken would die eight months later, in October 2018.) Meeting one of the giants of the implant revolution is a sobering honor, and after navigating a series of guard gates with electronic key pads (I'm armed with the codes) and driving along a paved road surrounded by volcanic rocks, I park under palm trees outside his office. Inside, Earl is seated on his scooter, flanked by bookshelves. There are awards and plaques here and there, but a toy Frankenstein doll up high on a shelf brings a smile to my face.

Earl Bakken recounts the old days, the threat of business failures along the way, his regrets and his triumphs, but the thing I am most impressed with is his belief in living a "full life." The mission statement of Medtronic genuinely beats in his heart, and our conversation is peppered with his lifelong sense of obligation to alleviate pain, restore health, and extend life.

Medtronic now has a traditional request of every patient who has benefited from one of their devices, a sense of obligation to "pass it on" in a meaningful way. Earl repeats this entreaty to me as our interview ends, imploring me to, "Live on! Give on!"

In a sense, that is the purpose of this book, to illuminate the contributions of the pioneers who have made modern life so less risky and human existence so much more enjoyable. We all benefit from the advances in medicine and surgery (as imperfect—even perilous—as it can be), and for those whose lives have been enriched and lengthened, there is a compulsion to "pay it forward."

The story of the pacemaker, is, of course, only one small part of the story of heart surgery. Prior to the heart-lung bypass machine, it was unfathomable that any cardiac defect or heart valve surgery could be considered. As noted earlier, it was the pioneering work of Minnesota surgeons that opened the door to the heart, and over the course of the 1950s, defect closure and valve repair became predictable and effective.

The first forays into valve surgery were rapid-fire incursions in which Mayo surgeons John Kirklin and Henry Ellis made small incisions in the side of a beating heart, working blindly with a specialized knife attached to the end of the surgeon's finger. The device was repeatedly and forcefully plunged into the diseased and constricted aortic valve.[15] The sobering reality is that the mortality rate in those early days was 20 percent, and was somehow deemed acceptable.

Dr. Lillehei achieved much greater success (in nearby Minneapolis) when he operated on diseased aortic and mitral valves while using a heart-lung machine. Instead of operating on a beating, blood-filled heart, Lillehei had the advantage of looking into the inner cavity of the heart and attempted to partially resect tightened and diseased valves, or to repair flimsy, incompetent valves.

The first artificial valve operations occurred in 1960, and with no device clearance needed, there was a "dizzying pace of progress" among cardiac surgeons in America.[16] The first valve replacement devices were silicon-covered Lucite balls contained in a stainless steel cage; the ball was designed to bob back-and-forth at the site of the resected valve. Although blood clots, arrhythmias, and sudden death were always a risk, lives were being saved and the quality of life for hundreds of patients was dramatically improving. "Open-heart surgery had evolved from an experimental procedure in 1955 to a standard treatment technique in less than a dozen years."[17] Any operation into the thorax was unthinkable prior to World War II, and by 1961, there were 303 hospitals in the United State fully equipped for open-heart operations and angiography.[18] Cardiac care had transitioned from the treatment of children with life-threatening cardiac anomalies to surgical management of cardiac valve disease. Completely unaddressed was the handling of coronary artery disease and heart attacks—an even more pressing issue—but it would take a fortuitous mistake to begin a critical revolution.

On October 30, 1958, Mason Sones, a cardiologist at the Cleveland Clinic, was performing a cardiac catheterization on a twenty-six-year-old male with valve disease as part of cardiac workup. At the time, a catheterization procedure consisted of inserting a thin, flexible catheter into the brachial artery (of the arm) and threading the catheter all the way to the root of the aorta, just above the aortic valve. (Today, catheterization is performed while watching massive, ceiling-mounted flat screen monitors, but from the 1950s and up into the 1990s, catheterization was captured on 35mm motion picture film and later viewed on a projector.) As Dr. Sones was sneaking the catheter tip across the aortic valve, an automated pressure syringe injected 50cc of contrast solution into the chamber.[19]

Almost all of the contrast material, instead of emptying into the aorta, filled the right coronary artery, resulting in "extremely heavy opacification" of the artery and temporary slowing of the heart. "Sones's fear that filling a coronary artery with so much contrast would cause a life-threatening ventricular arrhythmia gave way to a feeling of 'considerable satisfaction regarding the further diagnostic evolution of the technique.'"[20] His hopes buoyed by his experience, Sones soon collaborated with a company to produce custom, taper-tipped catheters to *intentionally* catheterize coronary arteries. Overnight, this resulted in the ability to image the coronary arteries, and more importantly, determine the degree and location of blockage. Doctors had always been impotent in determining the cause of chest pain or localizing the relevant area of vascular blockage—that is, until it was time for an autopsy. The grim reaper would now have to wait: physicians could now uncover the mystery of angina and heart attacks in real time, upon a beating heart.

The serendipitous discovery that visualization of the coronary arteries was possible—and not lethal—was seized upon by Sone's surgical colleagues at the Cleveland Clinic, realizing that they "had the best possible set-up for the surgical treatment in selected patients with coronary artery disease."[21] Indeed they did—and they do. Donald Effler, then chief of cardiothoracic surgery, made good use of the new diagnostic tool, performing the world's first coronary artery operation in January 1962. This would be the bailiwick of the Cleveland Clinic till this day, and more coronary artery bypass operations are performed there every year than anywhere else in the world. Coronary artery bypass grafting (CABG) was pioneered in Cleveland, and the technique of bypassing an area of blockage with vein harvested from the leg was the work of René Favaloro, an Argentinian surgeon on staff at the Cleveland Clinic. It is one of mankind's greatest operations, and is still performed today in every major hospital around the globe, with modifications.

The Cleveland Clinic is a world leader in multiple fields, ranking at the top of the list in multiple specialties, including most all surgical fields. This is no accident—the CEO of the clinic has traditionally been a surgeon, including the CEO at the time of this writing, Tomislav Mihaljevic, a cardiothoracic surgeon. The modus operandi in Cleveland is simply different than most hospitals, but it is no accident that some of the greatest institutions in the world, like the Hospital for Special Surgery, are also led by a surgeon.

Coronary artery bypass grafting was a decade old when the world's first angioplasty was performed in 1977 in Zurich, Switzerland, by Andreas Gruentzig. The development of angioplasty, and later coronary stenting, follows a typical path in medicine and surgery. Technological refinement begins with crude interventions, transitioning to less invasive and more sophisticated techniques, eventually leading to solutions that seemed impossible only a few years before. The first selective imaging of a coronary artery was in 1958, and less than twenty years later, angioplasty was developed to open up a clogged artery with a tiny inflatable balloon, followed shortly by the innovation of the cardiac stent in 1986.[22]

In the New York Public Library, there is a small globe (five inches in diameter) that is one of the earliest surviving cartographic spheres in existence. It is made of copper, and if you position yourself over the land mass of Asia, you can see the inscription, "*Hic sunt dracones*," Latin for "Here be dragons."[23] It might be the only globe (or map) that actually contains that expression, but has now become a popular saying for "no trespassing." The last major frontier of the human body was finally challenged, and mastered in the 1960s, and it was no accident that movies and television shows started portraying surgeons as heroes—unthinkable representations a century before. *Here be dragons* no longer applies in the human body.

We simply cannot fathom the extreme passivity of care given to the Senate majority leader Lyndon Johnson and President Dwight Eisenhower in 1955. Both men suffered heart attacks in the space of a couple months, and other than diagnostic EKGs, there was nothing to speed their recovery. In an era before angiography, cardiac stents, and coronary artery bypass surgery, it seems ridiculous that the President of the United States was given a pair of slippers and ferried about in a wheelchair, praying that his heart attack would respond to a program of six weeks' rest.[24] Today, every American undoubtedly expects a full cardiac resuscitation, with angiographic stenting or open-heart surgery following a heart attack. Cardiac valve repair or replacement, and aneurysm repair represent sobering surgical challenges, but it would not require undaunted courage to undergo such operations.

It is ironic that the heart was the last organ that yielded itself to the surgeon's scalpel, even though it was the first organ to be quantified through physiology by William Harvey. Denton Cooley, the famous Houston heart surgeon, said, "It's about the only organ in the body that you can really witness its function."[25] I had seen the heart of a dead horse in the necropsy room at my father's veterinary hospital when I was an adolescent, and it was confusing, lifeless, and smelly. But when I witnessed my first open-heart operation in college, I was dumbstruck, because there before me was a strikingly pulsating, wriggling, colorful, organ; replenishing and nourishing the entire frame of a human body.

No diagram or painting can possibly capture the dynamic function of the heart, and it was only when the hearthstone of the body was governed that surgeons could claim full ascendancy from a previously shameful trade to respectful—even glorious—profession. Think I'm too melodramatic? Position yourself in a hospital waiting room, and wait for a heart surgeon to meet with a small congregation of frightened family members, whose mother suffered a heart attack the day before, and witness the gratefulness for the ability, over the course of a few hours, to conquer the heart and sustain life. That's when the implant revolution feels like a miracle.

SIXTEEN

Specialization in Surgery

"As I see it, the outlook in the United States has never been so hopeful . . . The surgeons have had their day—and they know it! The American St. Cosmas and St. Damien—The Mayo Brothers—have their clinic today as important in medicine as it ever was in surgery. Wise men! They saw the pendulum was swinging."

—William Osler, 1919[1]

As a fourth-year medical student I am aware that I am of almost no practical use to the orthopedic service here at Denver General Hospital. What I can offer is enthusiasm, but more important, I must demonstrate my willingness to do almost anything to become an orthopedic surgeon. This is my "externship," a one-month rotation at another medical school that serves as a prolonged tryout. Because Denver is such a desirable location, the rotation at the University of Colorado and its sister hospital, Denver General, is chock-full of gung-ho medical students—gunners—who must pretend to be stellar team players while simultaneously outperforming the competition.

My first couple weeks here in Denver were at University Hospital, requiring daily diligence but no overnight fortitude. Now that I am at Denver General, known simply as DG, my grit will be tested, and I am eager to show that I am willing to work harder than anyone in my prospective class. I know my odds are long—there are hundreds of medical students in the class of 1995 across the country who want to be orthopedic surgery residents, and the vast majority of them will be crushingly disappointed. As my first night of call approaches, I am filled with a mixture of roller-coaster-ride anxiety and inquisitiveness over what I might find at Colorado's "Knife and Gun Club."

I accompany my supervising chief resident Joe down to the Emergency Room, where an elderly female has just been admitted with an injured hip. When we go to Bay 13, Joe, a sturdily built Chinese American, pulls back the dingy, beige curtain whose hooks screech along an overhead rail. Lying on the gurney is an emaciated woman who is in her late eighties, toothless and clueless about where she is. Her left leg is grotesquely askew and twisted, and Joe looks at me, asking, "What is your diagnosis?"

"Is her leg dislocated?"

"Are you asking me or telling me?" Joe queries.

"Her hip is dislocated," I conclude.

"Wrong!" Joe scolds.

Dang it. "Is her hip broken? I mean—her hip is broken."

"You're right," my boss affirms. "Let's get an X-ray to see how bad it is."

As we are awaiting the X-ray tech, the ER physician briskly walks down the cramped hallway, announcing to everyone in earshot that a gunshot victim will be arriving via ambulance in a few minutes. THIS is what I was hoping for, even though I am ashamed to admit it. I justify my morbid curiosity with the supposition that I don't want anyone to ever get hurt, but if they are going to get shot or stabbed, I hope that it happens on a night when I am on call.

Joe and I walk over to the trauma bay, a large room with supply shelves along one wall, and a row of sinks and cluttered countertops along another. Everything is messy, and the trash bins are already full from an earlier trauma. The cleaning crew is hurriedly trying to prepare the bay for the incoming patient, and with medical students and residents hovering in anticipation, this little space is jam-packed.

I walk out to the ambulance entrance with Joe, awaiting our GSW—gunshot wound—while trying not to look nervous. Facing Bannock Street on this warm October evening, I glance at the ER nurses and techs in light blue plastic aprons, and notice that everyone is lighthearted, even jocular. I think I'm the only one who is anxious. I've never seen a shooting victim, and although I am quite sure I will play no part in the resuscitation, my complete lack of expertise reinforces my apprehension. Behind me, the ER charge nurse approaches our crew of general surgery and orthopedic residents, informing us that a second ambulance has just radioed in, and will be arriving STAT with a stabbing victim. This pleases everyone.

The two ambulances arrive in synchrony, one from the north and the other from the south, lights flashing and sirens blaring. The doors of the first ambulance are flung open by our crew as I struggle to catch a glimpse of the patient. The paramedics slide the gurney out of the vehicle and its wheel mechanism deploys, stretching to meet the ground. The patient, a young Latino male, is conscious, and I am surprised how calm he appears, and for that matter, how relaxed the entire medical team is as well. His eyes dart around at most of the team members, but he remains silent as he is wheeled inside. An IV is in his right arm and a liter bag of clear fluids are running along the plastic tubing into his vein. There is a small amount of blood splattered on his white T-shirt, but certainly less than I was expecting from someone who was just shot in the abdomen.

The next ambulance is only seconds behind, but when its doors open, a vastly different scene plays out. An older African American man is sitting up on his gurney, a white blanket jumbled around his hands. Blood is everywhere—on his shoes, khaki pants, plaid, short-sleeved dress shirt, and on his face. He is in extreme pain, grimacing and writhing on his stretcher, and as he is wheeled into the ER, is pleading for something to deal with the pain.

"Find out what's going with that guy," Joe commands.

I follow the ambulance crew into a small ER exam room in time to hear their report.

"Henry Jones, sixty-eight-year-old male, who was accosted by a hitchhiker who pulled a machete on him on the side of the road after Henry offered to give him a ride. The asshole tried stabbing Henry in the gut, but the patient blocked the machete with his hands, nearly amputating both thumbs. No other injuries. One bag of LR in [lactated ringers IV solution], stable vitals, not up to date on tetanus, but has no other medical issues."

A veteran nurse, Stephanie, immediately takes charge once she realizes no residents or attendings are available. "Henry—we're gonna take care of you. No drug allergies, right?" He is gritting his teeth, almost holding his breath. His pain is white-hot.

"I'm not allergic to anything. Please give me something!"

Stephanie leaves to get some fentanyl from the pharmacy, and because all of the trauma team members are in the trauma bay, I am suddenly by myself with the patient and a young nurse. I grab a pair of gloves and introduce myself.

"Henry, I'm Dave Schneider, I'm a medical student on the orthopedic surgery team. I need to take a look at your hands. Is that okay?"

I begin unwrapping the makeshift dressing on Henry's hands, gently unfurling the blood-splattered blanket. When I untangle the last layer over the right hand, Henry shudders in pain. Lifting the white cotton blanket I see that the thumb is barely connected to the hand, dangling by a thin bridge of skin to a gaping wound of crimson, serrated muscles, shredded tendons, and pulsatile blood vessels.

Henry slams his head backward against the bed, screaming, "Oh my god!" I silently mouth the same sentiment. I grab some 4x4 sponges [pronounced "four by four"], vainly positioning the thumb back into position, and while holding the digit in its precarious station, wonder what will be my next move. I reinforce the temporary covering with a stouter ABD ["Army battle dressing"] sponge and stabilize it with layers of cloth tape. I like things to look tidy and organized; this looks like an 8th grade science project, and my lack of proficiency must be obvious to everyone.

Before I can report to Joe, I know I must also examine the left hand. Henry's eyes are closed, and I assume that the fentanyl is drenching his brain, finally relieving this poor man's agony. I repeat the process, carefully peeling open the onion of blanket layers, grimacing as I get down to the left hand, discovering another dangling thumb—this one really bleeding, obscuring my view of what is still holding the thumb to his body. It is cockeyed and sickeningly unhinged, and despite the powerful narcotics coursing through his body, Henry is snapped back into a hellish appreciation of his predicament because of my intrusion.

I finish dressing the left hand, and make my way to the trauma bay to update Joe.

In the ten minutes since we split up, another major trauma patient has arrived, this one a young man who was run down by a drug dealer driving a huge SUV. I glance into the room, and it's the craziest scene I have ever witnessed. Blood is everywhere. The young man is unconscious, and the anesthesiologist is hurriedly placing a breathing tube while the trauma team is cutting off his blue jeans with huge trauma shears. [Every medical student and surgery resident dutifully carry these industrial-strength, orange-handled scissors that can cut a penny and saw through any piece of clothing. Job one in a trauma bay is hacking off the clothes.] As a nurse is trying to take off the Timberland boots I can tell that his legs are a jumble of bones. Hideously, she twists the leg 360 degrees, and as the jeans get cut, it's apparent that the leg is completely amputated, only connected to the hip by the nerves, thick as a rope. Oh my god, I think again.

Joe scrambles to press on the thigh to stem the gush of blood. I think I am watching this beautiful young African American man die. He appears lifeless, and although these dozen or so lifesavers are scrambling in synchronized orchestrations, I can't imagine they can halt his demise.

While Joe waits for a tech to procure a tourniquet, he glances up at me, calmly asking, "What you got?"

I am mostly in shock, and not exactly sure what he is asking. Is he asking about this young man clinging to life?

"Dave—what's the deal with the knife wound . . . are his hands injured?"

"Yes, both his thumbs are almost completely cut off. I rewrapped both of his hands with sterile dressings."

Joe places the tourniquet around the demolished leg, stemming the flow of blood. He looks back at me and says, "Look, we're getting crushed here. I want you to numb up that guy's hands and wash them out in that ER room, and we'll try and sew them back on later tonight."

This is the reality of how a county hospital works (even today). There is simply too much work for a limited number of workers, and because all county and metropolitan hospitals are propped up by medical students and residents, patients often are cared for by unqualified practitioners. It is simply the truth.

I make my way back to Henry's exam room and tell him that our team is planning on reattaching his thumbs tonight, but because of the two crashing trauma patients, we can't go to the operating room now. My job is to wash the wounds now in hopes of preventing a terrible infection from the filthy machete. The problem is (although I don't tell him), I don't know how to do a regional anesthetic block.

When William Halsted pioneered the use of cocaine as a local anesthetic, he rapidly learned that targeting a specific nerve far away from the intended target could result in the entire limb becoming numb. However, this mandates extremely intricate knowledge of the three-dimensional anatomy of the nerves, and I simply do not yet have that type of comprehensive understanding. I am hoping that my injections into the hands themselves will work.

The nurse has pulled the lidocaine from the pharmacy and drawn it up into two large syringes. After removing my temporary dressings, I begin inserting the needle around the knife wounds, hastening bleeding and temporarily renewing pain. I keep injecting again and again in both thumbs and decide to wait a few minutes to let the medicine take effect.

I gather a large metal pan, resembling a baking sheet, to catch the irrigation water, and the nurse tells me that the orthopedic residents normally use a battery-powered pulsatile lavage device to blast water into traumatic wounds. I agree, although I have never used it. This is a bad mistake.

When I return to the room, I assess how numb Henry's hands are. The medicine seems to be working, and I decide to start washing the thumb wounds. Using the pulsatile device, I start spraying the raw wounds on his right hand. Henry shrieks in pain—he is clearly not numb enough to tolerate irrigation. I decide to inject more medicine, but I am afraid to cause damage to the nerves down his arms if I inject into the nerves themselves. Instead, I inject into the laceration sites themselves, and apologize to Henry that the medicine isn't working as well as I would like.

I understand that my role is to temporize Henry's situation, but I feel like all I am doing is making everything worse. I wait another few minutes, and I'm as frustrated as Henry is scared. I warn him that I need to try again to wash his wounds, and he positions his hands over the metal pan. Instead of using the pulsatile device, I use a simple bulb syringe to gently lavage his open wounds. At first contact, he cries out again, shuddering in pain, shaking the metal pan and spilling water over the edge of his gurney. Despite our repeated efforts and a change in technique, his pain is agonizing.

I stop using the bulb syringe, unable to continue subjecting Henry to what feels to me like torture. I am at my wits' end, and not sure what to do. I glance at this older gentleman, a man with kind eyes despite his predicament. I shake my head, and I am quite sure he knows I feel stunningly incompetent.

In a hushed voice, Henry intones, "Lordy, Lordy, help the doctor." This beautiful impulse on his part stuns me, and we lock eyes. Perhaps saying the only Bible passage he knows, Henry adds, "Jesus wept." I am stupefied.

I am at my breaking point, and by some miracle, my chief resident appears in the doorway. "How's it going, Dave?"

"Not well, Joe. I can't get his hands numb."

"Where are you injecting? Did you do a regional block?" Joe asks, sensing that I am in over my head.

"A regional block? Um, no."

Joe helps me save face, and realizing that we never discussed the technique of regional anesthesia, asks the nurse to draw up more numbing medicine. "I'll show you how," Joe affirms, and I'm starting to feel a rush of relief.

This is how medicine is taught—through a grueling series of trial-and-error assignments, failures and triumphs, overwhelming insecurity and occasional recognition. Oftentimes, when a medical student or resident fails, a singular tidbit of knowledge is missing, spelling doom to themselves, and more important, their patient. In a fully actualized training environment, *errors are detected early before harm occurs, and in this case, my chief resident intervened just in the nick of time to salvage the situation. Henry will receive the block he needs to properly cleanse his wounds and ready him for digit reattachment tonight.*

In 1540, by an Act of Parliament, the Company of Barbers and Surgeons of London was created. This codified their respective roles—only barbers could cut hair and shave beards, and only surgeons could engage in the primitive acts of lancing boils, setting bones, and suturing wounds. This was an uneasy marriage, particularly as the professionalization of surgery increased over the next two centuries. In 1745, the barbers and surgeons went their separate ways, with surgeons retaining the right to anatomize the bodies of executed criminals and to practice the more invasive aspects of their profession. Eventually, the Royal College of Surgeons of England was formed in 1800 by royal decree, in no small part due to the pioneering work of John Hunter, the first scientific surgeon.

Throughout the centuries, physicians continued to hold a much higher-privilege position in English society than surgeons; they were members of a socially elite caste, while surgeons were practitioners of a craft, and like apothecaries, were tradesmen. Physicians did not deign to work with their hands, and only "observed, speculated, and prescribed."[2] During the 18th century, the members of the Royal College of Physicians were exclusively graduates of Oxford and Cambridge, catering to aristocratic patients, and emulating the style and bearing of the upper class, "making every effort to attract attention to themselves by cultivating distinctive manners and fashionable dress."[3] The rigidly organized social strata in England did not apply in colonial America, where physicians, surgeons, and apothecaries existed together, oftentimes having attended crude medical schools, if any at all.

The Industrial Revolution broadened the middle class, expanding the ability to pay for medical services and decreasing the dominance of aristocratic patients propping up a small group of noble physicians. But it wasn't until the emergence of antiseptic surgery and the adoption of the German scientific mindset that the philosophical demolition of the hospital as a "death house" occurred. At the onset of the 20th century, an explosion in hospitals had transpired; from 1872 to 1910, the number of American hospitals had grown from 178 to more than 4,000.[4] American skepticism about the power of physicians to heal was well founded at the death of Abraham Lincoln in 1865, and even at the death of William McKinley in 1901, but medicine was soon to transform from incompetence to unimagined efficacy.

Cardiologist and medical historian Bruce Fye observed, "During the last third of the 19th century, an avalanche of new technologies transformed the United States. Some of the most impressive culture-changing inventions involved communication (telephones, typewriters, and phonographs) and transportation (expanding rail networks, turbine-powered steamships, bicycles, and automobiles). Electricity, another breakthrough technology that empowered machines and lit dark places, was also touted as a therapeutic tool."[5] In addition, modern building techniques utilizing newfangled Bessemer steel and George Fuller's internal load-bearing steel structure (as opposed to history's use of external load-bearing structures), along with the invention of the elevator (1883), propelled buildings higher and higher.[6] The first skyscraper was built in 1885, and it wouldn't take long for the new expanded building style to have its impact on the transformation of medicine.

On August 21, 1883, a cataclysmic tornado ripped through the tiny town of Rochester, Minnesota. The F5 tornado (winds greater than 260 miles per hour [7]) resulted in the deaths of thirty-seven and over two hundred serious injuries. The town of Rochester was a typical Midwestern farming community that had just become home to some twenty-four Franciscan Sisters six years earlier who had come to Rochester to establish a new congregation and to serve as teachers in the small southern Minnesota settlement. The cyclone devastated the town and highlighted the complete lack of emergency medical care. The town was fortunate to be served by an enterprising general practitioner—an immigrant from Lancashire, England, who had held more than a dozen job titles in his life, including newspaperman, riverboat captain, farmer, tailor, politician, and census taker, but as Rochester grew, his role as community physician became his identity. William Worrall Mayo (1819–1911) was a town father, but made his most important contribution to Rochester (and the world) in the rearing of his two sons, William James Mayo and Charles Horace Mayo.

Dr. Mayo and his wife Louise married when he was thirty-two years old, but Will Mayo wasn't born for another decade, when his father was forty-two. (Charles was born four years later.) Despite his advanced age, the elder Mayo was preternaturally energetic, even practicing with his sons for a decade upon their graduation from medical school.

Dr. Mayo graduated from the University of Missouri school of medicine in 1854, and upon graduation, his only option was to immediately enter practice—residency had not yet been invented. He eventually moved to Rochester in 1864, at a time when the town boasted only three thousand inhabitants.[8] After struggling to establish himself in Rochester, the ambitious "Little Doctor" (he stood 5' 4" tall) traveled to New York City and Philadelphia in 1869 to observe surgeons at work. Lister's technique of antiseptic surgery was brand new, and by the time Dr. Mayo returned to Rochester, his ideas about surgery had been transformed. The last time he had been in New York, in 1846, anesthesia had just been invented, and now for a second time he was witness to a scientific revolution.

For a man who had moved innumerable times and held countless jobs, his wanderlust was finally sated in 1874, and he would live in Rochester, as a surgeon, the rest of his life. By the time he became president of the Minnesota State Medical Society in 1873, Dr. Mayo had gained a regional reputation for his surgical skills. All of his medical care was provided in his tiny office and in the homes of his patients; there had simply been no demand for a hospital, and Dr. Mayo was likely one of the few people in town who had ever *seen* a hospital.

Will Mayo (1861–1939) was born in Le Sueur, Minnesota, within weeks of the Civil War's first major battle, and Charlie (1865–1939) was born in Rochester within weeks of the great conflagration's *last* major battle. Even as children, Will and Charlie accompanied the "horse and buggy" physician on house calls, even assisting in surgery. Will Mayo graduated from the University of Michigan in 1883, returning home to Rochester to join his father whose office was in "downtown" Rochester; Charlie was in medical school at Chicago Medical College at that time, and it was assumed he, too, would join the family business.

On August 21, 1883, all three Mayo men were in town the day the monstrous cyclone destroyed Rochester. Dozens perished and hundreds were injured, and the lack of any type of health care facility meant that the injured were triaged in the local school, small hotels, and the dance hall. It didn't take long for the leader of the Sisters of St. Francis, Mother Alfred Moes, to approach Dr. Mayo, offering to help build a hospital. Catholic sisters were partnering with town fathers across the country, and in the face of tragedy, Mother Alfred was inspired to build a small hospital on the plains.

Later, Dr. Mayo recalled, "I told her too that the erection of a hospital was a difficult undertaking and required a great deal of money, and moreover we had no assurance of its being a success even after a great deal of time and money had been put into it. 'Very true,' she persisted; 'but you just promise me to take charge of it and we will set that building before you at once. With our faith, hope, and energy, it will succeed.' I asked how much money the Sisters would be willing to put into it, and her reply was, 'How much do you want?' 'Would you be willing to risk forty thousand dollars?' I said. 'Yes,' she replied; 'and more if you want it. Draw up your plans. It will be there at once.'"[9]

In the five years it took for the Sisters to raise enough money to buy land and begin the construction process, Dr. Mayo and Dr. Will (as he became known) toured East Coast hospitals and met with architects. While the tornado had been a catastrophe, the timing was fortuitous. Thirty years before, hardly a soul lived in the area, but in the mid-1880s a small band of determined visionaries thought that building a local hospital was worth dreaming about. Constructing a hospital building with modern features was also possible; filling it with up-to-the-minute technology was a necessity—in fact a Mayo staple to this day—that helped the Mayo brothers leapfrog everyone in the region, and eventually, the world.

Six years after the ruinous tempest, Saint Marys Hospital opened its doors in 1889. For several more years, the three Mayo surgeons all operated together, and the numbers are nothing short of astounding.

Another longstanding hallmark of Mayo practice was the inclination to visit other surgeons. The "Grand Tour" tradition of Europeans (and particularly, the British) traveling throughout the Continent had been well established by the 19th century, and as travel became more reliable, Americans made the European voyage as well. David McCullough's book *The Greater Journey: Americans in Paris* details the pilgrimages of American painters, writers, sculptors, and doctors to Paris between 1830 and 1900. While the French had dominated medicine from the French Revolution till the last third of the 19th century, world leadership tilted east to Vienna and Berlin, and they became meccas for aspiring surgeons. First, however, the Mayo brothers would follow the lead of their father in traveling by train to New York, Philadelphia, Boston, and Baltimore, and only later traveling to Europe.

In 1889, the three-story, forty-five-bed hospital only contained one operating room, and in the first decade, the only surgeons were the Mayo brothers. Within five years, an addition was needed, and by 1894 there was a second operating room and a total of seventy-five beds. The same year that the original Saint Marys Hospital opened, Johns Hopkins Hospital had opened. In 1895, Will Mayo traveled to Baltimore to watch William Halsted and Howard Kelly operate, and recorded in his diary his impressions of the Hopkins surgeons: "Rubber sole shoes. Operating suits. Iodoform, Boric Acid & Bismuth. Sterilize by steam. Circle tables for instrument trays. Rubber gloves. Mattress stitch. Subcuticular stitch."[10]

Will Mayo would be the driving force behind the success of the Mayo Clinic for the next half century. Instead of tepid satisfaction with a small-town hospital, a series of hires was made that truly launched Rochester to the forefront of worldwide medicine. Instead of playing it safe and keeping their practice a two-man team of brothers, Will and Charlie hired additional diagnosticians who would screen the patients and perform testing with the latest laboratory techniques. By 1895, the brothers were performing an astonishing number of operations. Five years prior there was no hospital in Rochester; in 1895, there were 762 operations, including 95 intra-abdominal procedures.[11]

At the turn of the century, fewer than 2 percent of American physicians identified as surgeons,[12] but without apology, the Mayo brothers embraced the identity of surgical operators. By 1905, a dozen doctors had been added, all rallying around the Mayo surgeon-brothers. The greatest hire in Mayo Clinic history, it may be argued, was that of a twenty-nine-year-old physician who was also a Minnesota native.

In 1901, the Mayo brothers recruited Henry Plummer, a young man who had worked with his physician-father in nearby Racine, Minnesota. Plummer earned his medical degree from Northwestern in 1898, and had trained under Frank Billings, a professor of medicine in Chicago, who was a pioneering internist who had differentiated himself from the competition with advanced laboratory and diagnostic equipment. Moreover, Billings had predicted that the growing sophistication among physicians and surgeons would drive the development of specialization in medicine. Not coincidentally, it was Plummer who energized the growth of specialization at the Mayo Clinic over the next several decades.

Like William Welch at Johns Hopkins, Plummer championed bacteriology and laboratory medicine at Mayo. Whereas medicine had been a primitive art just a half-century before, now at the turn of the century, the small contingent of Rochester physicians and surgeons had a custom-designed office space, exam rooms, a clinical laboratory, and an X-ray machine.[13] In what sounds impossible, in 1903, Will and Charlie performed 2,640 operations, including 1,302 intra-abdominal procedures. "The number of patients who had died (sixty-nine) was remarkably low, and 84% of those who had died had undergone abdominal operations, acknowledged as a risky type of surgery."[14] There were thousands of small towns across America where groups of physicians had banded together on Main Street, hoping to serve their little communities. Why was the group in Rochester so radically successful in such a bewilderingly rapid fashion?

The achievement of establishing such dominance in a tiny town a long train ride away from everywhere was hard fought, but not inexplicable. The foundation for excellence was laid in the ambition of the father, William Worrall Mayo. The recipe for success is there at the beginning: the willingness to travel and be tutored by experts; the embrace of technology; the desire to hold elected office within medical societies; and the inclination to partner with others in the formation of a practical corporation. Dr. Mayo's sons continued this tradition and a scant decade following the construction of their hospital, had established their operation as the most advanced surgical enterprise in the world.

In 1906, Will Mayo was elected president of the American Medical Association. The forty-five-year-old surgeon stated in his presidential address that "the future will demand schools for the advanced training for those who desire to do special work." The Mayo Clinic School of Medicine was not founded until 1972, but Dr. Will was not advocating for a medical school. Instead, he was echoing surgical pioneers like William Halsted, who foresaw the need to spend additional years of training in the new specialties.

In the first decade of the new century, the Mayo emphasis on surgery continued. Not only were the Mayo brothers single-mindedly identifying as surgeons, the entire enterprise of the Mayo Clinic and Saint Marys Hospital was as well. For anyone who has visited Rochester, even today, it is hard to believe that it became one of the foremost medical complexes in the world. Wouldn't it be a guarantee that one of the East Coast port cities, or even Chicago, would be the leading surgical mecca?

In some ways, it is easier to believe that it didn't happen in a major metropolitan center; there, the competition was so keen that no single group of surgeons in those cities could commandeer a reliable cluster of patients to become expert in a general area of surgery, saying nothing of becoming a specialty surgeon. In Rochester, there was a "triumph of cooperation" among the disciplined troops of the Mayo Clinic.[15] In his commencement address in 1910 at Chicago's Rush Medical College, Will Mayo promoted multispecialty group practice, saying, "The sum-total of medical knowledge is now so great . . . that it would be futile for one man to attempt to acquire . . . a good working knowledge of any large part of the whole. The very necessities of the case are driving practitioners into cooperation. The best interest of the patient is the only interest to be considered, and in order that the sick may have the benefit of advancing knowledge, union of forces is necessary . . . [so] it became necessary to develop medicine as a cooperative science; the clinician, the specialist, and the laboratory workers uniting for the good of the patient, each assisting in the elucidation of the problem at hand, and each dependent upon the other for support."[16]

Henry Plummer continued to innovate, serving as the chief diagnostician and "systematizer" for decades. He and his colleagues invented the unified medical record, the hospital chart that is so common to us today. Prior to this ingenious idea, patients' information was handwritten in a daily log, making it impossible to retrieve patient information and track outcomes. Now, each patient was identified by a Mayo patient number, a numeral that stays with them for life—proudly, I have one—and an individual paper chart was created exclusively for that patient. Each subsequent hospitalization was recorded in that patient's chart, allowing for greatly enhanced continuity of care.

Plummer also played key roles in the construction of world-class medical buildings in Rochester, performed groundbreaking research, and served as one of the best diagnosticians in the world. His career spanned the decades of vastly improved medical technology, and his early adoption of X-rays and the EKG machine helped propel the Mayo Clinic to the vanguard of medical practice.

Both Mayo brothers died in 1939, by which time the Mayo Clinic had become the preeminent medical destination in the world. US presidents, CEOs, dignitaries, the wealthy and the destitute, and surgeons themselves had flocked to Rochester. It had become the world's oldest and largest multispecialty group practice, helping shape the very nature of specialization itself. Historian Rosemary Stevens states that "specialization is the fundamental theme for the organization of medicine in the 20th century."[17] By the end of World War I, most doctors were specialists, there was an outgrowth of technological expansion, the introduction of residency training, and there were improvements in the structure of medical schools following the Flexner Report.

The fact that the Mayo brothers had *not* completed surgical residencies, and yet helped transform the practice of surgery, the organization of hospitals, and the education of doctors—is amazing, to say nothing of the notion that it was worth traveling to tiny Rochester, Minnesota, to receive expert care.

The Mayo brothers, William Halsted in Baltimore, and the surgeons in London, Berlin, and Vienna were radically changing surgery at the turn of the 20th century. As generalists embraced specialization, additional years were required to inculcate the young physicians and surgeons with the special techniques that were being developed by the increasingly sophisticated practitioners. The ability to focus on an individual body system (or organ), boosted levels of sophistication that were unimaginable just half a century earlier.

Perhaps the most famous orthopedic institution in the world is New York's Hospital for Special Surgery. It was the not the first orthopedic hospital in America—that title belongs to Boston's "Orthopedique Infirmary," now defunct—but it is the oldest still in existence, founded in 1863.[18] Originally called the Hospital for the Ruptured and Crippled (R&C), the newfound hospital tailored its services to the deformed and the crippled filling the streets of New York City. At an age where epidemics were an enigma and tuberculosis was rampant, adults and children alike suffered the deformity of infectious disease and the catastrophe of trauma; once incapacitated, ruination was a certainty.

Despite its founding during the American Civil War, R&C did not have an operating room until 1889, the same year that Saint Marys Hospital opened in Rochester (with its single OR) and the same year that the Johns Hopkins Hospital opened. Though not a surgeon, James Knight (1810–1887) was the founder of the hospital, specializing in "surgiomechanics." Instead of operating, Dr. Knight used braces, bandages, and supports to treat scoliosis, hernias, varicose veins, and even hemorrhoids.[19] He remained skeptical of surgery his entire career, even as the R&C moved from its modest beginnings at his home on Second Avenue to a purpose-built hospital on the northwest corner of Lexington and 42nd Street (now the site of a Grand Hyatt Hotel). In retrospect, Knight's conservatism was well-founded, his career spanning an era before anesthetics, germ theory, and antiseptic surgery were known; and during a time that no one in the world could imagine antibiotics.

Virgil Gibney (1847–1927), a farm boy from Kentucky who attended medical school at University of Louisville, and later, Bellevue Hospital Medical College in New York, became the second leader at R&C in 1887. Despite losing the ring finger and little finger of his right hand as an adolescent, Gibney stubbornly persisted in his desire to become a surgeon, and fell under the tutelage of Bellevue's Lewis Sayre (1820–1900), the first orthopedic professor in the United States. Unlike James Knight, Dr. Sayre optimistically embraced the role of surgery when caring for musculoskeletal problems, and he pioneered the surgical resection of a tuberculous hip infection in 1854.

Dr. Gibney was surgeon-in-chief at R&C for almost forty years (1887–1925), and oversaw the creation of the first orthopedic residency in the United States in 1887, and the building of the first operating room at R&C in 1889. In the tumultuous last decade of the 19th century, the practice of orthopedics was upended at R&C under Gibney's leadership. Surgery took center stage, and the driving force for that change was the arrival of Royal Whitman in 1889.

The iconoclastic tendencies of Royal Whitman (1857–1946) perfectly matched Virgil Gibney's. A former student recalled, he "was always trying out new procedures—either those he initiated or those suggested by others. He had an insatiable curiosity about the pathogenesis of orthopaedic diseases and deformities, and an imagination which led him continuously to seek new methods of manipulative or surgical correction of musculoskeletal defects."[20] Whitman published a tour de force textbook, *A Treatise on Orthopaedic Surgery*, in 1901, revising it nine times over the next twenty years.

In the decades that Gibney and Whitman were active at the Hospital for the Ruptured and Crippled, yet another hospital was built (to accommodate the building of a new Grand Central Terminal), forcing a move east several blocks in 1912. By this time, there were multiple medical schools, religious and academic hospitals, and numerous surgical training programs in Manhattan. Orthopedics, however, was a clinical practice of casting and bracing, and a surgical discipline of excision and drainage. No one in the world was reliably implanting *any* metal, and, of course, plastics were yet to be invented.

The move to its present (and fourth) location along the East River at 70th and 71st Streets did not occur until 1955, but by that time the reputation of the Hospital for Special Surgery (the name had been changed in 1950) was sterling. Its association with the New York Hospital and Cornell University Medical College was critical, as was a deepening commitment to basic science research.

Around the world, orthopedics had become a specialty of bone, muscle, ligament, and tendon repair and reconstruction. Fracture care had transitioned from casting and bracing to "internal fixation" with plates and screws, and the treatment of hip fractures in the 1950s started to include partial hip replacement.

If, as noted earlier, "specialization is the fundamental theme for the organization of medicine in the 20th century,"[21] then hyper-specialization and the use of surgical implants is the fundamental theme of post–World War II medicine. If you seek proof of the explosion of surgery, consider that the Hospital for Special Surgery (HSS) in 1968, more than one hundred years after its creation, still only had *three operating rooms*. While Dr. Mark Coventry performed the first "FDA-approved total hip" at the Mayo Clinic in 1969, the first Charnley metal-on-ultra-high-molecular-weight-polyethylene hip had been performed at HSS in 1968[22] under the guidance of surgeon-in-chief Philip Wilson Jr. and Dr. Harlan Amstutz, later the chairman of orthopedics at UCLA.

Total hip replacement is arguably the most reliable, long-lasting, and patient-pleasing implant in the history of surgery. Sir John Charnley deserves the credit for innovating the use of polyethylene and bone cement in medicine, and his contribution to the world of medicine is still felt today by millions of people every year. The total knee replacement, however, was a product of HSS surgeons and engineers, and the "total condylar knee" (1974) was the world's first successful and widely utilized knee replacement.[23] The success of joint replacement and the improved reliability of spine surgery mandated that HSS increase the number of operating rooms to eight in 1972 and eleven in 1990, and double that today.

The slow expansion of the world's most renowned orthopedic hospital in its first fifty years had nothing to do with lack of leadership or commitment; HSS did not come into full bloom until the elements of the implant revolution had been conceived and refined. The postwar embrace of bioengineering and biotechnology fueled an orthopedic research program that is unmatched anywhere in the world, particularly in sports medicine. Like the Mayo Clinic, the use of high-tech diagnostic tools, the emphasis on resident education, and the yearning to create a center of excellence that attracted patients from around the world has made the HSS a beacon of hope.

Every surgical specialty has a short list of the most important clinics and hospitals where they began. Neurosurgeons acknowledge Henry Cushing's Johns Hopkins, urologists give the nod to Hugh Young and Johns Hopkins, and cardiothoracic surgeons embrace Minnesota, Cleveland, Philadelphia, Rochester, and Houston as seats of their profession, and so forth. Each one of these institutions was led by a visionary surgeon who, more often than not, was from a small town and who grew up working with his hands. Return to the pictures of Virgil Gibney and Lewis Sayre, the American giants of orthopedics, and consider that both of them were raised in the Lexington, Kentucky, area. Look closer at Dr. Gibney's right hand—he is missing two fingers from trauma suffered as an eleven-year-old.

There seems to be a secret behind being a dark horse among elite peers, a prototypical loner with the need to prove oneself, a practiced dexterity and a connection with manual work and innovative insight. Charlie Neer was all these things (although not a hobbyist or tinkerer, he was a horseman and tennis player). He changed the world of shoulder surgery by innovating the first shoulder replacement and refining the diagnosis and management of several shoulder conditions. The first post-residency fellowship in shoulder surgery was with him at Columbia University, and the fellows from that program are among the most famous shoulder surgeons who have ever lived. Dr. Neer was the founding president of the American Shoulder and Elbow Surgeons group, and was a force behind the establishment of the *Journal of Shoulder and Elbow Surgery*. All of this from a boy from Vinita, Oklahoma.

In the midst of the Normandy invasion, Charlie Neer was assigned, along with another medical "volunteer," to perform medical triage aboard a spartan hospital ship in the English Channel. Considered "expendable" because they were single, Neer and his colleague reached the ship by PT boat in rough seas, and quickly began assessing the battle injuries among the brave troops that landed on the French coastline. Charlie heard men crying out in the inky darkness, "Save me!" and "Help me!"—these entreaties would haunt Charlie for the rest of his life, with his patients' survival precariously hanging in the balance on converted warships, bobbing in murky waters.

Charlie Neer, now a twenty-six-year-old, would make land on D-Day "Plus 6," June 12, 1944. His army regiment established field hospitals in succession as the Allied forces made their way south and east, eventually establishing a more permanent venue in a schoolhouse in Arques-la-Bataille, a small village a few miles from the coastal village of Dieppe. Allied forces had suffered grave casualties in the opening days of the Continental invasion, and in the coming weeks would struggle to maintain a foothold in the north of France. Thousands of soldiers, marines, and airmen were killed in action, the victims of drowning, explosions, gunfire, and burns, while thousands more required medical attention in the temporary field hospitals that were established in parallel with regiment advancement. In Arques-la-Bataille, the newly minted medical officer, Lieutenant Neer, and his colleagues attended to traumatized Allied (and occasionally, German) troops, dressing wounds and performing surgery day and night.

Dr. Neer also served in the Pacific theater, working in the Philippines and in Japan. He saw the true horrors of war, including visiting the devastated Hiroshima bomb site. Charlie's experience in warfare medicine was never far from his mind; the ravages of war technology, the feebleness of 1940s medicine, and the agonizing cries of suffering, hopeless men. Decades later, he still remembered the beaches of Normandy.

Late in Dr. Neer's career, a phone call from Washington was patched through to his Columbia office. Secretary of State James Baker was urgently requesting a conversation with Dr. Neer. As a matter of patriotic duty, the secretary asked that Dr. Neer travel to Paris to attend to a wealthy individual who was an important ally of America (his identity not divulged here). Charlie Neer, the boy from Vinita, was being asked to provide medical consultation to one of the wealthiest people in the world at the behest of the president.

Acceding somewhat reluctantly, Dr. and Mrs. Neer were booked on a flight from John F. Kennedy Airport aboard Air France's supersonic Concorde jet. The harrowing flight at supersonic speed had begun with a terrifying take off, the plane tilting from side to side as it launched off the New York runway. The time savings did little to comfort the jittery jet-setters, and the distinction of traveling on the world's most famous airplane was lost in the terror of a rocket ship launch.

A handful of hours later, flying high enough to appreciate the circumference of the earth, Dr. Neer was in Paris, whisked to his patient's private residence in the most exclusive arrondissement in Paris. The shoulder examination was methodical and thorough; the discussion of the patient's diagnosis and prognosis was, as always, meticulous and detailed. The affluent patient understood that surgery was not warranted, and a simple plan of care was outlined.

Before departing, Dr. Neer paused to make a request: despite the generosity of his patron's travel provisions, Charlie was reluctant to return home on Concorde; the flight had simply been too vexing. His benefactor took no exception to his reaction to the flight, and promised that his staff would make different arrangements.

When Dr. and Mrs. Neer arrived at Charles de Gaulle Airport the next day, they were ushered to an aircraft that would be a charter flight just for them. Happily, there was no bone-chilling supersonic jet waiting for them, but to their great surprise, a gleaming Boeing 747 anticipated them on the tarmac. Boarding the behemoth, the Neers were shocked to learn that the entire 747 was theirs alone, compliments of their grateful patient. The pilot and staff greeted them, and inquired of Dr. Neer, "As we have entirely full gas tanks and no one else on board, is there anywhere else you would like to go on our way back to New York City?"

Charlie Neer, recalling his days as a young officer in the Army medical corps, landing on the Normandy shore among chaos and corpses, informed his aviator that he would love to see the English Channel, the coastal habitations of Northern France, and most important, the Omaha and Utah beaches.

The pilot was pleased to grant Dr. Neer's wish.

The private charter 747 jumbo jet, with two passengers, took off from Paris and traveled north to Normandy. Sweeping in from the east, the massive plane banked steeply to the left, the coastal villages of Normandy directly below. Memorial fields abounding with white crosses and broad, sandy beaches presented themselves to the boy from Vinita, now the world's most famous shoulder surgeon. His first trip to Normandy was forty-five years before, a scared and unsophisticated American who dreamed of becoming a surgeon one day. In that time span, antibiotics were developed, Medicare passed into law, joint replacements invented, and the major orthopedic device companies had been founded. And unbelievably, Charlie Neer was now returning to Normandy in his own 747. Commercial jet aviation was years away during World War II, and there is simply no way that he could ever have imagined returning to Omaha Beach in his own gargantuan airplane. The implant revolution, with its transistors, polymers, wires, biological materials, and modern alloys, was just as impossible.

SEVENTEEN

Implant Revolution

"The first industrial revolution used water and steam power to mechanize production. The second used electric power to create mass production. The third used electronics and information technology to automate production. Now a fourth industrial revolution is building on the third, the digital revolution that has been occurring since the middle of the last century. It is characterized by a fusion of technologies that is blurring the lines between the physical, digital, and biological spheres."

—Klaus Schwab, founder of the World Economic Forum

And now, our own John Charnley has become part of surgical history, and has taken his place in the gallery of the great master surgeons who have gone before . . . the Charnley prosthesis is in essence a biological design by a man who was also an artist. It is something which a Leonardo da Vinci might have envisaged. But today we are thinking about the man, the human person we knew and held in affection. He had so much to give to the world of surgery, both in fundamental knowledge and to the relief of human suffering."[1]

—Harry Platt

I'm performing shoulder surgery, but instead of wearing surgical scrubs, gown, cap, and mask, I'm sitting in my office chair in a white lab coat, dress shirt and tie. In fact, the patient is nowhere to be seen—only his scapula (shoulder blade) is here with me, a 3-D representation levitating on the computer screen in front of me. Just a few years ago, Stanley's case would have been unmanageable by any shoulder surgeon in the world, so significant is his bone loss and deformity, but now his case has become almost routine.

I first met Stanley four months ago in clinic, a man at the end of his rope. He had undergone total shoulder arthroplasty eight years before in the Midwest, and while the initial results were excellent, over the last few years his shoulder slowly became more painful. I reviewed his original X-rays and told him I would have been pleased had the implant placement been my own; however, in the intervening years the metal and plastic parts had started loosening. At the time of implantation, orthopedic prosthetic parts have to be extremely stable and secure. Any slight wobble in the bone dooms the replacement to eventual failure, but even ideal time-zero positioning is no guarantee for success.

Sandy-haired with a ready, gap-toothed smile and smoker's cough, Stanley had been evaluated by several surgeons prior to seeing me, and had been assured that his X-rays weren't worrisome. After performing a physical examination and scrutinizing the new X-rays I told Stanley that I thought his total shoulder replacement was loose, and possibly infected, which caught him by surprise. As a sixty-one-year-old male, he confided that he needed to work for several more years before retirement, and the realization that his replacement had failed was distressing. I recommended that we operate, remove the loosened implants, test for infection, and assess the degree of bone loss. In these types of cases, an unsteady glenoid component (shoulder socket implant made of polyethylene, resembling white candle wax) inflicts slow-motion destruction to the socket-portion of the shoulder blade. As the cobalt-chrome metal head of the replaced humeral head revolves over the glenoid prosthesis, the pegs of the polymer component can start to unseat, working away from the host bone like a rickety wooden post wobbles and fails to support a garden gate. Ignored for too long, the bone of the scapula impalpably fades away, leaving an eggshell of bone encasing tapioca-like fibrous tissue and an unmoored polymer implant.

In my first operation with Stanley a month ago, I found what I had feared: massive bone loss, and implants that were swimming in sloppy bone. After opening the deepest part of his shoulder, I found a combination of metal, plastic, synovial fluid (resembling apple juice), and fibrous tissue. All the foreign implants had to be removed, and cultures of the tissue and fluid were obtained to ensure that no occult infection was present. I then placed a cement spacer, comprised of acrylic cement, representing a marriage of liquid monomers and a powdered polymer mix, similar to an epoxy project of my childhood, with the appearance of blue Play-Doh. This spacer, which replicates the shape and function of a regular shoulder bone, also contains powdered antibiotics, fighting deep infection while we awaited the culture results from surgery. With the cement spacer in place, we then do something miraculous: obtain a CT (computed tomography), a sophisticated 3-D series of X-rays that facilitates preoperative planning for the revision total shoulder replacement.

One week after surgery, we obtained a high-resolution CT of his shoulder. The computer is able to assimilate all the X-ray information with visual-imaging software and construct virtual 3-D images. In the last decade, the imaging software has become so good that surgeons and radiologists are able to "subtract" all the surrounding tissues away (muscles, ligaments, and tendons), and "build" the bony structures on the computer screen. Using computer key strokes, the physician can spin or rotate the images to get a sense of what the skeletal bones look like. Imagine your car mechanic not lifting the hood of your vehicle; instead conjuring what bedevils your engine with a magical tool that could see through metal. In the last few years, technology has allowed us not only to see the bones in three dimensions, but now to plan the surgery, virtually implant the parts, and assess the placement. Even more mind-boggling, I can now work with engineers to create custom shoulder replacement implants that fit the specific defects of a particular shoulder.

My conference call with the implant designer from Quebec is about to start. Using a link from an email message from the implant company Zimmer Biomet, I connect to a Webex teleconference with an engineer in Montreal. With a few keystrokes, I am able to navigate the layers of protection to keep Stanley's information completely private. The engineer, Simone, speaks to me through my computer link, and she controls the images on the screen. On a light gray background, a 3-D image of Stanley's scapula is presented. Simone controls the imaging software with her mouse, although I can ask for different perspectives and orientations of the shoulder blade. We work together, imaging and imagining; Montreal and Denver suddenly don't seem very far away.

Simone cyber-manipulates the scapula, and it's like having a Gray's Anatomy *drawing come to life and pirouette in front of me. Instead of a stout glenoid, with bony integrity to support a regular implant, I am staring at a pockmarked shoulder socket with a cavernous central defect, a pothole unable to support the metal baseplate I need to implant. Here, Simone now transitions to remarkable modernity: we will build an implant together that will be a computer-aided manufacture (CAM) metal that will perfectly fit in Stanley's shoulder. His shoulder is so badly damaged that I would have never considered doing the case five years ago; today, I can work with Zimmer Biomet to custom-manufacture a one-of-a-kind implant that will fit hand-in-glove into the shell of bone, replete with the drill holes that will facilitate perfectly placed screws into the remaining bone. This system is a breakthrough that allows me to tackle a shoulder that five years ago I would have surrendered to. In less than an hour, we have completely designed the implant, and after signing off on the design, fabrication on the implant will begin in Warsaw, Indiana.*

Weeks have passed since we designed the implant, and the day of surgery has arrived. Everything else about the case is routine, including the preoperative interactions, prepping, positioning, surgical approach, and dissection. But once I get down to the deepest part of his shoulder, the banality ends. The specialized implant is separately boxed in layers of sterile packaging, awaiting implantation. Its doppelgänger is a 3-D printed white polymer stand-in of the exact same dimensions, along with a 3-D printed version of Stanley's shoulder blade. These life-size, lightweight, hard plastic models are identical to what I saw on the computer screen a few weeks ago, and help me to practice where I will place the real metal implant into Stanley's ramshackle shoulder socket.

After cleaning the fibrous scar from the cavitary defect, I am peering deeply into the shoulder socket, an eggshell of bone instead of a fortification capable of supporting an implant. Formerly, I would have quaked at such a finding, but we are prepared. Instead of trepidation, I am filled with pluck, even bravado, because I am armed with a tool that can transform this case from disaster to triumph. I carefully position the trial polymer implant in the defect and it perfectly clicks into place. I spend a few moments examining the fit, and satisfied, insert the actual implant into the deep cavity. The heavy alloy, odd-shaped implant seemingly seats itself into the crater, like a spaceship pod docking into a spaceport, with no less science fiction conviction.

I secure the custom-milled implant to the scapula with multiple screws, whose trajectory and length I determined weeks ago. What was previously impossible now seems mundane. Carefully drilling through the metal implant and blindly drilling into the compromised bone, I already know the length of the screws. One by one, the screws are exactly the same length that was predicted weeks ago when I was a cyber-surgeon. After implanting all the screws, Stanley now hosts an implant that was planned in Montreal, expertly milled by a team of skilled technicians in Indiana, overnighted to my partner company representative Jodi, who ferried it to me today. Engineering, commerce, bioresearch, computer imaging, satellite and fiber optic communication, advanced manufacturing, airline shipping, cooperative sales engagement, skillful surgery, and exceptional anesthesia, nursing, and tech support have all fused together to care for this man with a terrible problem that today seems like not that big of a deal. Dr. Neer would be astonished, and duly proud.

Charlie Neer's practice was similar to that of all orthopedic surgeons in the 1950s, when very few specialized in a particular joint. There were almost no specialty orthopedic clinics anywhere in the world, with the exception of hand surgery practices in San Francisco, New York, Chicago, and New York, under the guidance of the fathers of hand surgery, Sterling Bunnell and William Littler, among others. Dr. Neer continued to be a fracture doctor, even publishing a knee trauma paper in 1971, more than twenty years after finishing his residency. But as the pace of medical discovery quickened, surgeons like Charlie Neer trained their sights on particular joints. Just like the emergence of orthopedics as a specialty (separate from general surgery), the specialty domains of orthopedics were conceived by fanatics who were more narrowly obsessed.

War has always been a dastardly effective originator of advances in technology, transportation, communication, design, and medicine. With the advances in metallurgy and antibiotics during the 1940s, the orthopedic specialty was poised to launch into its most important era, but the treatment of arthritis had received little direct benefit. Charlie Neer had dedicated the first decade of his practice to fracture care, including surgical treatment of shoulder fractures.

For centuries, physicians and scientists have primarily communicated their discoveries to their colleagues via print journals. For academic university physicians, the credo of "publish or perish" mandated that a doctor actively conduct research and vie for journal acceptance. A young surgeon like Charlie Neer, imbued with optimism, energy, and a fresh set of eyes, was the perfect candidate to expose the inadequacy of orthopedics after the war. Dr. Neer's classic 1953 article, "Fracture of the neck of the humerus with dislocation of the head fragment," had highlighted the heretofore abysmal outcomes associated with nonoperatively and operatively treated severe fractures of the shoulder. On the last page of text in that article Dr. Neer included a photo of the shoulder implant he had designed, concluding, "replacement prosthesis presents logical possibilities and may prove of value in dealing with major injuries of the humeral head."[2] The world had its first peek at the future.

In 1955, Dr. Neer reported on a series of twelve patients who had undergone articular replacement of the humeral head.[3] Using the implant he had designed, he was able to show dramatically improved outcomes when treating trauma patients. Implantation of the Neer prosthesis following trauma was a major step forward, but in this paper, Charlie Neer hinted at another indication. Of the twelve patients, all but one had suffered from a fracture-dislocation of the shoulder. A seventy-year-old housewife (patient number eleven) was treated for "hypertrophic osteoarthritis," the world's first partial ("hemi") shoulder replacement for arthritis, on March 16, 1954. The patient returned to the Midwest, later writing Dr. Neer and telling him that she was "free of pain and leading a new life." Instead of limiting the use of the shoulder implant to patients who had shattered their shoulder, Dr. Neer was offering the device for treatment of shoulder arthritis.

As the postwar boom led to unprecedented growth and prosperity, physicians were sanguine that disease could be challenged in ways never imagined. Antibiotics opened the door to entering the abdomen and operating on abdominal organs and the bowels. Mechanical ventilation during and after surgery bolstered surgeons' ability to operate on ever-more critically ill patients. Pharmacological discoveries led to an explosion of medicines that made diseases like diabetes, malaria, gout, rheumatoid arthritis, and heart disease treatable. Finally, advances in chemistry and polymer sciences led to a handful of materials that are used millions of times every year in our world, including the world's most common plastic, polyethylene.

Dr. Neer performed forty-six *partial* shoulder replacements in the first ten years, since his 1953 breakthrough. Of those forty-six hemi-arthroplasties, seven were for osteoarthritis and not for fracture. In his 1963 paper in the journal *Surgical Clinics of North America*, Charlie Neer concluded: "The results of prosthetic replacement have been better in this group than any other." In the first decade, the father of shoulder surgery was performing less than one replacement per year for arthritis, but this number would quickly increase.

In Dr. Neer's next major shoulder replacement publication, "Replacement arthro- plasty for glenohumeral osteoarthritis," a twenty-year report on shoulder replacement, the New York City surgeon reported on forty-eight patients who had undergone arthroplasty surgery for arthritis. Therefore, in the first decade of application, seven patients were operated on for arthritis, but 41 patients were treated in the second decade, an almost sixfold increase. In the 1974 article, Dr. Neer, once again, gave a sneak peek into the future. Buried in data table 1 is patient number eighteen, a fifty- seven-year-old housewife who underwent *total shoulder arthroplasty*, with placement of a polyethylene implant on the glenoid, or shoulder socket side. Dr. Neer explained, "The technique was modified in this patient by inserting a high-density polyethylene glenoid, anchoring it with acrylic cement and using a slightly different humeral element."[4]

There is an illustration of the implant, but no X-ray. This "Neer II" Vitallium implant had been slightly modified by curving the edges and making the humeral head more spherical, and represents the primogenitor of all shoulder implants for the next several decades.

Every major orthopedic joint replacement developed since the 1960s has had three main characteristics in common: a plastic polyethylene cushion, a metal alloy articulating surface, and acrylic cement to hold the metal parts in place. Whether a shoulder, elbow, wrist, hip, knee, or ankle arthroplasty, every joint has been replaced with these three components. Newer developments have included cementless components where the texture of the implant causes bone ingrowth without the need for acrylic bone cement. This blueprint for joint replacement was drafted by Sir John Charnley.

Lancashire County lies in the northwest of England, and at one time was the world's most important industrial and commercial center and the locus of international capitalism. Lancashire's main centers are Liverpool and the world's first industrialized city, Manchester. Although an ancient Roman fortification site, Manchester came to prominence following completion of canals and river improvements that facilitated transport of coal and processed cotton from the surrounding countryside and shipping to the River Mersey, Liverpool, and the world. The Industrial Revolution started circa 1780, and although cotton has never been grown in England, by the 1830s, Lancashire was responsible for almost all of the world's cotton processing. The international influence exerted by Lancastrians would wane as the rest of the world emulated their steam engines, canals, factories, and trading centers, but arguably the most important Lancashire man of the last one hundred years came from the small town of Bury, outside Manchester.

John Charnley was born 1911 to a chemist father and nurse mother, and from a young age was noted to have a mechanical aptitude, building scale sailboats and tinkering with engines. John's sister would attend Cambridge, but he would go straight to medical school in Manchester after matriculating from the Bury Grammar School, winning science prizes and achieving high grades. It seems that Charnley was destined for life as a surgeon, even sitting for the Royal College of Surgeons of England examination while still a medical student, which he easily passed.

Charnley graduated MB, ChB (Latin: *Medicinae Baccalaureus, Baccalaureus Chirurgiae*, equivalent to the American MD degree) as a twenty-four-year-old in 1935. He began a career as a surgeon, working in London and later, back in Manchester, under the guidance of the esteemed Henry Platt, one of the early English orthopedic pioneers. His future career plans were shattered with the outbreak of war on September 1, 1939, and he was enlisted in the Royal Army Medical Corps

(RAMC) on May 1, 1940. At the same time German forces were sweeping across northern Europe (occupying Holland, Belgium, and France), Charnley was posted to Dover, across the English Channel from Dunkirk. He made multiple trips across the Channel to evacuate and care for the wounded, his life in grave danger during the miraculous evacuation of 370,000 troops from the French coast. He would later serve in the RAMC in Egypt and Palestine, all the while gaining invaluable experience in treating complex orthopedic injuries.

Once the war was over, Mr. Charnley (surgeons in England proudly retain the title of "Mr.") returned to Manchester, working part time at the Royal Infirmary. Needing additional hospital work, Charnley accepted a position at the Wrightington Hospital, twenty-five miles north of Manchester. Why would the young surgeon accept a post at a remote hospital in the countryside? And why was the hospital built there to begin with?

Tuberculosis sanitarium facilities had been built around the world in the 19th and 20th centuries, following a typical pattern of rural, purpose-built, single-story hospitals (as recommended by Florence Nightingale), where open walkways, large windows, and fresh country air was thought to help patients combat TB. After Robert Koch had identified *Mycobacterium tuberculosis* in 1882, scientists could only dream of a magic drug to kill the bacteria. Until that breakthrough, the contradiction of prepossessing hospitals in pastoral settings, housing diseased, coughing victims dying slow-motion deaths would persist. The Lancashire County Council purchased Wrightington Hall from a financially distressed titled family in 1920, converting it into a nurses' home with single-story hospital pavilions built to accommodate 226 chronic TB sufferers. Independent for decades, the hospital's authority was transferred to the National Health Service in 1948, about the same time that Charnley began making monthly visits to the bucolic outpost.

Most of the patients at Wrightington were suffering from bone and joint infections, rotting from the inside, with only palliative solutions to consider. Interestingly, the incidence of TB began to decline just as Charnley began to consult at Wrightington. As sanitation standards improved (including milk pasteurization) and living conditions advanced, fewer children were contracting TB, and with the introduction of streptomycin and para-amino-salicylic acid in the 1940s, a TB cure was possible. "Sanatoria and orthopedic hospitals all over the country were faced with the same predicament—how to use effectively that large number of beds which had been available for tubercular patients, and which were now no longer needed."[5] Patients could rightly expect *not* to die from tuberculosis, but the ravages of the disease had not disappeared: the joints of the afflicted were still wrecked. As Themistocles Gluck painfully learned in 1890, replacing an actively infected joint was no solution at all. But now, Charnley could consider surgically confronting diseased joints, TB or not, with the inclination that one of mankind's great burdens, arthritis, could be relieved, even cured.

While still working in Manchester (he still worked there part-time until 1958), Mr. Charnley evaluated a patient who had undergone a partial hip replacement with a Judet acrylic prosthesis (a clear plastic ball taking the place of the arthritic femoral head). The patient informed the sagacious surgeon that his replaced hip squeaked when he leaned forward. So severe was the screech that his wife could not tolerate his company. Instead of merely discounting the story (or even being amused by it), Charnley began to turn over in his mind why the noise was occurring. He observed that the sound rarely happened when the femoral head had been replaced following a femur fracture, where the cartilage from the hip socket was still intact (and ostensibly, still providing a slippery articulating surface), but these types of sounds only happened in arthritis cases where the hip socket had only roughened bone on both sides of the hip joint, squeaking when the replaced plastic ball interfaced with the arthritic hip socket. Importantly, *instead of primarily focusing on implants and gadgets, Charnley turned his scrutiny toward the organic*, considering the biomechanics ("the mechanical laws relating to the movement or structure of living organisms," according to the *Oxford English Dictionary*) of the vital and diseased tissues he was pondering. His mentor, Sir Harry Platt, described Charnley as a *surgeon-biologist*, rather than a surgeon-engineer. To generate a solution for hip arthritis, he would first need to understand the function of healthy articular cartilage. This would become the pattern for every implant ever invented: comprehending function before proposing a cure. It now seems laughable that Gluck was implanting ivory implants in 1890, before antibiotics, sterilization, modern biomechanics, and the presence of metal alloys and polymers.

The Industrial Revolution brought machines and engines, with their crankshafts, pistons, gears, and axles—all of which required lubrication. Engine grease and distillated viscous fluids, newly discovered from the nascent petroleum industry, were used to lubricate the metal interfacing machine parts. If man is a machine, it was a reasonable conclusion that our parts had similar biomechanical relationships. Reasonable, but wrong. Charnley began discussing his theories with engineering friends at the University of Manchester, and they all agreed that our joints were *not* lubricated under the same principles as metal mechanical parts, which used hydrodynamic lubrication, where a very thin film of fluid separates the articulating surfaces and the parts move rapidly. Charnley and his colleagues theorized that our joints, instead, used *boundary* lubrication where the lubricant (synovial fluid) had an affinity for the joint surface itself. To test his hypothesis, John Charnley and the engineers started to build testing apparatuses to evaluate the "slipperiness" of cartilage.

The coefficient of friction (represented by the Greek letter mu, "μ") is a mathematical ratio that expresses the friction between two surfaces. If the μ is very high, then a great deal of force is required to move an object against another. Rough sandpaper or a rubber tire has a μ value higher than 1. On the other hand, something very slippery, such as an ice skate sliding on ice, has a coefficient of friction of only 0.03, and it seems implausible that something could be more slippery than that. To determine the μ of cartilage, Charnley and his engineering cohorts built an apparatus that supported a platform, holding part of a human joint (a knee, and later, an ankle) in place. The upper half of the joint was positioned above, with a pendulum arm attached, which allowed the pioneering scientists to calculate just how slippery healthy cartilage is. What they discovered was astonishing. The coefficient of friction is 0.001, and is still the most slippery solid surface ever tested. It is (mathematically expressed) five hundred times more slippery than metal on bone and thirty times slicker than that ice skate on ice.

Mr. Charnley published his biological studies in non-surgical scientific publications; more important, he knew that the key to a good clinical outcome would be to design implants that had a low coefficient of friction. And he knew he could determine the μ with his testing apparatus while tinkering with the shape and size of the implants. The race was on to develop what he would always refer to as "low friction arthroplasty."

Surgeons had been replacing arthritic and fractured femoral heads for over a decade, oftentimes with acceptable results, but Charnley was striving for better outcomes and longevity. To achieve true *low friction arthroplasty*, a "slippery substance" was needed for the socket, and he began querying the newly trained polymer scientists in England about possible candidates. He was eventually guided to polytetrafluorethylene (PTFE), also known as Teflon. When we hear the name Teflon, an egg skillet comes to mind, but its original use was industrial, in the manufacturing of valve seatings and non-lubricated bearings. Charnley evaluated Teflon, finding it to be biologically inert, creating almost no local foreign-body reaction once implanted in a human (he conducted no animal trials). Teflon is white, semi-translucent, almost waxy in appearance, and able to be cut with a knife. Starting in 1956, Charnley started performing the world's first total hip arthroplasties, using Teflon cups that were pounded into the patient's own bony hip socket. The results were amazing.[6] Patients had wonderful range of motion and excellent pain relief. Charnley began reporting his results in the *British Medical Journal* and the *Lancet*, two of the most prestigious medical publications in the world.

One of the most dramatic changes Charnley made in hip replacement was the courage to change the size of the metal femoral head. All early hip pioneers, starting with Smith-Petersen and continuing with brothers Robert and Jean Judet and Austin Moore, had designed their partial joint replacements with a metal head that was the same size as the patient's own femoral head. With the introduction of a synthetic hip cup, Charnley made a genius decision to decrease the size of the metallic head. Again focusing on *low friction arthroplasty*, he determined that a smaller head would provide less friction, and the head was therefore changed from Moore's 42mm head (the size of a Ping-Pong ball) to 28mm, and finally to 22.25mm, about the size of an average toy marble. Many surgeons found John Charnley's hip design laughable, but he had math on his side.

In the beginning, Charnley was implanting the Austin Moore hip stem with the large femoral head with no acrylic bone cement. After a few years of implanting the Moore prosthesis and Teflon cup, he was looking for a more stable method of implanting the femoral component. In older patients with weaker bone, the slender metal stem of the Moore prosthesis could begin to wobble in the canal of the femur, leading to subsidence and pain. Charnley was used to consulting with the scientists at the University of Manchester, and having had initial success with Teflon, he turned to some of the chemists in the department of prosthetics in the University of Manchester's Dental School. Dentists were used to dealing with bony socket defects after the loss of a tooth; in England, following creation of the National Health Service in 1948, there were millions of patients who were seeking medical and dental treatment for the first time in their lives. This demand for healthcare led to a crusade to find better materials for dentures and tooth implants, and it was an organic chemist in Manchester, Dennis Smith, who recommended *polymethyl-methacrylate* to John Charnley.

Polymethylmethacrylate (PMMA), also known as acrylic cement, is a self-curing cement that is formed by the simple combination of a liquid monomer and a powdered polymer. The watery liquid monomer is stored in a vial (with an inhibitor chemical), while the powder (with the appearance of powdered sugar) is saved in a pouch. At the time of surgery, the surgical assistant combines the two ingredients in a mixing bowl, akin to making bread dough. The mixture is initially creamy, then becomes doughy, and in a few minutes is like fresh Play-Doh. Polymerization is the frantic race of smaller chemical molecules, the "monomers," linking with larger chain-like polymers to form a complex latticework of rigid matter. The chemical process is "exothermic," meaning that heat is given off as the molecular linkage occurs, starting as a viscous slurry, then progressing to a pliable plastic, and as it hardens, an elastic blob, before becoming a solid piece of polymer. Today, we see PMMA every day, in Plexiglas windows, display cases, eyeglasses, signs, bathtubs, and skylights. But Charnley saw PMMA as the ideal grout for holding hip stems in place. Without experimenting on animals, he first started using it in humans in 1958, and was immediately convinced of its potential. A half century later, Charnley's cement is used daily in every hospital in the world, with only slight chemical modifications.

Charnley left Manchester for good after 1958, initially working part-time in Wrightington, but eventually spending all his time at the previous TB hospital in the countryside. The local hospital board provided the funds for a biomechanical workshop and laboratory, and Charnley soon hired a laboratory technician, Harry Craven, a jack-of-all-trades who was at Charnley's side for many years during the pivotal moments of the 1960s. The lab opened in 1961, and with a dedicated staff and purpose-built operating rooms, Charnley was optimistic enough in his ongoing success to give his epicenter a name, the Centre for Hip Surgery—Wrightington Hospital.

Like all innovators of science and medicine, John Charnley was a tinkerer. He made things, fixed machines, fabricated models, and fashioned his own implants. He also had a workshop in his own house, including a lathe where he turned his own hip socket implants from blocks of Teflon. Craven assisted him in these endeavors, and the task of making the widgets themselves was key to solving the problem of hip arthritis.

Flush with the excitement of a Hip Centre and hundreds of hip replacement operations under his belt, Charnley was hopeful that an acrylic-cemented hip implant with a small head and Teflon cup was the long-term solution to hip arthritis. He had gone from one hundred operations a year in 1959 to over four hundred hip replacements in 1962. Of course, as a scientist, Charnley was interested in following his patients and confirming their ongoing success. It was late in 1962 that Charnley realized that something horrible was happening. Despite the robustness of patients' satisfaction and improved functional abilities, the three-year follow-up X-rays showed a disastrous change occurring in the Teflon cups. Charnley later explained:

It may seem strange that it took us some three hundred operations and between three and four years to arrive at this conclusion [that Teflon was unsuitable], but there were a number of different reasons. First, the results up to three years were so spectacular, and the patients so pitifully grateful, that we could not bring ourselves to face the suspicion that, in such highly successful results, the X-rays were showing incipient harmful evidence. Second, by its chemical nature PTFE [Teflon] was so extremely inert we felt that even if wear debris was present it would be harmless. Third, though we could see wear of the order of 1mm after about a year in the X-rays, I thought that this was not unexpected and could be explained by the "bedding-in" of the head in a socket that was deliberately machined to have an internal diameter larger than the head. It was only when the first year's wear was more than doubled in the second year and more than tripled in three years that the seriousness of the problem became evident.[7]

On the precipice of a world-changing revolution, Charnley had to wonder if all was lost. All the X-rays showed a similar pattern of superior erosion of the "roof" of the Teflon cup, with the impression that the metal femoral head was working its way into the plastic cup like a hot knife through butter, albeit over the course of a few years. Mr. Charnley began to re-operate, and to his horror, realized that wear of the Teflon cup was not the chief problem. Worse, "wear debris" was found around the hip joints of patients who had suffered failure of the Teflon cups. He discovered globs of fibrous tissue surrounding the Teflon particles within the hip capsule. The adverse tissue reaction to a material he had initially concluded was "inert" reinforced the idea that Teflon, despite its early promise, was completely unsuitable for human use. To further confirm his hunch, Charnley prepared specimens of finely ground PTFE and *injected them into his own thigh with a large bore needle*. After waiting nine months, he dissected out the nodules underneath his skin, and examining the blobs of PTFE surrounded by an amalgam of fibrous tissue, knew that he would never use Teflon again. A block of Teflon in the body *is* inert; particles of Teflon are not. We are amazed that Charnley would inject Teflon particles into his thigh, but considering the action of his English forebear, John Hunter, the father of scientific surgery, who applied syphilitic pus onto a self-inflicted scratch on his own penis as an experiment, Charnley doesn't seem as unhinged.

John Charnley, staring defeat in the face, was racked with guilt and was disconsolate. For weeks he lived in total despair, his wife (he had finally married at age forty-six) finding him awake at night, sitting up in bed with his head in his hands. She felt that "everything was gray and there was an all-pervading gloom."[8] The anguish lasted for weeks, when a chance discovery set him back on track.

In May, 1962, an industrial salesman representing a German plastics company arrived in Wrightington and asked to speak with Charnley, or his assistant John Craven. The salesman was selling plastic gear parts, which were being used in the Lancashire weaving trade (still substantial in the 1960s), and surmised that Charnley would need mechanical parts for his lab. Craven met with the salesman, and at a glance, saw the raw physical similarity to Teflon. He obtained a four-inch sample block of the material, high molecular weight polyethylene (HMWP), intent on showing Charnley.

Craven showed the block of polyethylene to Charnley, who handled it, dug his thumbnail into it, and realizing that he was able to score the material with his nail, concluded that the "poly" would be a similar disappointment as Teflon, and told Craven he was wasting his time. Undeterred, Craven kept the sample of the new polymer and planned on analyzing it in a custom-made testing apparatus he had devised. The initial results were astounding and incomparably better than Teflon had performed on the same Wrightington machine. Charnley left the country for a meeting in Copenhagen, and unknown to him, the machine churned on, oscillating the stainless steel heads over the HMWP block.

No doubt Charnley had left for Copenhagen empty of optimism, not even aware that HMWP was being tested in his lab. He had gained notoriety for hip arthroplasty, but as the horror show unfolded before him, Charnley had to wonder if he was a fool. When he returned from vacation, he later recalled:

> My office door opened to reveal Craven who asked me to come down
> to the lab . . . Down I went to see the HMWP. After running day and
> night for three weeks, this new material, which very few people even in

engineering circles had heard about at that time, had not worn as much as PTFE would have worn in twenty-four hours under the same conditions. There was no doubt about it: *we were on.*

Charnley obtained more material from the German company, Ruhrchemie (later purchased by Hoechst AG), and like his previous experiment with Teflon, injected polyethylene into his thigh. After six months, there was no nodule formation in his leg. He wrote a letter, quickly published in the British journal *Lancet* on December 28, 1963, and reported his worrisome reaction to Teflon and encouraging response to the finely ground HMWP, partially motivated to warn surgeons about Teflon because he had heard surgeons were contemplating using it in a rudimentary knee replacement.

It only took a few months for Charnley to begin performing total hip replacements with polyethylene cups, starting in November 1962. Many of these early poly total hips were revision operations, taking out the failed Teflon cups and cementing in the new poly cups. Initially, all the cups were made by Charnley himself, and sterilized chemically by soaking them in Cidex (glutaraldehyde) overnight. Later, his manufacturing partner, Thackray, manufactured the cups and irradiated the poly with gamma irradiation (others recommended immersion in ethylene oxide—still controversial). He continued to implant the new HMWP cups by the hundreds, never using systemic antibiotics and only using stainless steel stems (and not cobalt chrome, like we use today). Charnley waited to publish his results, fearful of an unexpected Teflon disaster repeating itself, but it never happened.

In fact, John Charnley hardly changed a thing over the next twenty years, performing thousands of hip replacements in Wrightington (and later, in Midhurst as well), operating almost till the day he died, suffering a heart attack at age seventy. Even today, with more modern manufacturing processes, advances in metallurgy and polymer sciences, improvements in surgical techniques, and educational innovations, no one has ever demonstrated superior results compared to Charnley. The surgeon-biologist from a tiny town in rural England not only changed orthopedics, introducing materials that would be implanted millions of times every year in our world, but, as much as any person in the world, helped change the mindset that receiving a foreign substance into your body is extraordinary.

I should know: I am one of the millions of people who have had their hip replaced. And I'm extremely grateful to Sir John Charnley for immeasurably improving my life and relieving my previous suffering.

❖

Charlie Neer first reported on the use of a polyethylene shoulder socket component in his 1974 article. His first use of the component was in 1973, and in a 1982 *Journal of*

Bone and Joint Surgery article, Neer reported on 273 patients who had undergone total shoulder arthroplasty by him over a nine-year period.[9] A decade of use of polyethylene in the hip had transpired before Neer attempted its use in the shoulder, in part because fixation in the shoulder was much trickier in the unforgiving, small shoulder socket. It boggles the mind that Charlie Neer was one of the few surgeons in the world performing shoulder replacement even three decades ago, especially in light of the fact that over 100,000 shoulders are replaced in America every year. This number pales in comparison to the half-million hip replacements and roughly one million knee replacements performed every year in the United States.

Joint replacement arthroplasty is the most commonly performed implant operation in the world. It is not only pain-relieving, it dramatically improves function and the ability to work and play. It is one of mankind's greatest innovations, and the credit for discovering the combination of the right metals, plastics, and cement goes to John Charnley, an industrialist working with his own hands in an otherwise unremarkable countryside hospital.

EIGHTEEN

The Birth of Sports Medicine

I am sitting in the executive office suite at Arthrex in Naples, Florida, waiting to meet with company founder and president, Reinhold Schmieding. Today is a very busy day for Reinhold, with hundreds of young surgeons in attendance for the annual resident symposium, and his impending trip to Munich, Germany, where he spends every summer, overseeing the European branch of Arthrex. Despite all of the commotion, the American-born son of German-American dentists has agreed to the unthinkable: a sit-down interview. Quite possibly, it's only because I am a busy shoulder and elbow surgeon with the proper pedigree of having trained at the famous Kerlan-Jobe Orthopedic Clinic in Los Angeles, but whatever his reasons for allowing me to pepper him with questions, I am pleased to sit down with one of the world's great entrepreneurs.

As I review my research papers while seated on a modern, black leather chair, I glance down at a simple white table with tubed, aluminum alloy A-frame legs, and a thought occurs to me: is this the drafting board of legend, where Reinhold Schmieding designed the Arthrex logo that has endured almost forty years? While living in a small apartment in the Olympic Village in Munich in 1981, the twenty-six-year-old simultaneously invented his company's name and logo while perched over a $50 drafting table purchased at a home improvement store, and the thought of a "small beginnings" nostalgia is appealing to me.

Reinhold Schmieding is sixty now, still very fit and impressively energetic. After a few moments in his presence a visitor realizes his pride of ownership of Arthrex, which is now the world's most successful sports medicine implant company, and you realize that his fierce competitiveness and loyalty are shared among all his employees. The company got its start at the very beginning of the development of arthroscopy, and Schmieding's "better mousetrap" surgical instruments gave his empire firm footing.

As a surgeon, I've implanted literally thousands and thousands of Arthrex sutures, suture anchors, screws, and various other devices, and I am well aware of how Arthrex has changed the face of sports medicine around the world. But I am here to learn more about its founder and how he has achieved so much in one lifetime.

I have come to Naples armed with a few observations, and I am eager to learn how Reinhold persevered and has landed himself on the Forbes 400, poised to become one of the richest one hundred people in America. [1]

My first question: is this the selfsame drafting table? Yes, he confirms with a warm smile. (As I have gotten to know him better, it's not a surprise he still has the table. He's an interesting mix of thriftiness and extravagance, measured analysis and gut instinct.) I think the table is a talisman of sorts that memorializes simple origins and the power of self-invention.

Just as the light-powered arthroscope was becoming practical after years of development by Masaki Watanabe in Japan, surgeons in North America and Europe turned from merely peering into a joint to performing work in a joint. Watanabe's mentor in Tokyo had initiated the development of a tool, a pencil-slim metal gadget with an ocular opening that allowed surgeons to squint into the eyepiece while bent over, face close to the operative joint. The first model was created in 1931, and subsequent models were sequentially numbered. On the twenty-first attempt, in 1958, Watanabe delivered the model upon which all subsequent arthroscopic lenses were designed. The "Watanabe No. 21" was still powered by a tiny incandescent light and had to be held to the surgeon's face, but it led the revolution in minimally invasive surgery around the globe. Over the next two decades, arthroscopy turned from a novelty to a powerful tool, particularly with the application of a flexible, fiber optic light source and the attachment of a small video camera to the lens. No longer was the surgeon consigned to bending over and banging (and contaminating) the lens against his eyeball.

By 1981, surgeons were relieved to be standing up, manipulating the camera with their hands, and miraculously investigating the miniature world of any joint in the body. The quandary was the lack of tools to reliably accomplish anything while looking.

Enter Reinhold Schmieding and Arthrex.

Although Reinhold was born and raised in Michigan (graduating from Michigan State in 1976), he relocated to the Black Forest region of Germany to become an international manager for the US-based orthopedics company, Richards. After three years of selling orthopedic implants to surgeons in and around Germany, he became fascinated with the new developments in the field of arthroscopic surgery. While viewing the inside of the knee arthroscopically was becoming more commonplace, there was a glaring lack of dependable instruments to manipulate the tissues; namely, to grasp, cut away, and remove pieces of cartilage, bone, and meniscus. Fortunately, for the budding entrepreneur, he was stationed in southwestern Germany, long a home to the finest craftsmen in the world for the manufacturing of surgical instruments. Sensing an opportunity, Reinhold began designing and developing instruments for minimally invasive surgery, and within months, decided to leave Richards and form his own company. Naming his new enterprise "Arthroscopy Excision Instruments," shortened to "Arthrex," he sat at his economical drafting table and sketched out the Arthrex logo, still used today for the company that is worth billions.

Arthrex struggled to survive in its opening years, much as Medtronic had in the late 1950s. Cash flow problems, travel costs, surgeon resistance, and initial limited product offerings had Arthrex on life support. A few key early surgeons ordered surgical instruments and provided cash, just as the breakthrough guide for arthroscopic ACL reconstruction was designed by Schmieding. With Arthrex in its infancy, Reinhold Schmieding traveled to Zurich, Switzerland to exhibit his small set of knee instruments to a highly-regarded surgeon. After scrutinizing the tools, he turned to the young upstart, and questioned, "Herr Schmieding, davon wollen Sie leben?"[2] He was asking, "You want to live *from this?" The small set of instruments didn't seem like an empire in its infancy.*

As has been documented numerous times in this book, most of the great pioneers in medicine and surgery had great facility with their hands and an unquenchable urge to tinker with gadgets and machines. Reinhold Schmieding loved art as a child, and sensed, even in his early school years, that he processed and thought about things differently than his peers.

Sitting in front of the drafting table, I ask him for an example.

"A really important day for me was in high school, when my history teacher gave us an assignment. Placing a candle in front of us, he asked us to describe it with one word descriptions, giving us only five minutes to come up with as many words as possible. The words just kept coming to me, and I was writing them as fast as I could. When the time was up, my classmates revealed that they had come up with 15, 20, or 25 words. The teacher looked at me, and I told him 225 words. That's when I realized I have a different way of thinking about things."[3]

This example of "ideaphoria," or the high flow of ideas, is characteristic of the creative problem-solvers in all branches of business and medicine. Combined with Schmieding's advanced spatial reasoning and artistic bent, Arthrex has been on the cutting edge of surgeon education from its inception. Because Arthrex was born at the advent of arthroscopy, practicing surgeons had not been trained in arthroscopy during their residencies. What was needed was a company that could provide practical teaching to surgeons who were reluctant to look awkward with the completely new technology in their hands.

What may not be obvious to the lay-reader is that arthroscopic surgery is a wholly different set of technical skills, with angled mirrors on the end of lenses that tilt the perspective on the television screen. It's like backing up a car using the rearview camera on your dashboard the first time—things work in reverse, and a lifetime of using your eyes is of no use. In time, we all become accustomed to using the technology, but no proud surgeon wants to look like a fool, or worse, put their patient at risk, while conquering a new skill.

Arthrex, while still a young company, began hosting skills labs and producing artwork that was vastly superior to its competitors. When I was a sports medicine fellow in Los Angeles in 2002, I was gifted an entire collection of animated surgical DVDs from Arthrex, years before any other orthopedic implant company could try and match them. The artwork has always come at great expense, but directly stems from Reinhold's artistic sensibilities, and in my estimation, has paid great dividends for our patients. It also has been a great business decision, and I have always been mystified when other surgical companies skimp on their educational budgets. Reinhold tells me, "What made it necessary is what made it great."

Arthrex survived its first decade with home offices in Munich and eventually a world headquarters in Naples, Florida. More than twelve years after its founding, Arthrex began producing its first medical implant, the specialized titanium "interference" screw used in knee ACL reconstruction. This ushered in a tsunami of implants since 1993, and in the last twenty-five years a dazzling lineup of implants for every joint in the human body has been developed by Arthrex. Rotator cuff repairs, ACL reconstructions, shoulder stabilizations, fracture management, ankle ligament repairs, and over ten thousand other implants and techniques are all addressed with Arthrex devices.

So dominant is Arthrex's place in the sports medicine world, it has become obvious to me that it is now impossible to turn on any NFL, NBA, MLB, or NHL game and not see an athlete who has an Arthrex implant in his body. I ask Reinhold how he feels about this, and he tells me, "I feel really lucky that I came along when I did. It is obvious to me that I was perfectly prepared for this, for the founding and nurturing of Arthrex. I enjoy strategy, leadership, service, and medicine . . . and when presented with a challenge, can rapidly see the solutions in my head. (In the early days) when we struggled, it was my sense of survival and tenacity that kept us afloat, and I am humbled more than ever to work hard to help surgeons treat their patients better."

I am eager to tell Reinhold Schmieding my big observation, based on years of analysis and calculations. While there are medical companies with higher annual revenues, like General Electric's healthcare division (which does not develop implantable devices), most medical implant companies build revenue through costly, big ticket items like total hip implants and pacemakers. Additionally, among all of the largest orthopedic implant companies, the reigning CEOs have been at the helm less than five years. There is no other major medical device company that is privately held, and certainly not by the founder of a company that is almost forty years old.

Because Arthrex has so many thousands of implants, almost all of which are permanent, and not degradable, and due to Schmieding's unprecedented tenure, including the last twenty-five years of incredibly high-volume implant manufacturing, there is one stunning conclusion. There is not another individual on earth who has personally overseen the design, manufacturing, and distribution of so many medical devices that exist in the bodies of our fellow earthlings. With Arthrex's distribution in 150 countries, there is probably nowhere on our planet that Reinhold can go without bumping into a person who has some type of Arthrex implant in her body. In this book about the implant revolution, there is one man who has physically touched the lives of more of his fellow men and women than any other, and it seems that Reinhold Schmieding is just getting started.

On a chilly, overcast day[4] the last Saturday of November 1888, the Yale squad is struggling to defeat Princeton to preserve a perfect record. The Yale Bulldogs football team has arrived at the Polo Grounds in New York City, having won the previous twelve games by an *average of fifty-seven points*, having not allowed a single score all year. The roster includes William Heffelfinger (who will eventually become the first professional football player) and Amos Alonzo Stagg, who will go on to become a legendary coach in multiple sports.

Walter Camp is standing on the sidelines, but he doesn't know that history will regard him as the Father of American Football, or that he's watching one of the most dominant teams of all time. Like any coach, all Camp cares about in this moment is the game before him. Football is not even two decades old, but it is clearly different from the game that preceded it—rugby. Many of the sport's early changes are Camp's, including the line of scrimmage, the position of quarterback, and the system of downs.

Determined to achieve a perfect 13–0 season, the game is uncomfortably close. Although Yale has waltzed through every match this year, this tilt against Princeton is savagery among gentlemen. In fact, the Harvard graduates who are officiating the game kick Hector Cowan, the Princeton captain, out of the game for rough play.[5] They consider it good fortune to win 10–0, and retreat to New Haven as the mythical national champions.

Now, put yourself in Walter Camp's shoes: what if someone had been seriously injured? Would you rely on a team physician to examine one of the young seminarians and diagnose a critical injury?

Of course, in 1888 there was no team physician. What many in the crowd may not have known is that a few years earlier, in 1882, Walter Camp had been just two classes short of graduating from Yale Medical School when he dropped out. Camp would later note in a biographical questionnaire "the death of a surgeon with whom I had expected to practice medicine caused me to leave the medical school and go into business."[6] He had earned his bachelor's degree from Yale College in 1880, and then continued at the Medical School in New Haven while playing for the football team, acting as the team captain.

Another setback had occurred in 1882: *Camp injured his knee in practice, thus ending his playing career.* Camp had likely suffered an ACL or meniscus tear, and in the late 19th century, there was not a single operation for any type of sports knee injury. Even a relatively minor knee injury resulted in the termination of an active life.

In the Cro-Magnon days of collegiate football, injuries were common and deaths occurred with shocking regularity, and by 1905, some Ivy League schools suspended their football operations. That same year, *eighteen* college students died and there were 149 serious accidents.[7] President Roosevelt himself summoned Walter Camp and representatives from Harvard and Princeton to the White House to respond to demands for the abolition of football in America.[8] Within a year, the Intercollegiate Athletic Association of the United States (forerunner to the NCAA) was formed, with Walter Camp at the helm of the rules committee.

The Yale-Princeton game was 125 years ago. What medical technology was at Mr. Camp's disposal?

Nothing.

Prior to our modern age, sports medicine just a century ago almost exactly mirrored gladiatorial times, with sophistication no better than an emphasis on eating meat, taking a cold bath after a competition, and have a rubdown in the training room. It shouldn't be too surprising that eighteen college students died in 1905 while playing a violent, largely unregulated sport with essentially no advanced medical treatment available.

When a player suffered an *open* ("compound") ankle fracture, he faced a potential death sentence. In the Franco-Prussian War the mortality rate of a lower leg fracture was 50 percent; in World War I, the mortality rate for an open femur fracture was a startling 80 percent. Nobody worried that Joe Theismann might die on that fateful Monday night game in 1985, nor did any viewer contemplate Ed McCaffrey's mortality when he suffered a similar open tibia fracture in a game broadcast on *Monday Night Football* September 10, 2001 (just hours before the 9/11 terror attacks).

Following the first public demonstration of surgical anesthesia in the Ether Dome at Massachusetts General Hospital in 1846, Koch's experiments to prove that bacteria were real, and the development of antisepsis by Lister, surgery became vastly safer. Coupled with the introduction of antibiotics during World War II, and the transition of medicine from a merely observational science to an investigational one, sports medicine was finally able to alter the lives of athletes.

Los Angeles was doubling in size every decade for a century up till the 1950s. As transcontinental travel was becoming more practical, the Cleveland Rams shockingly celebrated their 1946 NFL championship by moving to Los Angeles, the first move west by any major sports team. Waiting for the Rams was a gregarious orthopedic surgeon with a tragic medical secret.

Robert Kerlan was the son of a general practitioner in the small town of Aitkin, Minnesota, about an hour's drive west of Duluth. An all-star high school athlete, Kerlan first arrived in Los Angeles as a sixteen-year-old to play basketball at the University of California in Los Angeles. After one year at UCLA, he transferred to USC, matriculating both undergrad and medical school. Like so many medical students with an athletic past, Bob Kerlan gravitated toward orthopedics, and after finishing his surgical training, became one of the earliest orthopedic surgeons who steadfastly served as a team physician for professional sporting franchises. In Los Angeles, he had the best seat in the house for every sporting event in the booming fifties and sixties.

Dr. Kerlan was hired just the day before the opening day of baseball for the new Los Angeles Dodgers in 1958. (Having won the World Series in 1955, the Brooklyn Dodgers broke fans' hearts and ventured west prior to the 1958 season, as did the New York baseball Giants.) He had volunteered for a minor league baseball team for a few years, but this was different. The Los Angeles Dodgers became a dominant team for a decade, and their bold venture signaled the emergence of sports as big business. Dr. Kerlan's good fortune overshadowed the fact that he was stricken with *ankylosing spondylitis.*

Ankylosing spondylitis, an inflammatory condition of the spinal column, propels its victims into a forward thrust, turning the flexible human spine into one long, fused, rigid piece of ratcheted bamboo. Slowly, painfully, the patient is turned into a spectacle. Once fully entombed, the sufferer cannot raise his head to look forward. The worst cases result in a vulture's posture; I've seen patients who were forced to walk backward to see where they were walking, similar to the gaze of a jockey looking over his shoulder. Bob Kerlan battled ankylosing spondylitis (AS) his entire professional life, and yet famously maintained his good humor and positive outlook.

Because AS is an inflammatory condition, the treatment requires anti-inflammatory medications, along with physical therapy and exercise. The most commonly used anti-inflammatory in the 1960s was Butazolidin, known around racetracks as "Bute." In a 1969 *Sports Illustrated* article, Dr. Kerlan described the use of Bute on Elgin Baylor, Jerry West, Wilt Chamberlain, and racehorse jockeys. Bob Kerlan, himself, was taking Bute and aspirin by the handful, and had another famous Southern Californian taking Butazolidin as well: Sandy Koufax.

Dr. Kerlan continued operating for a decade and a half, but by the early 1970s, his disability was so severe it left him unable to safely navigate the operating room and to manipulate surgical tools, eventually forcing a reluctant submission to his jailer. Undaunted, he persisted in his care of the Dodgers, the Lakers (who had arrived in 1960), the Rams, the Kings, and the jockeys at Hollywood Park. And all the while he cultivated friendships with Hollywood actors like Walter Matthau and Danny Kaye, and the giants of sports from Willie Shoemaker to Wilt Chamberlain. Watching him soldier on, no one bothered complaining to Dr. Kerlan about their aches and pains. In a 1969 *Sports Illustrated* article, Sandy Koufax said, "His own physical problems are far more serious than most of those he treats, and yet he is always having a good time—telling jokes, kidding people, and getting kidded in return. I always liked him as a doctor, but more than that I liked him as a man."

Overwhelmed in his unique practice, Bob Kerlan sought the help of a partner. Before the orthopedic surgery community authentically recognized sports medicine as a specialty, in 1965, Dr. Kerlan was able to recruit a genuine and honest young man originally from the small town of Boone, North Carolina—Frank Jobe.

Although he would dramatically change sports by co-creating sports medicine, Dr. Jobe was never particularly athletic, telling me in response to a query about his own baseball prowess, "I was never a very good ballplayer. I realized later that my talents lay elsewhere." Dr. Frank Jobe had enlisted as an army medic in World War II, triggering a lifelong interest in medicine. After the war, he received all his medical training in Southern California and initially started his own private practice in metro Los Angeles.

When Dr. Kerlan and Dr. Jobe joined forces in 1965, they combined two very different and extreme talent sets. Dr. Kerlan's tremendous interpersonal skills married to Dr. Jobe's incredible innovative surgical insights; the art of medicine matched with scientific research; a body ravaged and limited by disease joined with superlative, gifted hands. The Kerlan-Jobe Orthopedic Clinic was born, and in the City of Angels, it was a match made in heaven. The two men joined together to form a working unit that superseded any one man's gifting.

Dr. Kerlan could work a room. I have never met a person who, when questioned about the man, didn't tell me a side-splitter. (Dr. James Andrews is the unquestioned king of sports medicine in the world today. While still at the Hughston Clinic in Georgia, Dr. Andrews came to Los Angeles for visits in the '70s with Dr. Kerlan. These visits left a deep impression on Dr. Andrews, and he told me, "Dr. Kerlan was *THE* MAN!" He held me by both elbows when he said it, so important was it for me to understand that statement.) Frank Jobe was the down-to-earth, matter-of-fact foil to Bob Kerlan—his straight man. No less dedicated to his patients or his craft, Dr. Jobe was a once-in-a-lifetime surgical innovator, scientist, and visionary. While Dr. Kerlan was a jester and bon vivant, Dr. Jobe was more literal and less hands-on (ironic, in that it would be Dr. Jobe's hands that would change the baseball world).

Tommy John is a person, a patient, a ballplayer. Baseball fans remember him primarily as a player for the LA Dodgers, Chicago White Sox, and New York Yankees, who underwent the most famous elbow operation in history on September 25, 1974. In the field of medicine, syndromes and conditions are almost always named for the physician who first described them (infrequently, a disease is named for the location in which the disease occurred, such as Ebola and Lyme). Thus, we know the names Parkinson, Huntington, Hodgkin, and Marfan—the physicians who describe the disease and not the names of the patients who suffered the maladies. Almost without exception, the names of the victims are lost to history, with the notable exception of Lou Gehrig. Every baseball fan knows the name Tommy John: the name attached to the triumphal solution to a failed elbow.

Dodger fans had watched another phenomenal lefthander, Sandy Koufax, surrender to a failed elbow a decade before. Neither Sandy Koufax nor Tommy John was able to undergo an MRI of their elbows—MRIs were not commonplace until the 1980s. A torn elbow ligament (the medial collateral ligament) was only a clinical ("hands-on exam") diagnosis. Following a traumatic injury in the midst of the 1974 season, Tommy knew that he couldn't throw anymore—his injury was not the subtle presentation that most baseball docs see today, but was instead a grossly unstable, floppy elbow.

Dr. Jobe examined John's elbow at Dodger Stadium and initially recommended placing the southpaw's arm in a cast for a trial period. No doubt Dr. Jobe was thinking about Sandy Koufax and the collapse of his career in 1966. Two years earlier Dr. Jobe had operated on Tommy's left elbow, removing bone chips, and following dedicated rehabilitation, Tommy returned to excellent form. Now, Tommy John was facing certain doom and the end of his career. Realizing that splinting was a failure, and sitting together in Dr. Jobe's office, the two men discussed reconstructive surgery. After a "night to think on it" Tommy said, "let's do it."

Tommy John's willingness to undergo surgery underscores the change in mindset away from surgery as a last resort. No surgeon, however, had operated on the ulnar collateral ligament of the elbow. While a professional athlete often equates a career-ending injury with death, most sports medicine operations are entirely elective; life could easily go on without surgery. Tommy was sailing into uncharted waters, placing his limb and his career entirely in Dr. Jobe's hands, trusting that Dr. Jobe had the creativity and skill to accomplish what no other surgeon had.

What makes a great surgeon? From William Halsted to today's notables, a requisite set of characteristics sets apart the heroes. Many patients don't realize that many, perhaps even most, surgeons are not superlative. A high percentage of surgeons have adequate hand control, but not special. Very few practitioners are true innovators who have breakthrough creativity when thinking about injuries and diseases. They have a three-dimensional understanding of anatomy—the kind you can "feel." It's a concept that is difficult to describe, but is most like someone who has a great sense of direction, driving in the dark in an unfamiliar part of town, and yet, still knows the way home.

A phenomenal surgeon has vision, poise, insight, skilled hands and (surprisingly) humility. *Poise* is what the "clutch" player has who actually wants the ball at the end of the game. When the airline pilot Chesley "Sully" Sullenberger realized his aircraft was doomed as his first officer was taking off from LaGuardia, he calmly said, "My aircraft." His first officer, following protocol, said, "Your aircraft." Every pilot, surgeon, and true leader who heard that story nodded to him or herself in knowing agreement. Surgeons who throw massive tantrums in the OR (there are many) usually betray the fact that they do not do well under pressure, and the "macho" display is actually an admission that their nerves also do not do well under pressure.

Insight is a way of incorporating all your teaching from a very diverse group of mentors and specialists in many different disciplines, and accessing it, sometimes on the fly, to critically problem solve. *Skilled hands*—truly gifted hands—are characterized by naturally tremor-free control, powerful yet dexterous, elegant and sensitive to touch yet able to make rapid, precise moves. A cross between Itzhak Perlman and Norm Abrams. *Humility* rounds it out. Knowing when you are at the limits of your powers, when another surgeon is better at a particular procedure, when you were wrong—and admitting it.

Finally, the best sports medicine practitioners today all have one thing in common: an overwhelming ability to communicate to their patients "all will be well." This is usually accompanied by physical touch: a comforting pat, a solemn reassurance.

To those who were blessed to be in his sphere, Dr. Jobe was all things.

The groundbreaking elbow operation was performed at Centinela Hospital on September 25, 1974, but it can be hard to know what happened forty-five years ago in an operating room in Inglewood, California. I was fortunate to know both Dr. Jobe and Tommy John, but we also have a detailed report in a 1986 publication. *The Journal of Bone and Joint Surgery (JBJS)* is the bible of orthopedics, and although it is surprising that it took over a decade to publish "Reconstruction of the Ulnar Collateral Ligament in Athletes," Drs. Jobe, H. Stark, and S. J. Lombardo give us interesting tidbits about the important operation in 1974. Typically, medical journal articles are exceptionally dry reading, and even case reports contain only bland information about nameless patients. But in the *JBJS* article, we find Case I, a twenty-nine-year-old professional baseball player. This is obviously Tommy John.

The article presents the clinical information about his initial injury, suffered as a twelve-year-old Little League pitcher. Amazingly, Tommy had been injected with steroids approximately twenty-five times over an eight-year time span as a professional. After having his elbow debrided (cleaned up) in 1972, he returned to form, and was actually having a great 1974 season, starting out 13–3. In July, while throwing a hard slider, he felt severe pain and actually felt his elbow snap and give way. We have a stress radiograph labeled "1974," also obviously Tommy.

Years ago, Dr. Jobe confided to me that the groundbreaking operation was performed with no prior practice on cadaver arms, and with no biomechanical testing to scientifically prove its merits. This is a startling revelation to a younger surgeon, further buttressing the notion that Dr. Jobe was a courageous innovator.

Pioneering surgery of the hand had demonstrated the utility of transferring a tendon from one part of the hand to another when treating polio. Thus, a partially paralyzed leg or arm could regain function. One of the trailblazers in this type of surgery was Dr. Jacquelin Perry, who worked for decades at Rancho Los Amigos in metro Los Angeles, and was a lifelong friend of Frank Jobe. It made sense to Dr. Jobe to harvest one of these forearm tendons, the *palmaris longus*, which is of little functional importance, from Tommy's right arm to his left elbow. Amazingly, the body senses that this newly placed tendon from another place belongs in its new home, and quickly begins to vascularize it and bring it to life (in a biological and biomechanical sense).

In the article there is a series of drawings that explain the revolutionary operation, and details the surgical technique. The drill holes are strategically placed in the humerus and the ulna, and if precisely placed (to the millimeter) the new tendon set into the drill holes will replicate the function of the torn ligament. The collateral ligament of the elbow is only about an inch long, and less than the diameter of a pencil. Careful biomechanical studies have (now) shown which part of the ulnar collateral ligament (UCL) is important in throwers: the anterior bundle of the anterior band of the UCL.

When Tommy awakened from surgery, he groggily attempted to feel his right arm. Having already endured left elbow surgery two years before, Dr. Jobe had told

Tommy that this new "reconstruction" would entail use of the tendon from his *right* arm.[9] Dr. Jobe had told Tommy there were two eventualities: another simple cleanup, or the new reconstructive operation. Once the Dodger ace felt bandages on his right arm, he knew history had been made.

Dr. Jobe had told Tommy that his chances of returning to play were one in a hundred. After venturing into the sports equivalent of an untethered spacewalk, Tommy came back and famously won more games after his surgery than before. Tommy John surgery is now one of the most reliable operations in all of sports, with a quoted return to play rate of at least *80 percent*. A stroke of genius has saved hundreds of baseball careers, and there is not a single major league baseball team without numerous Tommy John "survivors."[10] And because the Tommy John operation is almost entirely elective, only reserved for elite baseball players, in some ways it represents the ultimate expression of the implant revolution, where surgeons now perform operative reconstructions on athletes who play for our entertainment and vast sums of money.

If you were taking a walk on the windswept beach at Kill Devil Hills, North Carolina (just south of Kitty Hawk), on December 17, 1903, you might stumble upon the Wright brothers making history with their Wright Flyer. You likely wouldn't believe your eyes, but the meaning of manned flight wouldn't resonate until you saw an airplane full of passengers making a long-distance journey. Similarly, if you were in Los Angeles at the Centinela Hospital operating room on September 25, 1974, you might be intrigued to see Tommy John on the operating table. The deeper meaning of what that moment meant for baseball and sports medicine would be impossible to know until hundreds of baseball players and elite "overhead athletes" would have their careers (and lives) changed by Tommy John surgery. Although sports medicine has many birthplaces, only then would you appreciate that you were at one of those special moments, witnessing a master at work, a humble pioneer helping birth sports medicine.

NINETEEN

Calculating the Impact

What is the impact of the implant revolution? Comprehending the reach of medical devices in our world requires a full accounting of the costs, a consideration of the number of operations, and a tabulation of the total number of devices implanted every year.

Turn on a televised political debate and you will hear about the "massively inefficient" US medical system, so described because of its "runaway expenses." This focus on costs is critical, especially because Medicare spending is part of the enormous direct, or *mandatory*, spending that is considered an *entitlement* for US citizens. The US government spent $692 billion of our dollars on Medicare in 2016, which was a 9 percent increase over 2015.[1]

Importantly, "inefficient" can be different than "costly." Do we decry the costs associated with building (and rebuilding) municipal airports because modern jet airplanes need longer runways? Do we complain about the new costs of visually stunning LED flat panel televisions in relation to our grandparents' radios? Of course, we all complain about the expenditures in our public and private lives, but we gladly pay for home Wi-Fi, smartphones, and drive-through coffee shops—modern conveniences we can't imagine living without.

The real question is, how much are we willing to pay for healthcare? As was outlined in chapter 12, no member of Congress could have possibly guessed what was brewing in the minds of scientists and physicians in 1965. In 1967, the first full year of Medicare and Medicaid implementation, total federal spending on Health Services and supplies was less than $10 billion, with $5.5 billion allocated on hospital care[2] ($38.4 billion in 2015 dollars).[3] Who could have dreamed that annual federal spending on health care could increase 1,700 percent in half a century? Outcomes in the treatment of cancer, heart disease, and arthritis have dramatically improved in the last fifty years, but the question remains, how much are we willing to pay?

Some healthcare critics have cited a doomsday event of sorts wherein one day we might spend more money on healthcare than our mortgages. Even acknowledging the tragedy of uninsured Americans and poor healthcare outcomes among disadvantaged families, shouldn't we prioritize the health of our bodies above the status of our houses? Hopefully, we can avoid that mathematical reality, but a bit of the sting of sticker shock is placated by an appreciation of how very far we have come in the last seventy-five years.

A tabulation of implantable medical devices will now be offered. The simplest way of accomplishing this will be by specialty, but because there is no national registry this is a daunting calculation. The methodology is based upon a combination of US federal government information and industry reports. Both require the purchasing of reports and the consultation of experts.

A review of implants based upon their longevity (temporary versus permanent) and structural makeup was proposed (organic, biological, plastic, metal, and electronic). Every implant that exists in a human can be described as some combination of those descriptors.

Another way of categorizing implants is based upon their function. Implants are used for: *repair, reconstruction, substitution, stabilization, restoration, augmentation,* and *electrical stimulation.* This classification is mine, and is presented, not as a factual representation but as a conceptual approach.

Repair implies the realignment of our own tissues, like sewing lacerated skin edges together or attaching tendon edges to a bone, like rotator cuff repair. Repair almost always entails the use of suture, and more often than not, permanent suture that is intended to stay forever. Open-heart surgery and repair of the mitral valve (between the left atrium and left ventricle) involves a great deal of permanent sutures and, oftentimes, a permanent reinforcing mesh ring for annuloplasty.

Reconstruction involves implanting a new tissue into a specific area with the expectation that the body will (miraculously) respond by: 1) not rejecting the tissue, 2) microscopically responding by accepting and incorporating the tissues to nearby structures, and 3) functionally adapting the tissues to adopt the role of previous tissues. The most common reconstruction operation every year in the world is anterior cruciate ligament (ACL) reconstruction, in which tendon from a patient (or cadaver) is implanted in a patient's knee, and after healing, functions as a stabilizing and critical knee ligament.

Substitution surgery involves complete replacement of a worn-out or diseased body part with an implant that is meant to restore function. Examples include joint replacement, wherein the arthritic bony surfaces, with their worn-out cartilage, are resurfaced with metal or ceramic parts, or heart valve replacement surgery, in which diseased or misshapen valves are replaced with animal (pig or cow) or metal implants. Substitution surgery can not only restore previously lost function, it can prolong life and dramatically improve a person's existence.

Stabilization surgery involves the implantation of devices that bolster the body during a restorative healing process. Stabilization differs from repair in that an additional structure, like a metal plate and screws, is configured adjacent to a biological zone for healing. Severe displaced fractures are internally stabilized with trauma hardware, which allows the bone ends to slowly knit together. Similarly, spinal fusion is achieved by linking vertebrae together with large screws and rods, attaining a fusion of neighboring structures. Under the right conditions, this type of fusion is advantageous, but only occurs when the stabilization is so secure that the cells can accomplish solid healing across the chasm.

Restoration surgery reinstates function when the body cannot achieve it on its own. When a heart has lost its own rhythmic pacing ability due to diseased electrical conducting nerves, there is no medicine or nonsurgical treatment that can restore normalcy. Implantation of a pacemaker restores regular pacing of the heart—a profound and modern miracle. Another, more recent marvel is deep-brain stimulation, in which precisely placed electrical leads are positioned in tiny areas of the brain to affect a mood, an inhibition, or excitation. Tremors or seizures are decreased (even abolished); memory is enhanced, depression is lifted, and perhaps, Alzheimer's is held at bay.

Augmentation surgery and other types of plastic and aesthetic surgery do not enhance function, but instead alter appearances with permanent implants. In the case of augmentation breast surgery following mastectomy, the line is blurred between augmentation and restoration.

In fact, the lines are often blurred among categories. For instance, a cochlear implant is used to restore hearing among the deaf, but is this restoration or substitution? When a cardiac stent is precisely snaked into a coronary artery, is this repair or restoration? Whatever the case, implants powerfully facilitate healing at the hands of physicians and surgeons. As Francis Bacon dreamed four hundred years ago, "Let us hope . . . there may spring helps to man, and a line and race of inventions that may in some degree subdue and overcome the necessities and miseries of humanity." Here, then, is a computation of the lines of invention.

Orthopedic surgeons facilitate the implantation of more devices than any other specialty. Joint arthroplasty of shoulders, elbows, wrists, fingers, hips, knees, ankles, and toes, combined with spinal fusion, fracture management, and tendon and ligament repair result in millions of operations every year in America. Prior to Smith-Petersen's cup arthroplasty in Boston in 1938, almost no implantation of metal occurred in America, and for another decade, a slow uptick in the rate of implantation occurred. During the 1950s, orthopedic surgeons in America, Europe, and Japan began pioneering joint arthroplasty, and by the 1960s the revolution was in full bloom.

JOINT REPLACEMENT

As discussed elsewhere in this book, there is no US joint registry, so appraisals about joint replacements are based upon commercial and governmental estimates. The most reliable estimates are contained in the National Inpatient Sample (NIS), the "largest all-payer inpatient care database in the United States, containing data on more than seven million hospital stays."[4] NIS is the largest database compiled by the Healthcare Cost and Utilization Project (HCUP), and is sponsored by the Agency for Healthcare Research and Quality (AHRQ), itself a part of the US Department of Health and Human Services (HHS), with a tiny budget compared to other divisions like the NIH, CDC, CMS, and FDA. Information in the NIS captures about 20 perent of all hospital discharges, so extrapolation of the data must be performed to come up with national estimates.

The first FDA-approved total hip replacement in the United States was performed by Dr. Mark Coventry at the Mayo Clinic on March 10, 1969. John Charnley had been performing "modern" total hip replacements for almost a decade when Coventry officially implanted the first hip in America. FDA approval for medical devices would become much more stringent after the Dalkon Shield fiasco in 1976, and the Medical Device Amendments of 1976 (to the Food, Drug, and Cosmetic Act of 1938) greatly strengthened the federal government's oversight of devices.

At the passage of the 1965 Medicare Act, there had been no total hip replacements performed in the United States. Because the AHRQ was only established in 1989, there is scant evidence about the volume of hip arthroplasty volumes prior to the 1990s. In the Mayo Database (formed in 1969 with the insertion of the first hip), there is precise information about joint replacement volume at the Mayo Clinic. At one of the world's busiest joint replacement hospitals there was a total of 35,167 hip replacement operations from 1969 to 2000.[5] Even though that number started slowly and ballooned, it only represents a little over one thousand hips per year. What has happened over the last few decades, then, is remarkable.

From the AHRQ's NIS, there were 290,700 hip replacements in 1997 in America. By the year 2000, there were over 300,000 hips replaced in America; in 2005 there were 383,500. In 2007, in the *JBJS*, an investigation on the numbers of hip and knee replacements was conducted, and based upon the volumes and the changing demographics of America (i.e., the aging of the Baby Boomers), the authors predicted that by 2020 there would be 384,000 primary totals hip replacements and 67,600 revision hip operations.[6] That number swells to 572,000 primary and 96,700 revision hip replacements per year by 2030. Those 2030 predictions represent a 137 percent increase over 2005. The striking observation in reading that 2007 article now is the degree to which the 2014 data blows away the predictions.

A December 2017 Statistical Brief by the AHRQ entitled, "Overview of Operating Room Procedures During Inpatient Stays in U.S. Hospitals, 2014" reports that in 2014 there were 522,800 hip replacement operations (not counting the nearly 300,000 operations for hip fractures, which are often treated with partial or total hip replacements).[7] Therefore, in 2014, the 522,800 hip replacements is 16 percent more than the 451,600 predicted by 2020 in the *JBJS*. Even savvy statisticians miss the mark, consistently underestimating the impact of even commonly performed operations.

In patients sixty-five and over (almost 100 percent of whom have Medicare insurance), there were 315,400 hip replacement operations in 2014, representing the second most common operating room procedure among that age group.[8] Hip replacements in 2014 were the third most costly inpatient stays in America, averaging over $17,000 per stay, which is over 5 percent of all costs associated with admission to US hospitals. As will be seen, pacemaker operations are twice as expensive, and heart valve operations are three times as expensive as hip replacement, but because they are less commonly performed are less costly overall than hip surgery. The aggregate costs for hip replacement hospitalization (not counting outpatient therapy and nursing) is over $8 billion dollars, which is, all by itself, more than the entire 1967 Medicare budget for hospital care. What started in a tiny workshop in Lancashire, northwest England, at the hands of Sir John Charnley, has grown into one of mankind's most effective interventions, occurring almost one million times per year in America as you read this book, if you combine its use for arthritis and fracture. The old saying in orthopedics that "man enters life through the womb but exists through the hip" is not nearly as true as it used to be, and we can thank Charnley for that. It was cheaper to fold an afghan over the incapacitated legs of an invalid with a fracture—but not very effective or humane.

Total knee replacement numbers are more sobering. Arthroplasty of the knee is the second most expensive procedure in America, costing almost $12 billion in 2014.[9] In fact, the six most commonly performed musculoskeletal operations account for one-quarter of aggregate costs of all inpatient stays in America, totaling $41.2 billion in 2014.[10] In the year 2014, there were 723,100 total knee replacements performed in US operating rooms.[11] The same 2007 *JBJS* article referenced above forecast 1,641,00 knee replacement operations by 2020 (7 percent of which are predicted to be revision operations), and 3,749,00 total knees by 2030 (of which 268,000 are estimated to be revision operations).[12] Based on 2014 AHRQ data, wherein the average cost of knee arthroplasty is $16,300, treatment of knee arthritis alone in 2030 would be an astounding $61 billion dollars.

Shoulder replacement arthroplasty has ballooned at a rate over the last fifteen years unlike any other joint over the last half century. Charlie Neer's original hemiarthroplasty underwent few changes throughout the 1950s and 1960s, and total shoulder replacement was still practiced by a small subset of orthopedic surgeons in the 1980s and into the 1990s. Throughout the 1990s, glenoid implant design had changed very little, but at the turn of the century a major change in the shape and style of fixation was proposed at several orthopedic implant manufacturers, both in Europe and in America. By 2006, total shoulder replacement finally surpassed partial shoulder replacement, and this trend has never reversed.

The number of total shoulder replacements doubled in a short few years around the time that the FDA approved *reverse* total shoulder replacements, in March 2004. This third segment has dramatically changed (and improved) the way orthopedic surgeons address arthritis, rotator cuff tears, fractures, and previously failed shoulder operations. Not surprisingly, the number of reverse replacements has exponentially grown since introduction just fifteen years ago. Combined, there were 66,485 shoulder implant operations in 2011, growing another 50 percent as this book goes to press.[13]

Elbow replacement entails either complete replacement of every bearing surface, or simple replacement of the radial head alone. In the United States, the annual rate of total elbow arthroplasty is 5,800 patients, while the rate of radial head replacement is 9,200. Combined, 15,000 patients undergo some type of major implantation of metal in their elbow every year in the United States.

Wrist and ankle replacement operations are relatively rare compared to hip and knee arthroplasty. Instead of hundreds of thousands of cases per year, there might be less than four hundred wrist replacement operations[14] per year in the United States. The combined total of partial and total wrist replacement operations in 2014 was only two thousand cases.[15] Total ankle arthroplasty is much more common, with an estimated 13,145 cases performed over an eleven-year span, from 2000–2010.[16] There was an increase in total ankle procedures after 2006, but the number is still likely less than two thousand per year in the United States. Ankle fusion is much more common (perhaps six times), and requires a substantial number of metal screws, plates, and rods. It is reasonable to conclude that more than ten thousand patients a year in America require major implant surgery for serious ankle problems.

In the United States, there were approximately sixteen thousand thumb and finger replacement operations and twelve thousand toe replacement operations in 2014, according to SmartTRAK, an orthopedics industry analyst.[17]

In sum total, in the year 2014 (the most recent year for which we have the most reliable data), there were 522,800 total hip replacements, 723,100 total knee arthroplasties, 90,000 total shoulders, 15,000 elbow replacements, 16,000 thumb and finger replacements, 12,000 toe replacements, 2,000 ankle replacements, and 2,000 wrist replacements. *The grand total of all joint replacements in America was 1,381,300 in 2014.* There is a small subset of patients who received more than one joint replacement in a single calendar year, so it is not accurate to conclude that 1,381,300 unique Americans had a joint replaced in 2014, but until the United States. has a joint registry, this tabulation is as close to accurate as we may produce. This number will swell to roughly two million joint operations *every year* by the time this book goes to press and *four million every year by the year 2030.*

Spinal fusion surgery is the costliest inpatient operation in America. Although spine surgery cases are about half the volume of knee arthroplasty operations, the mean costs are almost double. In 2014, there were 413,200 inpatient stays for spinal fusion surgery, with nearly all requiring metal screws, plates, and/or rods.[18] Although laminectomy or disc removal spine surgery is the fifteenth most expensive operation (with aggregate costs of $2.3 billion), these operations, which do not involve device insertion, are not central to this implant-focused tabulation. Nonetheless, spine fusion surgery is massively expensive, representing over 7 percent of all operating room costs in 2014.[19] Because many spine operations are now performed on an outpatient basis, the above noted number of 413,200 spine operations vastly underestimates the actual number of operations performed. Cooperative industry tracking estimates compute the number of *instrumented spine operations in 2014 as 778,180 operations.*[20] Roughly a half-million Americans have spinal implants inserted into their bodies annually, about a third of which (155,900) are between sixty-five and eighty-four, representing somewhere in the neighborhood of $4 billion of Medicare spending on spine fusion operations, saying nothing of the astronomical costs of nonoperative spine care and the sobering loss of productivity among American workers who suffer from a "bad back."[21]

TRAUMA

Bone fractures are treated either with nonoperative intervention (like casts, splints, and slings) or with open reduction and internal fixation. "Internal fixation" requires plates, screws, pins, and anchors. Every single bone in the body has certain types of fracture patterns that are best treated with implanted devices—to not have surgery is to sacrifice the eventual function of that limb. A wise surgeon knows which fractures to treat with a cast, and which ones demand an operation. The tabulation below is therefore exclusively a compilation of surgically treated fractures that have been addressed with implanted devices.

Fracture fixation of the upper extremities, including clavicle, shoulder, elbow, wrist, and finger device implantations, totaled 350,388 procedures in 2016.[22] With an increase of about 25,000 such operations per year, a retrospective extrapolation of 300,000 internal fixation implantations in 2014 seems reasonable. A total of 1,862,134 internal fixation procedures were performed on the lower extremities in 2016, and assuming a 6 percent growth in internal fixations over the last decade, a total of 1,627,924 lower extremity implant-related cases were performed in 2014.[23] These include tibia and femur fractures, ankle fractures, pelvis and hip fractures, and foot fractures. Combined, *there were approximately 1,928,000 extremity fracture operations in the United States in 2014*, not counting spine stabilization operations that were previously included above. By the year 2020, this is expected to pass three million internal fixation cases per year, due in large part to age-related hip and leg fractures among the population.

SPORTS MEDICINE

Sports medicine as a specialty started as knee reconstruction on football players. Today, sports medicine operations typically entail arthroscopically performed, minimally invasive operations on joints and their ligaments. Oftentimes these operations include multiple procedures performed together, most notably knee meniscus repair and ACL reconstruction. Therefore, the number of procedures is greater than the number of operations. Because the focus of this book is upon the impact of the implant revolution, and because I am driving toward an overall number of patients in America who have a device implanted in their bodies in a given year, combined procedures are counted as a single operation. This complicates an already daunting task of calculating the reach of the device industry, particularly in orthopedic sports medicine. No one truly knows the precise figure, but by cross-checking multiple sources, like industry reports, medical literature publications, insurance databases, and state and federal agency databases, the numbers get more reliable.

ACL reconstruction stabilizes the knee, and is typically performed on non-Medicare patients in an outpatient setting. Recent studies show a significant increase in the rate of ACL surgery over the last fifteen years. In 1994, the rate of surgery was 33 per 100,000 capita,[24] increasing to 40.9 cases in 2004,[25] and further increasing to 45.1 per 100,000 capita in 2006.[26] That ten-year-old data equates to 134,421 ACL reconstruction operations per year. The US Census Bureau estimates that there were 318,646,275 people in America on July 4, 2014,[27] which I will use to calculate numbers of procedures in this book. Even if the rate of 45.1 per 100,00 capita did not increase (unlikely), a total of 143,689 ACL reconstructions were estimated to have been performed in 2014. However, that number is a vast underestimate if compared to the market analysis performed by those "in the know." In 2016, there were 493,328 knee ligament reconstructions (including 34,005 multi-ligament reconstruction operations).[28] With a compound annual growth rate of 3.93 percent in knee ligament repair, an extrapolated number of 455,600 knee ligament reconstructions were performed in 2014, triple the number that would have been predicted from the medical literature a decade earlier.

There are over half a million knee arthroscopic operations per year in which a portion, or all, of the knee meniscus is removed. In 2014, there were 416,400 knee meniscus repair operations and 2,200 meniscus allograft (transplantation) operations.[29] Surprisingly, with a total of 418,600 operations involving implantable devices, meniscus repair was almost as common as cuff repair.

Rotator cuff repair is an even more commonly performed operation. The rate of cuff repair was estimated to be 98 per 100,000 capita in 2006,[30] equaling an estimated 312,228 cuff repairs in 2014, if we believe the rate has not increased. However, recent studies show a stunning growth in the rate of rotator cuff repair, with a 353 percent increase in cuff repair in Florida from 2000 to 2007, and a 238 percent rate increase from 1995 to 2009 in New York state. With Baby Boomers just now reaching Medicare age, and with comprehensive proof that rotator cuffs tear around that age, there is certain to be an explosion of cuff tear operations (all of which entail the use of implants).

A (relatively) modest 60 percent increase since 2006 would equate to 500,000 cuff repairs in the year 2014, a figure that everyone in the orthopedic industry agrees that we have already achieved.

Shoulder stabilization surgery is usually performed on an outpatient basis, so NIS data is unusable. Utilizing commercial databases, authors calculated numbers of open and arthroscopic shoulder stabilization operations in large populations (more than a tenth of the US population). Extrapolating from these numbers, there were 30.7 shoulder stabilization operations per 100,000 capita in 2012, or an estimated 97,928 operations in the United States.[31] With rates growing around one patient per 100,000 in that five-year span (2008–12), there were likely 100,000 shoulder stabilization operations in 2014, all of which require multiple surgical implants.

Arthroscopic repair of soft tissue elements of the hip was virtually nonexistent prior to 1990, and has been the fastest growing segment in sports medicine since 2000. In 2014, there were approximately 100,000 hip repair operations in the United States, all of which required permanent device implantation.[32]

Cartilage implant surgery involves implantation of whole or morcelized pieces of cartilage from one part of a patient's body to an injured part, or from another patient—so called "allograft cartilage" implantation. In 2016, there were 15,452 cartilage replacement operations from one (deceased) donor to another patient.[33] With an estimated 10 percent annual growth in operations, it is reasonable to estimate that 12,500 such cartilage implantations occurred in 2014.

Today there are numerous soft-tissue operations performed on every particular bit of real estate in the human body—from the little toe to the sternoclavicular joint at the breastbone. Combined, these cases approach several hundred thousand operations. Totaling these together with ACL reconstruction, cuff repair, meniscus repair, shoulder stabilization, hip surgery, and cartilage implantation, *the US market was approximately two million procedures in 2014.*

ORTHOPEDIC SUMMARY

In 2014, the total number of operations by subject were:

Sports Medicine: 2,000,000
Fracture operations: 1,928,000
Joint replacement: 1,381,300
Spine instrumentation: 778,180

Combined, there were 6,087,480 orthopedic and spine device implantation operations in 2014.

CARDIOVASCULAR

The heart, lungs and great vessels are addressed by cardiothoracic surgeons, cardiologists, and vascular surgeons. Prior to angiography, cardiologists performed few procedures, but as minimally invasive techniques improve, cardiologists perform more interventions every year, some of which are nothing short of unbelievable.

Cardiothoracic surgeons specialize in performing "open-heart" surgery, with coronary artery bypass grafting (CABG), valve repair or replacement, lung resections, and treatment of the great vessels. While many lung operations are performed "thorascopically," or with the use of endoscopic equipment and small incisions, most heart operations are performed via a thoracotomy, where the sternum, or chest bone, is split with a saw and the rib cage is (gruesomely) spread open with a metal cranking device. The heart and its neighbors lie in full view, accessible to surgeon's hands and devices.

Valve surgery is most commonly performed on the aortic valve, followed by the mitral valve.[34] Recent analysis shows that isolated valve surgery (operating on one valve in a single operations) occurs 89 percent of the time, while 11 percent of open-heart valve operations involve a combination of valve repair or replacements (most commonly the aortic and mitral valve together).[35]

Tricuspid valve surgery, like all valve surgery, entails either repair or replacement. Tricuspid valve replacement surgery is performed one thousand times per year in the United States, about half the time in concert with another heart procedure.[36]

There is a significant trend in repairing diseased mitral valves, instead of replacing them. For those undergoing replacement, there has been a definite trend in replacing the mitral valve with a xenograph (so called "bioprosthetic" animal valves) tissue graft instead of a mechanical heart valve.[37] In cases of repair, permanent sutures and reinforcing implants are used; therefore, device implantation occurs with repair *and* replacement. In 2005 there were a total of 16,997 isolated mitral valve operations in the United States, with almost an equal distribution of repair and replacement.

Alternatively, in 2005, there were 28,360 isolated aortic valve operations, with 97 percent valve *replacement* operations and only 3 percent repair. Almost all the replaced valves are bioprosthetic, with fewer and fewer mechanical (metal) valve operations every year. In addition, like other valve operations, where there is a trend in multiple structures treated, there is an increase in the percentage of patients whose thoracic aorta is also addressed at the time that their aortic valve is replaced. A recent study showed 28.5 percent of patients with a bicuspid (congenitally malformed) aortic valve underwent surgery on their thoracic aorta. This is a threefold increase from 1998 to 2008, and represents a 7.5-fold increase in costs from $156 million to $1.2 billion in the same time span.[38] The Harvard physicians who published this paper were able to demonstrate a meaningful change in mortality, but the study highlights how a simple change in attitude among surgeons results in a rapid, and expensive, ballooning in costs and device usage.

Attempting to decipher the number of valve operations per year based upon recent publications limits this author to patients at least a decade ago. The most recent clinical volumes among US cardiac surgeons represent the year 2007. In a 2011 publication, using the Society of Thoracic Surgeons (STS) database, a total of 292,543 valve operations were performed over the five-year period of 2003–07. Simple arithmetic yields an average of 58,509 operations per year. That squares with data from above-mentioned subtotals. There were about 45,000 isolated valve operations per year in the early 2000s, with at least ten thousand more combined valve operations. Unbelievably, in just one decade, that number has tripled to *143,500 valve operations in 2014*,[39] and with the Baby Boomers just reaching the average age of valve surgery patients (sixty-seven years),[40] valve surgery promises to rapidly grow.

CABG is the most commonly performed open-heart operation, with *201,600 operations in the year 2014*.[41] There has been a trend over the last fifteen years of decreasing CABG numbers, in large part due to the ability of cardiologists of addressing coronary vessel disease "percutaneously," or via a catheter in the femoral artery in the groin. It is likely that cardiologists will continue to address more heart conditions with minimally invasive technology, sometimes in a fashion that seems daring, impossible, and perhaps risky, as will be seen. CABG surgery is performed by "jumping" a stenotic (clogged) artery with a patient's own vessel graft (vein from a leg or small artery from the chest wall). It is sewn in place with a polymer suture that is permanent. Although the suture is not the device around which the operation revolves, CABG cannot be reliably performed without modern polymer implants like polypropylene suture.

Pacemakers were first used sixty years ago, and have become the eleventh most expensive inpatient operation. In 2014, there were $2.8 billion in aggregate costs for inpatient stays, but a significant portion of pacemakers are placed on an outpatient basis.[42] Recent analysis has shown the rate of pacemaker placement in 2009 had risen to 61.6 implantations per 100,000 capita, equating to 188,700 patients. Assuming no further growth in pacemaker placement (unlikely), pacemaker placement in 2014 would equal 194,346 patients. Steven Kurtz and his coworkers have shown that implantable cardioverter defibrillators (a device that senses an arrhythmia and shocks your heart back into normal rhythm) have approached a rate of about 40 percent that of pacemakers.[43] Therefore, the 2014 volume of defibrillators could be estimated at 77,738. The total US volume of implantable electronic cardiac devices in *2014 is estimated to have been 272,000 devices*. (A study by the European Society of Cardiology in 2015 showed 500,411 pacemakers and 85,289 implantable cardioverter defibrillators were implanted in 2013 in the greater European area.[44] All taken together, the worldwide market for implantable electronic cardiac devices is set to exceed one million per year.)

Coronary artery stents are small, expandable cylindrical devices made of metal (and more recently, polymer) that open up coronary arteries. They have been in use almost a half century, at times facing intense scrutiny over their potential overuse.

Initially made of bare metal, and now commonly coated with drugs that inhibit the formation of clot and scar tissue on the stent itself, stents are placed either following an acute heart attack or in cases of ischemic chest pain. Trying to determine the number of stenting operations per year with a semblance of accuracy may be one of the most challenging computations of this book. The *New York Times* reports that there are 500,000 such procedures per year,[45] and *USA Today* quotes a Vanderbilt cardiologist estimating one million operations per year.[46] Other estimates claim 700,000 stenting operations annually.[47]

The crux of the problem in analyzing stent implantations is the dual nature of inpatient and outpatient procedures and the large numbers of patients who are not of Medicare age. A recent report analyzed data from the Centers for Medicare & Medicaid Services from 2001 to 2008, concluding that there were 319,567 procedures in 2008.[48] Alternatively, a *Journal of the American Medical Association* paper evaluated the NIS, including all adults, and concluded that 809,400 adults had stent placement in 2008.[49] No more recent data analysis exists in the medical literature, and with the population aging, and stenting remaining popular, it is entirely reasonable that the real number is *one million patients per year* undergoing stenting per year in the United States. With over seventy companies manufacturing stents, the worldwide market is estimated at $12,000,000,000 (yes, $12 billion) annually.[50]

Device-related repair of large vessel aneurysms has dramatically decreased morbidity and mortality of the dreaded condition. After leaving the heart, the aorta is still as big as a garden hose, and if the wall of the great vessel weakens and balloons out, a patient can be on the precipice of death. Aneurysms can also occur in other vessels as they divide into the lower limbs. While only 2,000 such repairs occur every year,[51] there are about 30,000 abdominal aortic aneurysms repaired each year, averaging the values of an estimated 26,257 cases in a survey of surgeons in 2013[52] and a computed number of 32,464 procedures among Medicare recipients in 2006.[53] Combined, there were *32,000 aneurysm repairs in 2014.*

The combined financial impact of stent placement, bypass grafting, heart valve, and pacemaker procedures was $23.4 billion dollars, in large part due to the high per case costs (e.g., $52,000 for heart valve surgery, $41,900 for bypass surgery, and $35,000 for pacemaker insertion).

CARDIOVASCULAR SUMMARY

In 2014, the total number of procedures per subsection were:

Coronary stents: 1,000,000
Pacemakers: 272,000

Coronary artery bypass graft: 201,600
Valve operations: 143,500
Aneurysm repair: 32,000

Combined, there were about 1,650,000 cases in 2014 involving the cardiovascular system. Therefore, for musculoskeletal and cardiovascular systems, there were 7,737,000 operations in 2014.

NEUROSURGERY

A ventricular-peritoneal (VP) shunt is used to treat hydrocephalus, a condition of excess cerebrospinal fluid accumulation. A VP shunt is a thin piece of plastic tubing that is positioned deep in the brain and is tunneled under the skin, down the neck and into the abdominal cavity, where the excess fluid drains. There were 27,870 shunt-related procedures in 2000.[54] A more recent study estimates *30,000 VP shunt procedures per year*, and this will be used as a 2014 figure.[55]

The treatment of a brain aneurysm is either by applying a clip from outside the vessel via brain surgery, or by floating a tiny coil of tangled metal through the brain artery. By 2010, this was occurring six procedures per 100,000 Medicare enrollees,[56] which in that year was 47.7 million people.[57] That equates to *2,862 Medicare aneurysm treatments in a single year, but is estimated at 12,000 Americans per year* in more recent literature.[58]

Neuromodulation devices target the brain, the spinal cord, or peripheral nerves. Neurostimulation devices were developed in the 1980s, based upon cardiac pacemaker-inspired prototypes, and not surprisingly, were first developed in Minnesota.[59] Implantable devices deliver small amperage electrical pulses via tiny wires to the brain or to the spinal cord or peripheral nerves. Implantable drug pumps deliver small molecules to the central nervous system, allowing for neuromodulation of brain function.

Neuromodulation of brain function is called deep-brain stimulation (DBS) and is principally used to treat Parkinson's disease, essential tremor, and epilepsy. Spinal cord stimulation (SCS) is primarily used to treat failed back syndrome, chronic regional pain, and peripheral nerve diseases. Peripheral nerve stimulation (PNS) is used to treat an interesting combination of diseases, like incontinence, migraines, obesity, obstructive sleep apnea, and abdominal problems.

DBS was approved by the FDA in 1997 for essential tremor, and in 2002, was indicated for Parkinson's disease. As with other medical devices, the FDA allows for case-by-case use under the humanitarian device exemption (HDE), and with the demonstrated efficacy of DBS for Parkinson's disease, the FDA granted exemptions for neurosurgeons to implant DBS devices for dystonia and obsessive-compulsive disorder.[60] There is a growing list of off-label uses of DBS, including major depression, Tourette syndrome, anorexia, and even dementia. The most recent academic paper on the US trend of DBS devices estimated 5,385 implant operations in 2011, and based upon growth trends in the preceding years, an estimated *6,596 DBS devices were implanted in 2014 in the United States.*

SCSs are typically placed by physicians who specialize in pain control, and in recent years those implantations occur in an outpatient setting, which, as always, complicates the computation of devices used. A scholarly article in *Neuromodulation* in 2009 concluded that 4,000 SCS systems are implanted each year in the United States; this was rebutted by a thoughtful letter by the immediate past president of the North American Neuromodulation Society, who concluded the 2007 number to be over 27,000 SCS implant operations—a stunning difference.[61] The only recent credible source of neuromodulation device implantation rates comes from industry sources. One such report shows that, in 2014, there were approximately three times more spinal cord stimulators placed than deep-brain stimulators[62]; thus, *in 2014, there were an estimated 20,000 SCSs placed in the United States.* Similarly, there were *7,000 sacral nerve stimulator and 2,000 vagus nerve stimulator operations in 2014. All told, there were approximately 35,000 neuromodulation operations performed in America in 2014.*

Combined, there were 77,000 central and peripheral nervous system implant operations in 2014.

OTOLARYNGOLOGY

The cochlear implant is perhaps the greatest success story among medical implantable devices. The only device to restore one of the five senses, the cochlear implant arguably achieves the most satisfying emotional results. The worldwide market for cochlear implants is about 50,000 devices per year, and as of December 2012, 28 percent of all cochlear devices have been implanted in the United States.[63, 64] Therefore, an estimated *14,000 cochlear implants were inserted in the United States in 2014.* There were also a small number of middle ear implants performed in that year, but this number is privately guarded by industry.

Sinus surgery, although common, entails the use of a temporary device, and therefore will not be included in this permanent implant calculation. Similarly, tympanic ear tubes are also temporary implants and not counted here.

TRANSPLANT SURGERY

The first successful kidney transplant was fifty years ago, but it wasn't until the advent of powerful anti-rejection medicines like *cyclosporine* in the 1980s that organ transplantation in the United States and around the world was unleashed. America has, by far, the highest rates of organ transplantation of any country in the world. In the year *2014, there were 29,539 organ transplant operations in the United States.*[65] As opposed to every other data source in this chapter, this is an *exact* number, due to the supremacy of the United Network for Organ Sharing (UNOS) and its stellar record keeping and oversight powers. In 2014, there were 17,108 kidney transplants, 6,730 liver transplants, 2,655 heart transplant operations, and 1,925 lung transplant operations. All heart transplant operations are from deceased donors, but overall, about one-fifth of organ transplant operations are performed from a living donor, including one-third of all kidney operations.[66]

In vitro fertilization (IVF) is arguably the greatest life-giving operation known to mankind, as it is truly the creation of a human being in a laboratory setting. Beginner biology students learn about asexual reproduction among spores, bacteria, and many single-cell organisms. Except for rare examples of *asexual reproduction* among fish, amphibians, and even birds, animals develop from fertilized eggs; sexual fertilization (whatever form of copulation occurs) demands an interaction between male and female. Anonymous fertilization of numerous eggs by, say, a disinterested king salmon is the norm among lower invertebrates, but the creation of a mammal requires an intimate interaction among male and female—until now.

According to the CDC, there are about *72,000 live born infants every year* in America arising from assisted reproductive technology (ART).[67] Two centuries ago John Hunter achieved groundbreaking insight by pickling chicken eggs at varying stages of development, leading him to believe that humans also developed from eggs in utero. Now, in the last several decades, IVF technology has given modern man the ability to achieve a sort of asexual reproduction, resulting in 1.5 percent of American births[68] and one million American "test tube" babies.[69] One could argue that IVF embryos are temporary implants, but with the appreciation that fetal cells do cross the placental barrier, and bizarrely, independently exist in the host mother's body, it is fair to say that ART does result in a permanent cellular implant, saying nothing of the millions of ART humans now inhabiting planet earth (perhaps even you, Dear Reader).

UROLOGY

The inflatable penile implant is used in erectile dysfunction, but its use is actually decreasing, as more precise surgery is perhaps contributing to a lower incidence of surgery-associated impotence, even as the overall number of men suffering from erectile dysfunction has increased. Even with an aging population, the *surgical volume of inflatable penile implants is only 5,000 per year in the United States.*[70]

Urinary incontinence is much less common among males than females. Operations for men include placement of an artificial silicone sphincter or insertion of a sling implant to internally check the flow of urine. A recent study estimates approximately *2,500 such operations every year in the United States*, which is slightly more than 1 percent of the national total of similar operations for women.[71]

OPHTHALMOLOGY:

In 2014, there were 1,428,800 intraocular operations performed at community hospitals in the United States.[72] Essentially all lens replacement operations are performed in an ambulatory surgery setting, even if the operation is performed in a hospital. That million-and-a-half number is misleading, however, because many states do not require a full accounting of non-hospital owned surgery centers. Therefore, the data sources are incomplete, and the aggregate numbers included in NIS calculations are a vast underestimate of the number of cataract operations.

Like almost all implant trends in the United States, the rate of implantation of intraocular lenses has steadily increased since their introduction. Curiously, a 1986 article predicted that the number of lens implantations per year would decrease and level off to less than one million operations per year.[73] In essentially all cases of device implantation, it is always a safer bet that the number of operations will increase, not decrease.

The *true number of intraocular lens implantation operations per year is approximately 3,000,000 per year.*[74] More than 99 percent of cataract operations are on a single eye, with the second operation scheduled weeks later.[75] Therefore, the number of American receiving lens implants on an annual basis is roughly 1,500,000.

GENERAL SURGERY

Insulin pumps, (and the blood glucose monitors with which they interface) are not true implants, existing outside the body and connecting to the insides via infusion setups, small devices that are perched on the skin and have a tiny needle that crosses into the subcutaneous fat.

Prosthetic mesh is the most commonly implanted foreign material in general surgery, used as a permanent reinforcement in hernia surgery. There were approximately 190,000 inpatient abdominal wall surgeries in 2012, but the number of outpatient operations is vastly greater.[76] It has been estimated that more than 80 percent of hernia repairs are performed with mesh,[77] and there are more than 800,000 inguinal hernia operations per year in the United States.[78] Commercial reports indicate that there are over 500,000 abdominal wall hernia operations per year,[79] with a combined total of 1,300,000 hernia operations per year, of which at least *one million hernia operations per year in the United States entail the use of permanent implanted mesh.*

Gastric bypass surgery has changed and grown over the last several decades. Surgery is nothing short of a miracle for the nearly 200,000 people treated annually. The American Society for Metabolic and Bariatric Surgery estimates that 193,000 operations for weight loss were performed in 2014.[80] More than 75 percent of bariatric operations involve nothing more than ingenious (crazy?) surgical transpositions of intestines and stomach. A small percentage of large metal and polymer devices are inserted into the belly of an obese patient, but in *2014 there were 193,000 bariatric operations,* either using simple permanent surgery while sewing guts together, or using a band or other mechanical device to slow food down from being transferred or digested.[81]

GYNECOLOGY

The most common use of implants in gynecology is transvaginal mesh (TVM), used to treat pelvic organ prolapse (POP), a condition that most frequently affects women after pregnancy. There are an estimated *200,000 TVM operations performed every year in the United States.*[82] While POP is a common postpartum condition (approximately 10 percent of pregnancies),[83] and TVM is ordinarily performed, the FDA in 2011 issued a safety communication stating that "the use of transvaginal mesh may put women at a higher risk of complications without increased benefit to their quality of life."[84]

Urinary incontinence treatment is now most commonly treated with the use of a synthetic device that functions as a sling.[85] Among American women aged eighteen to sixty-four, the rate of sling repair is 198.3 per 100,000 women; the rate for women older than sixty-five is even higher, perhaps 60 percent more.[86] The total of *sling operations for urinary incontinence is estimated at 215,000 women per year in the United States.*

PLASTIC SURGERY:

The American Society of Plastic Surgeons reported *286,254 breast augmentation operations in 2014.*[87] Most women choose silicone gel-filled implants, which were provisionally approved in 2006 by the FDA.[88] Most other plastic surgery operations do not incorporate implants, including face eyelid surgery, breast reduction, and liposuction. Many plastic surgery operations entail the use of injected fillers that get degraded and removed by the body, or transfer of fat cells from other parts of one's own body. Neither qualify as permanent implants

ORAL SURGERY DENTAL IMPLANTS

It has been estimated that *450,000 osseointegrated dental implants are placed every year in the United States.* [89] Gone are the days of exclusively relying upon dentures and dental adhesives. Instead, millions of patients have artificial teeth that are anchored to the bones of the skull with metal posts that cross the gums and bond with the bones.

In the United States, in the year 2014, there were:

6,087,000 orthopedic and spine device implantation operations
1,650,000 cases involving the cardiovascular system
77,000 central and peripheral nervous system implant operations
14,000 cochlear implants were inserted
29,539 organ transplant operations
72,000 live born infants every year from IVF
7,500 urological implants
3,000,000 lens operations
1,000,000 hernia operations entailing the use of permanent implanted mesh
193,000 bariatric operations
200,000 TVM operations performed
215,000 sling operations for female urinary incontinence
286,254 breast augmentation operations
450,000 osseointegrated dental implants

Therefore, in the United States in 2014, the total number of implant operations was about 13,280,000. Assuming a compound annualized growth rate of 4.5 percent (a very conservative growth rate), the number of implant-associated operations per year by the year 2020 would be 17,294,000.

As has been emphasized in this chapter, these calculations are occasionally educated guesses, but are far more scientific than anything that has ever been proposed. At the time of this writing, as we approach the year 2020, we, American citizens, politicians, employers, medical device manufacturers, hospital administrators, and healthcare workers, must shake off our somnolence and deal with the fact that implant-oriented surgery is expensive, particularly when things go wrong. I am fortunate to have a very busy total shoulder replacement practice, and I delight in the implant revolution, humbly recognizing the pioneers who, with great insight and courage, imagined the synthesis of metals, drugs, plastics, and dexterous technique that allows surgeons to so powerfully free our patients from the necessities and miseries of life. But our conversation about healthcare reform and the future of medicine must be based upon facts—and the fact is that *by 2020 there will be over 17,000,000 implant-oriented operations in America.*

TWENTY

Brain Implants

"To do things right, first you need love, then technique."

—Antoni Gaudí

"Neither physiochemical concepts of the body machine nor hopes for technological breakthroughs are of use in defining the ideal man or the proper environment unless they take into consideration the elements of the past that have become progressively incarnated in human nature and in human societies, and that determine the limitations and the potentialities of human life. The past is not dead history; it is the living material out of which man makes and builds the future."

—René Dubos, *So Human an Animal* (1968)

With my head titled heavenwards, I am mesmerized by the carved wooden figures adorning the ceiling overhead. Standing in one of the oldest academic buildings in the world, completely encircled by spruce wood—the floors, walls and ceilings are lined with honey-stained planks—it is the central carving above me that commands the room. The University of Bologna is the Alma Mater Studiorum *to every school in the Western world, claiming 1088 as its founding year, but it is this four-hundred-year-old anatomy theater that has drawn me to this central Italian mecca. Intricately hand-tooled panels overhead exhibit fourteen representations of the constellations and signs of the zodiac; a cursory study of the symbols reveals Leo, Virgo, the Gemini, and all the others. But the fully carved naked life-size figure floating forty feet above my head summons me to stop in my tracks.*

My private tour guide Luca, in a Bolognese-English accent, remarks, "Apollo." Suspended in space, Apollo gracefully points directly at the white marble table next to me. Here is the Greek and Roman god of music, truth, healing, and light, the son of Zeus and the father of Asclepius, the most important god of medicine and healing. (Asclepius himself was the father of several important goddesses associated with the medical arts, including Hygieia and Panacea.) I strain to see what object is there behind Apollo, and with some prompting from the guide I am delighted to comprehend it is a lyre, appropriate equipment for our god of music.

Apollo swims overhead, and at first, I don't understand the message of the ancients. My wife and I are alone in the theater with the guide (who was kind enough to not kick us out at closing time), and preoccupied with trying to remember the astrological signs, I am missing the whole point. Looking back at Luca, he gestures in a grand circle with his arm, "All of these constellations are around Apollo, the god of light and the sun. Imagine this room hundreds of years ago, with the light only coming in through the windows. And Apollo is bringing light, or understanding, to the room, which is why he points to the dissection table."

I position my body until Apollo's finger is pointing right at my face.

I want understanding.

Luca starts again, "While Apollo is at the center, and is the god of light, he is surrounded by the main constellations, which in the 17th century had the power over life and death, illness and disease."

The power. Ancient man yearning to understand, to explain, if not control the forces of disease. And then Luca says it. "The constellations and Apollo, exert their influenza, our word for 'influence.' That is how we got our word for the enigmatic disease, influenza, that was so inexplicable to man since the dawn of modernity. What causes plagues, epidemics, and pandemics? The stars, the gods. Influenza."

Now my eyes trace the same circle as Luca's arm, my personal Apollo, as I hurriedly gaze at the dizzying astrological divinities. Not that long ago, in this hallowed room, the continent's most promising students gathered to understand the enigmas of the human body and what forces bedeviled man. And in the most obvious demonstration, the esteemed professors communicated to the students, it was the heavens that dominated their existence. Swirling above, the constellations reigned over the learned, while humankind was confusingly vulnerable.

As Luca closes the shutters to the eastward-facing windows, I notice inscriptions burned into the panels surrounding Apollo. On the gently curved, shield-like panels, surrounded by graceful curlicues, are Latin phrases. The most poignant axiom proclaims, ET CUNCTORUM SUBIECTA POTENTIA NOBIS, meaning, "To us, and before all, their power."

For the ten millennia that man has domesticated crops and animals, organized towns and cities, computed math on clay tablets and paper, bartered in money, sailed on the seas, conducted warfare with fabricated weapons, brewed alcohol and cured cheese, spun cotton and weaved wool blankets, created roads, formulated concrete, dredged canals, constructed dams, diverted rivers, siphoned away sewage, and conveyed fresh water, he could make no progress on understanding life's most pressing inquiry: Why do we get sick?

With no possible way of explaining, our forefathers turned their gaze to the heavens to disentangle themselves from their snare. It was the influenza.

After eons of bewilderment, our present generation has (mostly) grown comfortable with the explanations of the scientific community about the causes and treatment of disease. The implant revolution has been so successful that we are nearing the point that a lack of diagnosis or the failure to completely restore function is entirely unacceptable. Additionally (in America at least), it is becoming challenging to find an individual above the age of fifty who doesn't have some type of permanent implant in his mouth or body, and downright impossible to find a young person who can't name an implant recipient. This transformation has occurred in the last generation, and will only get more profound.

What will the next few decades bring? My guessing is perhaps true folly (if the last few hundred pages are instructive), but surgeons aren't timid, so here goes. But first, a story about one of mankind's most remarkable interventions.

In July 1982, a gentleman in his early forties was admitted to the Santa Clara Valley Medical Center, in San Jose, California, in a state of frozen wakefulness. He was motionless and stiff as a board, and while he was almost totally (physically) unresponsive, the admitting neurologist, William Langston, sensed that he might be normally alert. A true medical mystery, Dr. Langston had to determine how this patient had, overnight, become "paralyzed" without losing cognitive ability. [1]

A quick physical exam ruled out stroke and catatonic schizophrenia. Although the patient was incapable of moving his limbs, his body was not flaccid, but was instead rigid. In fact, the patient was described as having "waxy flexibility," wherein his arm could be raised overhead, and once the examiner's supporting hand was removed, the limb would stay in place. No, this was not an infection, nor was it a brain hemorrhage, nor was it insanity. Talking with those who knew him revealed that these symptoms had come on literally overnight. It was as if the patient had become the first person in history to come down with a case of profound Parkinson's disease (PD) in the span of one day.

The mystery was rapidly solved through careful detective work and a "trail of ironies,"[2] piecing together ER admission stories from neighboring cities, police reports, news media alerts, and what is ironically considered "a bit of luck." Dr. Langston and other medical investigators identified six additional Bay Area cases of sudden-onset PD, and all seven patients had one striking feature in common: they had used a new "synthetic heroin" that had recently been pushed on the streets of northern California during the "designer drug phenomenon" of the early 1980s.

The question remained, what was the causative agent behind these catastrophic afflictions? The scientists, working with law enforcement and "cooperative" drug dealers, obtained samples of synthetic heroin, eventually hitting the jackpot and discovering a batch of adulterated drugs that was comprised almost entirely of MPTP, an unwanted molecular byproduct of botched kitchen (or Winnebago) chemistry. It turns out that while manufacturing synthetic heroin, temperature is paramount, and if a (not so) trustworthy drug dealer is breaking bad and sabotages the cooking of heroin, a nasty byproduct results that is toxic to a very particular part of one's brain.

◆

Langston and colleagues published their findings in *Science* in 1983, suggesting that MPTP was toxic to a specific part of the brain that was implicated as the locus of destruction in PD. Never before had an animal model been discovered that mimicked PD, and researchers around the country pounced on MPTP as an agent they could use to artificially create PD in lab animals. It would later be discovered that MPTP was broken down into another, more toxic, molecule (MPP+) that was aggressively taken up in the cells that use dopamine as their primary "neurotransmitter"—in essence sending little chemical "smart bombs" into the depths of the brain where movement is coordinated.

Giovanni Morgagni's breakthrough observation about the seats and causes of disease had changed physicians' understanding of the role of individual organs in disease. The component parts of the brain were not understood by scientists into the 19th century, but a curiosity was building about the role of the shape and size of the skull in determining a subject's personality and functional ability. German physicians developed *phrenology* as a pseudo-science, believing that bony prominences and unique head contours could give an alert examiner a leg up on diagnosing psychological issues. Like most forms of quackery (even today), there was no disproving that phrenology was false, but it did open the door to considering that *different parts of the brain had singular functions*.

Phineas Gage was a Vermont railroad worker who suffered a macabre injury to his head when a railroad tamping iron was blasted out of a hole beneath him, rocketing through his head and landing some eighty feet away. This was 1848, with little medical care to speak of, and death a certainty. But somehow, he lived. The local doctor, for his day, was well trained and understood the value of wound debridement and gentle care of the tissues near the blast injury site.

The trajectory of the javelin-sized tamping iron had started below his left eye and continued out the top of his head. Amazingly, Phineas was initially conversant, but within days, became semi-comatose and hovered at the brink of death. The local physician performed removal of clot and abscess. Nearly a century before the advent of antibiotics, Gage would live or die based solely upon the power of his own immune system. He lived, losing the function of his left eye, but more important, losing his old personality and temperament.

Phineas Gage lived another decade, but was transformed from a man of mild disposition and normal social interactions to a desperate patient who indulged in the "greatest profanity (which was not previously his custom), manifesting little deference for his fellows, impatient of restraint or advise when it conflicts with his desires, at times pertinaciously obstinate, yet capricious and vacillating, devising plans of future operations, which are now sooner arranged than they are abandoned in turns for others appearing more feasible," wrote his physician in a Massachusetts medical publication.[3] Here was a patient, who at least in the short term, suffered a dramatic change in personality, without losing his ability to move his limbs, speak, or process information. The obvious conclusion was that the "frontal lobes," the portion of the brain that lies over the eyes, had no role in moving the limbs, controlling speech, or controlling facial functions. More importantly, the case of Phineas Gage is one of the earliest "lesion case studies" in which loss of a specific portion of the brain reveals the function of that part. Although crudely understood for decades, the concept of *cerebral localization* would flourish as scientists developed more precise ways of delving into the mind.

Pierre Paul Broca (1824–1880) was a French physician who had trained with some of the most esteemed medical practitioners throughout Paris in the mid–19th century, and then alternately practiced pathology, surgery, and anatomy. Broca was an inveterate investigator, broadly curious, and deeply committed to medical research and publication. In 1861, Broca was summoned to the Bicêtre Hospital to examine a patient who had lost the ability to speak. The Bicêtre Hospital was a suburban Paris hospital that specialized in mental illness, but Broca arrived *not* to evaluate a mentally ill patient, but a fifty-one-year-old man who had not spoken for twenty-one years—other than the word "tan." So well-known was his singular vocabulary word that he was not called by his name, Louis Victor Leborgne, but by the moniker "Tan."

During the first decade of his stay at Bicêtre, Tan's only limitation was his inability to speak. Broca later reported that his "intelligence seemed unaffected, his mental and physical faculties, intact and responsive . . . he never stopped trying to communicate."[4] In the years before his encounter with Broca, his right side had become paralyzed, and gangrene of the limbs was hastening death. Even in a near-death state, Broca and the patient Leborgne were able to interact. As always, his only utterance was "tan," but Broca determined that he was able to comprehend the language of others, follow commands, and track numbers.

Paul Broca had just attended a lecture at the Société d'Anthropologie de Paris, where Ernest Aubertin had presented the case of Monsieur Cullerier, a patient who attempted suicide by shooting himself in the forehead. Tragically, the bullet smashed open the front of the victim's skull, exposing the brain, but remarkably, did not kill the patient. Admitted to the Saint-Louis Hospital, Cullerier was conversant for the last few hours of his life. Aubertin seized upon the opportunity to examine the patient and to perform an outlandish experiment. He later wrote, ". . . curious to know what effect it would have on speech if the brain were compressed, we applied to the exposed part a large spatula, pressing from above downward and a little from front to back. With moderate pressure speech seemed to die on his lips; pressing harder and more sharply, speech not only failed but a few words were cut off suddenly."[5] Aubertin argued before his Parisian colleagues that *brain functionality was local*; Paul Broca pondered if the facility for speech was indeed housed near the front of the cerebrum.

A week later, Louis Victor Leborgne (Tan) died, and an autopsy was performed, including a dissection of his brain. An isolated syphilitic lesion in the frontal lobe, near the lateral sulcus (the large fissure between the frontal and temporal lobes), was found, and Broca surmised that this must be the location where speech is generated.

He was right. For the first time in human history, a scientist had identified a specific region of function in the human brain. Even today, this region is still called "Broca's area." The curious doctor from Paris set the stage for *cognitive neuroscience* and jump-started the true understanding of localization and lateralization of brain function.

Wilhelm Röntgen discovered X-rays in 1895, heralding mankind's original capacity to peer inside a live person's body. But X-rays are completely incapable of revealing internal damage to the brain. While *pneumoencephalography* was invented in 1919 by Johns Hopkins's Walter Dandy (wherein air was pumped into the hollow chambers deep inside the brain and follow-up X-rays of the cranium revealed a ghostlike silhouette of the brain), it wasn't until the 1970s that CT scans and MRI scans could painlessly and accurately show the internal structures of the cranial vault. Therefore, from the 1860s until the 1960s, the process of determining the functional areas of the human brain was entirely based upon postmortem examination of a brain-injured patient. To truly unlock the secrets of the brain, a breakthrough technique was required to turn the gaze of neuroscientists down to individual brain cells. Only then could doctors understand the unfathomable tangle of cells that is the brain.

Microscopy significantly improved with the innovation of the achromatic lens by Joseph Jackson Lister (1786–1869), but more significantly, histology (the study of cells) dramatically improved with modern chemistry and the use of dyes to bring the tissues to life. But nerve tissue was stubbornly impervious to staining, and controversy fulminated over how the brain cells contacted and communicated with each other. In fact, there were experts who believed that the brain was a singular organ comprised of a single cell that interconnected with itself in a massive jumble of hair-like fibers—the "reticular theory." Every attempt at staining neural tissue was met with a microscopic rat's nest of nerve fibers, and it is easy to see how those mid–19th century histological pioneers were bamboozled by the brain.

William Perkin's discovery of mauveine in 1856 (described in chapter 8, Antibiotics), launched the synthetic dye industry, altering clothing manufacturing and revolutionizing modern chemistry. Soon, German universities and corporations were dominating world markets—in pharmacology, chemistry, and manufacturing—and it's no surprise that is was often German scientists who were identifying the next histological staining technique to bring tissues to life under the microscope. One of the first people to ever see (and know he was seeing) an individual nerve cell was Otto Deiters (1834–1863), a young German neuroanatomist, who (at age twenty-six) invented a technique of staining nerve cells and teasing away the individual cell under extremely high magnification (300x) and with remarkable dexterity. Deiters exhibited his findings with drawings by his own hand, as photomicroscopy had not yet been invented, and the presentation was obvious: the brain and spinal cord were comprised of individual cells that had serpentine tentacles to facilitate interaction. He had succeeded in isolating a singular cell, unbelievably, by hand with tiny needles,[6] but to make sense of the jangle of cells in the brain, a miracle of science was necessary. Looking at a cross section of brain tissue under the microscope was like looking at a bowl of spaghetti, and trying to identify a single nerve cell is tantamount to following a single spaghetti strand in that bowl. Perhaps Deiters would have deduced the answer, but sadly he died from a typhoid infection at age twenty-nine, even before his work was fully published, and it would be up to another researcher to unlock the secrets of nerves.

Camillo Golgi (1843–1926) was born and raised in the Lombardy village of Corteno, nestled near the Swiss border. Golgi's father was the local physician, and he followed in his father's footsteps when he traveled south to the University of Pavia in 1860. While his academic record during his first decade is unremarkable, Golgi did study under early pioneers in psychology and histology, pricking his interest in the microscopic study of nerves. One thousand kilometers away, Robert Koch was busily innovating the use of microscopy in studying bacteria, and like Koch and Robert Lister, Golgi later performed his most important research in a makeshift lab in a kitchen.

Golgi left the comfortable confines of Pavia in 1872, where he had acquired the analytical skills of a serious histopathologist, and made the short trip to nearby Abbiategrasso, itself on the outskirts of Milan. For the next three years, Golgi innovated a new way of staining nerve cells, stubbornly altering the timing and sequence of reagents until one of science's great eureka moments occurred. Playing with his concoctions in a kitchen lab, Golgi entombed a sample of a dog's brain in a block of paraffin wax. Earlier, he had "fixed" the tissue with formalin so that it would not decay, and once the wax had hardened, the thirty-year-old scientist sliced incredibly thin, almost transparent sections of the animal's olfactory bulb. This time, in 1873, Camillo Golgi first exposed the sample to potassium dichromate ($K_2Cr_2O_7$), and then silver nitrate. Mysteriously, this resulted in the staining of only a few neurons on the slide, which were ink-black from the silver, while the rest of the field was a saffron yellow. To this day, it is unknown why only an isolated few neurons pick up the silver nitrate, but the result was the unveiling of the identity of a single nerve within its brainy habitat. With repeated stainings of adjacent slices, the architecture of the sample could be elucidated, and here, Golgi put his artistic skills to great use.[7]

An incredible piece of medical artwork was published by Golgi in 1875, exhibiting the columnar organization of the nerve cells of the mammalian olfactory bulb. Picasso or Dali could not have been more inventive than Golgi when he accurately captured the botanical layout of the nerve bodies and their dendritic appendages. This "black reaction," now called "Golgi's method," is still the standard method of staining nerve tissue,[8] and enabled Golgi to publish the world's first drawings of neurons. Ironically, Golgi concluded that the complicated tangle of nerve fibers was all part of a single nerve unit, reinforcing the reticular theory of brain organization—wrong, of course. Fortunately, an imaginative Spanish physician saw Golgi's incredible artistic representation a decade later, inspiring a Nobel Prize–winning research program on the structure of the brain.

Santiago Ramón y Cajal (1852–1934) was thirty-five when he first learned about Golgi's method of staining nerve cells. His father had been an anatomy teacher in Zaragoza, but Santiago was reluctant to follow in his medical footsteps. A turning point was a legendary adventure to a cemetery, where his father implored him to put his artistic skills to use and sketch skeletons.[9] The gifted artist (self-described as a shy, unsociable, secretive student),[10] found his niche as a painter and scientific illustrator, attending medical school where his father was an anatomy teacher. Having contracted malaria during army medical service as a young doctor, Cajal was too weak to practice medicine. Instead, he turned to histology, which further suited his natural introverted personality. "I finally chose the cautious path of histology, the way of tranquil enjoyments . . . [so that] I should feel myself happy in contemplating the captivating spectacle of life in my forgotten corner . . ."[11]

In 1887, as a middle-aged physician who had not left his mark yet, Cajal encountered the drawings of Golgi, and there seems little doubt that a technical and artistic fascination overcame the Spaniard. He would spend the next half-century detailing the visual appearance of the brain and spinal cord, and more important, unlocking the secrets behind the complex organization of the neural structures.

The artistic merit of Cajal's work is indisputable. His scientific drawings have traveled the continents, exhibited both for their artistic merits and intellectual significance. Similar to the visual brilliance of Vesalius's *Fabrica*, Cajal's works are beautifully presented, but because he was representing an *idea* about the structure of the neurons, his illustrations conveyed a reality that could never be found in any single microscopic slide. Here, artistic ability and imagination were more important than even a photograph could capture.

While Cajal's representations are gorgeous, his scientific discoveries opened the door to the new field of neuroscience. Golgi was the pioneer of neural staining—Cajal was the innovator who carried the field to unimaginable heights. Golgi believed in the reticular theory of the brain, but Cajal was able to show that the brain was instead comprised of billions of nerve cells. (It is widely accepted that the brain is formed by one hundred billion cells, and that each brain cell can connect to tens of thousands of other cells.) Cajal is the unquestioned father of neuroscience: the titan who paved the way in demonstrating the stunning pathways and conduits that nerves travel throughout the brain, spinal cord, and body.

Quite opposite of seeing the brain as a blob of gelatinous material, Cajal envisioned the tissue as "the hum of the restless beehive which we all have within us."[12] Even the most naïve student of medicine can recognize that cells are infinitesimally small, of course not visible to the human eye. But Cajal, and those who followed in his footsteps, were able to point to an astounding fact: between the outermost layer of the brain, the cortex, where the impulse to move is initiated, and the muscle it moves, say the thumb flexor muscle, *there are only two nerve cells.* Microscopically tiny, the *upper motor neuron* (UMN) starts in the brain cortex, and sends its axon, the spiny appendage that carries the electrical signal, down through the brain and toward the brain stem. Here, it crosses ("decussates") and descends down the spinal cord. This same axon fiber tendril continues on the opposite side of the spinal cord, until it synapses with the *lower motor neuron* (LMN) in the neck. This nerve cell travels from the spinal cord, out the nerve root between neck vertebrae, and down the arm to the thumb muscle in the forearm. The axon for this LMN is *over two feet long* in a regular-sized adult! It simply boggles the mind that something so unimaginably slender can be so long. And it's inconceivable that any scientist figured out that the thinner-than-spider-silk axon fiber runs the entire length of the connection from the spinal cord to the muscle. (It also explains the near-impossible task of reconnecting every axonal gossamer strand following a spinal cord injury, and why there is no surgery to recouple the shredded ends . . . yet.)

"When we look at his [Cajal's] drawings today, we see not diagrams or arguments, but the first clear picture of that remote frontier, drawn by the man who traveled farthest into its endless reaches."[13] Santiago Ramón y Cajal ignited an investigation into the labyrinth of the brain and the mind. He, like Golgi, lived to eighty-two years of age, still exploring, even on his deathbed. Decades would pass before CT scans and MRI imaging elucidated the live function of the brain, but Cajal was the voyager who endured years of tedious exertion, squinting over the eyepiece of his microscope, encapsulating the landscapes of our microscopic constitution. He was, and will always be, the greatest cartographer of the mind.

Edward Evarts (1926–1985) was born in New York City and attended both Harvard College and Harvard Medical School, earning his MD in 1948. He immediately began a life of psychoneurological investigation, briefly completing two years of psychiatric training before returning to the neurophysiology laboratory at the National Institute of Mental Health (NIMH) in Bethesda, Maryland, for more than three decades. Evarts spent his professional career delving into the functional pathways of the brain, but instead of relying upon the histological staining of dead brain matter, innovated a way of testing the electrical transmissions of the brain. His patients were not humans with psychological problems, but a small number of cats and monkeys.

Evarts made several breakthrough discoveries in the 1960s, particularly in inventing a way of tracking single cortical neurons in animals. In 1962 and 1964, Evarts published articles detailing the use of glass-insulated platinum-iridium microelectrodes in unanesthetized, unrestrained cats[14] and monkeys,[15] respectively, in their waking and sleeping states. Later, Evarts was able to track activity of single neurons during operantly conditioned movements in monkeys.[16] These studies were based on decades of research by trailblazing neurophysiologists, following the predictable pattern seen in every branch of medicine: anatomy, physiology, and eventually, pathology. Evarts wasn't the first to use implantable electrodes in lab animals, but his own "brilliant perfection of the method of single unit recording"[17] led to its widespread use, and potentiated the ability to track even more complex circuitry in the brain.

Evarts' ultimate goal was understanding the physical basis of mental activity, but concluded that the initial program would be to comprehend the patterns of firing during limb movement. He held that "one must understand movement before one can understand the mind behind it."[18] Evarts was able to decode the timing and sequencing of neural firing over the next two decades, but in addition to his lab techniques, his greatest contribution to neurophysiology was his mentorship of leading neuroscientists, including a young Harvard medical resident who was doomed to be shipped to the theater of war in Vietnam if he didn't find a job working for Uncle Sam in the public health system. Fortunately for Mahlon DeLong, MD (and humanity), the National Institute of Health (and its NIMH), qualified as an acceptable stopover.

Mahlon DeLong (b. 1938) attended Stanford University, spending time with a physiologist who studied the nervous system of the crayfish. That professor was Donald Kennedy, the future president of Stanford and editor in chief of *Science*.[19] Not exactly the typical preparatory material for medical school, but DeLong's interest in biological systems was piqued, and he headed east to Harvard Medical School, graduating in 1966. He stayed in Boston to begin his medical residency, but the logistical demands of the Vietnam War gave rise to the infamous 1969 Vietnam draft lottery. Rather than risk being deployed in the medical corps overseas (as young physicians were), DeLong accepted a position as a research associate in the NIMH laboratory of Edward Evarts. For the next five years, DeLong was part of the team that spearheaded the inquiry into the circuit board of the mind.

When DeLong arrived, he soon realized that most of the easily approached areas of the brain had been claimed by other research associates of Evarts's. The out-of-the-way center of the brain was poorly understood, and for DeLong it "was like exploring and charting unmapped Africa or the Amazon."[20] Since all the "good stuff" (such as the motor cortex and cerebellum) was relatively well understood, Mahlon DeLong turned his attention to the *basal ganglia*, where the normal anatomy and physiology were almost completely unknown. The young researcher, with no PhD in the field, quickly discovered an astounding fact: just as there was specificity in the function of cortex, there was localization for the nerve pathways that encode movement in the basal ganglia.

As Paul Broca had predicted, brain functionality was local; there are extremely specific areas of function in the brain, such as the motor cortex. Not only is all muscle function controlled in the "motor strip," there is a bizarrely composed plan of motor responsibility in this ribbon of wrinkly brain, and can be represented to the reader with the "homunculus," or "little man."

As one can see in the diagram, the localization for sensation is preposterously predictable. For instance, sensation for the knee lies near the top of the head, where the hemisphere of the brain takes a right angle and the split between brain halves occurs. Similarly, the movement of the facial muscles is initiated from the motor strip in the mid-portion of the hemisphere—Dear Reader, if you touch your scalp above your ear, the bit of brain about an inch deep from your finger is what controls your face on the opposite side of where you are touching. DeLong's big finding was that there is a similar map of uniquely active cells in the basal ganglia; these nerve cells are active with specific face, arm, and leg movements. His 1971 publication[21] was "very much a landmark,"[22] and greatly challenged scientists' preconceived notions about the role of the basal ganglia in movement.

No sooner had DeLong published this classic paper than he returned to academic clinical medicine, but instead of returning to Boston, he transferred to Johns Hopkins, where he completed a three-year residency in neurology. Donning the white lab coat of a resident, there can be no doubt there were times of awkward role reversal when Dr. DeLong was schooling his superiors on the function of the basal ganglia. Not surprisingly, DeLong continued his research as a resident, and then stayed on at Hopkins until 1989. In these years, DeLong and his team showed that the basal ganglia "structures were not a funnel of diverse influences to the motor cortex but rather components of a series of independent parallel circuits, receiving from and sending information to specific cerebral cortex. And, another big surprise, the circuits were not only involved in movement but also in cognition and emotion."[23] In short, Mahlon DeLong had decoded the deepest parts of the brain, finally understanding why a malady like Parkinson's disease simultaneously causes shaking and spasmodic tremors of the hands while freezing someone's legs from moving.

The brilliant scientists of the 19th century had positively no explanation whatsoever for tremors, seizures, migraines, and brain infections. Santiago Ramón y Cajal mapped the neural pathways he could see. The succeeding wave of neuroscientists used electrodes to determine the firing patterns of the neural cells, but what was needed was a mammalian model of PD that would allow scientists to further illuminate the complex pathways of movement disorders, and perhaps, miraculously, intervene surgically. No one was thinking of a brain implant in 1982 . . . that was the stuff of science fiction. But Mahlon DeLong continued to hope for an animal model to test his hypotheses, when he stumbled upon a report in *Science* that detailed the bizarre story of heroin users in the Bay Area who, overnight, became PD sufferers.

Dr. DeLong stayed at Johns Hopkins following his neurology residency, forming his own lab to continue his exploration into the basic anatomy and function of the basal ganglia. The news of MPTP-afflicted patients inspired DeLong to see if he could create an animal model for PD. His hunch was that PD's damage was far more complicated than theorists had proposed. To decipher the cause and cure of the disease would demand great imagination, and because DeLong "was very good at putting things in their perspective and thinking conceptually,"[24] he was the right man to crack the complex code of the black box in the deep part of the brain.

DeLong's earliest work on the lab animals with artificially created PD turned the world of brain research upside down. While most of us associate a tremulous hand with PD, the other major symptoms are stiffness of limbs, slowing of movement, the loss of facial expression, and the slurring of speech. All told, most of the symptoms imply an inhibition, or "dumbing down," of the nerve transmissions from the motor cortex to the muscles; therefore, scientists had concluded that, conceptually, PD could be summarized as a slowing of nerve transmissions among the neurons that control and coordinate movement. *Instead*, Mahon DeLong recalls, "There were skeptics, but on Day One we saw that the activity patterns and firing rates were quite altered, and that the output from the basal ganglia was *increased* rather than decreased."[25]

DeLong and his colleagues realized that the Gordian knot of the basal ganglia was the key in understanding movement disorders. As experiments progressed, a Rube Goldberg contraption was revealed, where neural circuitry between the cortex of the brain and the elements of the basal ganglia came into focus. The breakthrough observation was that some of the neural pathways triggered an inhibitory signal to other nests of cells, so that increased firing along an "inhibitory" conduit would result in decreased signaling along the next stopover.

This is a critical concept in physiology and medicine: there are functional elements in our DNA, in our cells, between glands and organs, and along nerve pathways, where an *increase* in the signaling molecule or nerve transmission results in a *decrease* in a corresponding effect. A classic example is the tumor suppression gene, p53. When active, the "anti-oncogene" p53 helps repair DNA defects and stabilizes cellular integrity, thus checking the devolution of a cell's status from healthy to cancerous. Many of our hormones act in a similar fashion, where an increase in the secretion of a protein hormone results in the decrease of the presence of another protein or ion. Looked at another way, the loss of inhibition is *net stimulation*. DeLong was able to demonstrate that the basal ganglia was a complex set of "independent parallel circuits," with separate pathways to and from various areas of the basal ganglia to specific areas of the cerebral cortex, some with excitatory potentials, others with inhibitory. And just as surprisingly, the circuits were also involved in emotion and cognition, not just movement.

The deepest part of the basal ganglia is the *subthalamic nucleus*, and DeLong concluded by the late 1980s that this clutch of cells was critical in PD. While it was obvious that the substantia nigra was the area of degeneration that triggered abnormal physiological output throughout the entire system, DeLong's MTPT animal models showed increased excitatory impulses from the subthalamic nucleus to other parts of the basal ganglia. He had an idea: how about intentionally destroying (or "lesioning") the subthalamic nucleus?

"It was quite a remarkable result that Mahlon would suggest making a lesion in this region to restore the balance of activity in Parkinson's disease, and lo and behold he got that result. I can't overemphasize what a significant leap this was, because the one constant that people had known for more than fifty years is that if you make a lesion in the subthalamic nucleus, you produce an abnormal movement disorder, and here Mahlon proposed treating Parkinson's disease by putting a lesion just in that region and in fact got the results he predicted,"[26] observed his former lab associate. It would be like treating a skin cancer patient with prolonged sunlight.

Mahlon DeLong published his findings in a 1990 article of *Science*.[27] In laboratory monkeys who were suffering from experimental parkinsonism, lesioning of the subthalamic nucleus reduced all of the "major motor disturbances in the contralateral limbs, including akinesia, rigidity, and tremor."[28] In medicine, our preliminary instincts are often wrong. And DeLong's suspicion that causing additional nerve damage in another part of the brain could ameliorate parkinsonism was the counterintuitive jackpot.

With the success of experimental subthalamic lesioning, the obvious next step was to begin precise radiation therapy treatment to the same area *in humans*. To accomplish this, DeLong turned to his neurosurgical partners, particularly as he transferred his practice and lab to Emory University. Immediately, the results were positive.

Neurosurgeons had been performing "functional surgical intervention" for epilepsy patients in the few years before the first "pallidotomy" (selective lesioning of a part of the basal ganglia) at Emory in 1992. Previous attempts at thermally damaging parts of the brain, had, on occasion, led to disastrous results, leaving patients drooling, depressed, and barely speaking. And because these interventions had been performed with intentional permanent destruction, there was no going back. DeLong had reasons to be optimistic in 1992, however, including the assurance that the precision of brain imaging and electrophysiological mapping techniques were considerably better than they had been in the decades before.

Experimental surgery is at once defined by bold and visionary practitioners, desperate patients, courageous families, and the (necessary) passage of time to prove its efficacy. The initial slow trickle of patients in the preliminary years turns into an ever-growing influx of patients who are unwilling to be victimized by their disease. Pioneering surgeons mandate that the original cohort of patients be followed for a few years before it seems prudent to conduct a comparative study, which then requires a few years to organize and enroll enough patients to draft a manuscript. Many months of editorial jousting are consumed before publication is finally achieved. Thus, it is no surprise that a *decade* passes before a randomized study of an experimental operation (versus medical treatment alone) appears in a medical journal, which is the timeline that occurred in the case of pallidotomy. The Emory group published their article in 2003, revealing the superiority of surgical intervention over medical treatment in PD.[29]

During the 1980s, DeLong and his partners became expert in passing tiny electrodes into the deepest parts of the brain. These wires were meant for monitoring the firing pattern of the neurons at the tip of the electrode. What if a surgeon intentionally left the electrode in place, and then, boldly, experimented with the application of an electrical current at various frequencies? Probably impossible to perform in America, the land of billboards with earnest-faced injury lawyers, but an intrepid surgeon in another country could be the one to perform the breakthrough operation.

Alim Louis Benabid (1942–)is a French neurosurgeon in the small city of Grenoble, nestled at the base of the Alps near the Italian border. Grenoble has become one of Europe's most important research and tech centers, and its most important science and engineering schools is Université Joseph Fourier (UJF). Benabid had the unusual distinction of holding a PhD degree in physics and an MD from UJF, ideally preparing him for a biophysics revolution.

Although the drug levodopa had significantly improved the lives of hundreds of thousands of PD patients, it tended to lose its efficacy after about five years, and even in patients who benefited from the medicine, there were meaningful side-effects. Dr. Benabid was still performing lesioning operations throughout the 1980s, always targeting the thalamus in the brains of severely affected PD patients. The usual technique involved carefully positioning an *awake* patient on an operating table, her head carefully secured with clamps that would hold the skull perfectly still. A small drill hole was made at the top of the head, and a long needle stealthily inserted deep into the brain under the guidance of live X-rays. With the complex anatomy of the brain held in consideration, a small electrical impulse was then delivered to the deeply positioned needle electrode. Aiming for the central thalamus, "If you are just a little bit posterior, you are in the sensory thalamus and the patient feels a tingling; if you are too lateral you are in the pyramidal tract and the patient exhibits contractions of the hand or face."[30] Placing the electrode in the wrong location could paralyze a patient, so the mimicking of neural firing with low frequency impulses was critical.

One day in 1987, in his Grenoble operating room, Benabid asked a question, not surprising for a physicist: what would happen if I changed the electrical frequency of the probe? Benabid conducted an experiment on an elderly patient who suffered from a severe tremor. Once the probe was in the correct location, he began altering the frequency. "I explored the effect from very low frequency—1, 5, 10 Hz, etc.—up to 100 Hz, and this is how I got lucky." All of a sudden, without losing other neurologic function, merely by changing the electric frequency of the probe, the Parkinson's victim had complete cessation of his tremors. "At 100 Hz, we had suppression of the tremor . . . I thought, aha, this might be the solution."[31]

There can little doubt that Benabid's discovery must rank as one of the most spectacular moments in the history of surgery. Had it been a haphazard, accidental finding, it would still have been amazing. But in context, where the surgeon-scientist dared to believe that he might affect a change, and then witness an instantaneous cure of tremor, an ancient malady with religious insinuations, Benabid's breakthrough is a marvel, a eureka moment of the highest order.

For an encore, Benabid left the wire, with four metal contacts at its tip, in place in the patient's thalamus. Benabid connected the wire to an external battery source, and then programmed the device using a small box with buttons and archaic-looking switches. This allowed him to individualize the settings and continue to experiment with the frequency delivered. Benabid immediately reported his miraculous results[32] when implanting the electrode in the thalamus. It doesn't seem possible, but in 1987, there was still no Institutional Review Board (IRB) in Grenoble, and no French equivalent of the FDA (the Agence du Médicament was established in 1992). With permission from his neurosurgical boss at the UJF, Benabid continued placing the world's first deep-brain stimulation implants.

Concurrently, across the Atlantic, DeLong was finalizing the map of the "functionally segregated parallel circuits"[33] and homing in on the importance of the previously ignored subthalamic nucleus. After the successful reversal of parkinsonism by the Emory team in 1992, Benabid began altering his technique in Parkinson's patients, placing the DBS wire implant in the subthalamic nucleus. Shortly before DeLong's 2003 publication of comparative subthalamic lesioning, the FDA approved DBS implants for PD.

I have been fortunate to be in the operating room for the neurosurgical placement of DBS implants for PD. The experience of watching six-inch-long rigid needles being coaxed to the deepest parts of the brain utilizing intra-operative CT scan guidance is phenomenal. But this pales in comparison to the postoperative appointment a couple weeks later, when the neurologist and the assisting technician program the transistor-filled electronic unit that has been implanted in a patient. In one moment, the patient sits rigidly in her chair, hands tremoring with every attempted movement, and the act of raising a glass of water to the mouth an embarrassing impossibility. However, with an instantaneous initiation of the electrical signal from the pulse generator unit, the tremors stop. For the first time in years, taking a drink of water is a possibility, and the patient (and her family) burst into tears. I defy you—do an internet search and watch a few videos of DBS units being turned on and try to not weep.

DBS has been used over 150,000 times around the world, with about 10,000 such operations per year in the United States. The last year with complete United States implant data is 2014, and there were an estimated 20,000 SCSs placed. Similarly, there were 7,000 sacral nerve stimulators and 2,000 vagus nerve stimulator operations in 2014. All told, there were approximately 35,000 neuromodulation operations performed in America in 2014.

"A recent study estimates the global neuromodulation market—including deep-brain stimulation along with technologies such as spinal cord stimulation, and transcranial magnetic stimulation—at €3.31 billion ($3.65 billion) in 2015 (*MarketsandMarkets*). The market is projected to grow at a CAGR of 11.2% to reach €5.62 billion ($6.20 billion) by 2020."[34] Because DBS placement is more effective than medicine in early stage Parkinson's, and because its effectiveness continues for years, it is a powerful weapon in the battle against movement disorders.

It is striking that DBS is also being used now for chronic depression, dystonia, epilepsy, obsessive-compulsive disorder, Tourette's syndrome, and even Alzheimer's disease. In a perspective article in the *New England Journal of Medicine*, Michael Okun concluded, "DBS therapy is usually considered only after all other treatments have been exhausted, but becoming 'bionic' has provided many patients with a new lease on life. Thanks in large part to the contributions of two extraordinary scientists [DeLong and Benabid], we have entered the era of human neural-network modulation."[35]

Deeply placed wire electrodes with low frequency current have now been used around the world for three decades. Interestingly, scientists are not exactly sure of the mechanism of action, but then again, we still don't fully understand how many anesthetics work during surgery, and how many of our disease-modifying drugs function. We just know *that* they work.

The implantable devices of the future will not just deliver a haphazard electrical pulse, but will instead function as recording and transmitting machines. The general title for these types of implants is Brain Machine Interfaces (BMI). While today's FDA-approved brain implants merely generate an electrical field within a specifically identified location, BMIs will record and stimulate individual neurons. As Edward Evarts pioneered the ability to perform "single unit testing" in the lab, scientists today are trailblazing the path of recording single units in humans, with the goal of transmitting neurons' signals to a machine.

Our brain, the central processing unit (CPU) of our body, takes in all of life's sensory information (sight, sound, touch, smell, feel, balance), and then processes that information, both subconsciously and consciously. Oftentimes, the inputs are wordless, requiring reasoning and computation of meaning. Language, whether written or spoken, similarly demands cognition, analysis, memory formation, and response.

Our outputs include motor movements of our face and limbs; it is interesting that we can gather information without moving a muscle, but cannot communicate without either moving our mouth and vocal cords or using our hands to type, write, or sign. At least, for now. A BMI is the ultimate implant. Up till now, implants have functioned as replacement parts—stenting open coronary arteries, substituting for arthritic joints, standing in for aged lenses, and reinforcing weakened abdominal walls. In the near future, BMIs will propel society into a bionic future, where men and women will not just be cured of disease, but be made better, stronger, and faster (to quote *The Six Million Dollar Man*).

As biological and genetic treatments improve, there can be no doubt that the focus of medicine a century from now will not be on cancer (it will be cured), and it will not be on chronic disease (diabetes, auto-immune disorders, degenerative conditions, arthritis, and cardiac disease will be a distant memory); medicine in the 22nd century will concern itself with superior humanoids, where implant operations are not used to treat disease, but to make better cyborgs. Sound chilling? Perhaps, but it's more inevitable than you think.

TWENTY-ONE

Cyborg *Future and* Homo Electrus

"In days gone by, surgery was all about blood and guts. In the future, surgery will be about bits and bytes."

—*Bleeding Edge*

"The robot will see you now."

—*Bleeding Edge*

I am running late on my way to class here in my second year of medical school. Having moved from a suburban apartment to a three-bedroom bungalow in a tumbledown neighborhood close to the medical center, I can easily walk to my morning lecture along cracked and uneven sidewalks. Backpack slung over my shoulder (following convention of only one shoulder strap and avoiding Poindexter associations of using both straps), I am preoccupied with thoughts of my upcoming bacteriology exam, the infamous ultimate flunk-out challenge in my school.

And now I encounter an unusual scene—an elderly couple standing on the curb in the middle of the block, hand in hand. They are facing the road, motionless, and appear to be waiting for something. I slow down, glancing across the street to see where they are looking. It then dawns on me that they are blind, as their heads are cocked slightly downward and their gazes are unfocused. I come to a stop, realizing that they are aiming to cross the street, and I surmise that they are listening for traffic on this one-way street.

My block is precisely lined with two- and three-bedroom craftsman bungalow houses built in the 1930s, where most of the front yards are decorated with dandelions and have a sunken pit where excavated trees formerly found anchorage. Generations of medical students and residents have populated these neighborhoods, and like those denizens before me, I have no money to improve my abode and no energy to care. Living in our neighborhood is the occasional older couple who has lived here for decades; this blind twosome must have crisscrossed these streets for years.

A dilapidated pickup truck rumbles by, and the standoff continues. I am across the street from the pair of careful travelers, and with no approaching traffic, I wonder what they are waiting for. The man appears to be about sixty years old, with slicked-back dark hair and the build and clothing style of Archie Bunker; the woman is in a colorful dress and sensible shoes, her white wavy hair adorned with a practical headband. Abruptly, the man jerks her hand, and they rapidly scoot across the roadway. Once on my side of the street, they continue along the same sidewalk, but given my potential tardiness, I can't afford to indulge my curiosities about this couple. I wordlessly pass by, wishing I could initiate a conversation, but am left hoping that I encounter them some other time.

A few months later, during my physical examination class (in which we learn to incorporate all our senses and formulate a diagnosis), I mention my encounter to my professor. In an instant, he knows the couple I'm talking about. "They're the Dowdys. They are not just blind, they are both deaf." BOTH DEAF? Now I think back to the moment I saw them on the road, exercising extreme care before they zipped across the street. They weren't listening for an oncoming vehicle—they were feeling for it.

I learn more about the Dowdys from my professor. This married couple has been together for years, Leonard having lost his vision and hearing as a five-year-old from bacterial meningitis in 1932 (prior to the 1940s, everyone in the world was completely at the mercy of a bacterial infection of the brain), and Betty being born deaf, and later losing her eyesight to retinitis pigmentosa as a young adult. They communicate with each other by using sign language in each other's hands ("fingerspelling"). How do they converse with someone who doesn't know sign language? My professor isn't completely sure of the mechanics, but tells me it has something to do with touching the speaker's face.

Now, I am even more eager to see the Dowdys and interact with them, hopefully without threatening them.

A week later I see Mr. Dowdy at the medical center. He is waiting in the lobby, seated by himself in a waiting area. Here is my chance.

I approach him slowly, wondering if he can sense my advance. I gently place my hand on his shoulder, and grab the empty plastic seat next to his.

Leonard quickly turns his head my direction and a broad, open-mouth smile appears in an instant. As I reach to shake his hand, his whole body is turning my way, and after a brief touching of our hands, his right hand tactfully is drawn to my face. His thumb is immediately on my lips, and his index and middle fingers are on the angle of my jaw. The meaty part of his palm now presses against my voice box. I guess I've always had a phobia of people touching my face, but the suddenness of our coupling overwhelms my social qualms.

An energetic smile still electrifying our first meeting, he mouths, "Hello, I'm Leonard," with an understandable phonation that is nonetheless typical for a deaf person.

"My name is David," I intone, intensely reading his face. Can we cross the chasm?

"Hello David!" Leonard exclaims with gusto. A new friend.

I glance around, a smattering of patients in wheelchairs, teenage mothers with newborn babies, a disheveled homeless guy in a windbreaker, and a smartly dressed pharmaceutical rep all staring at us. I explain that we live in the same neighborhood, that I have seen them walking to the medical center. Leonard explains that Betty works here at the medical center in the laundry department. With a little bit of difficulty, we are communicating, and I'm getting used to my face being handled and interrogated. A few minutes in, Betty appears on the arm of one of her coworkers.

Leonard and Betty grasp hands, immediately fingerspelling together in a cacophony of blurred silence. Betty tilts my way, and in almost normal speech, says, "Nice to meet you, David."

I am dumbfounded, really. Over the course of a few minutes we are able to acquaint ourselves, and I'm hopeful that in the future they'll remember me. I scramble back to one of the academic buildings on campus, inspired by the Dowdys resolve and good-natured charm. I later learn of their trips to Europe, his job in a machine shop, his woodworking hobby, and other wonders. I also learn more about Leonard's ability to use his hands to understand people. It's called "Tadoma," and utilizes the feeling of the breath, the shape of the lips, the movement of the jaw, and tactile vibrations of the voice box to synthesize an understanding of speech.

From time to time I encounter the Dowdys around our neighborhood, and while I find their communication ability a marvel, it is their capacity for transit that I find a miracle. I observe that short jaunts around our local streets do not necessitate Leonard's use of a blind person's cane, so comprehensive is his knowledge of every curb and obstacle. I follow them one day to investigate the technique, and it's like a mortal-combat three-legged race of Pin the Tail on the Donkey. Coming to an intersection near our houses, I see their united pause, with motionless hyperacuity, as a 1970s Thunderbird slops along. We are only a block from Rainbow Boulevard, a city thoroughfare heaving with trucks and cars, and I wonder how they perceive which cars are harmlessly traveling north and south, and which vehicles are threateningly barreling east and west here on 41st Street. When the coast is clear, with no vehicle in sight, the speed-walking race commences, hand-in-hand in synchronized gait to the other side, feet knowingly raised at the opposite curb. They silently slow to a gentle pace, safely back on the crumbling sidewalk.

As my third year of medical school commences, my rotations bring me to the Harry S. Truman Veterans Administration Hospital, and it's been a few months since I've seen the Dowdys. Now that my Psychiatry rotation has brought me back to the University hospital, I am hopeful that I see them soon.

On call tonight for ER admissions for Psych patients, my chief resident and I are hastened downstairs for a patient who is suffering from a life-threatening manic episode. She has been awake for three days, is hoarse and out of her mind. I have cared for a few patients in the grip of mania, but I am a bit frightened by this emaciated patient in a padded cell, free of furniture and loose objects.

My face at the window of the room, I stare at a woman who is pacing in a frenzy, ges- ticulating with a tiny nub of a cigarette. It's not lit, but she uses the prop as a pointing device, a baton, and as a writing instrument. Ignoring me, but addressing an invisible throng, she gives an impassioned plea about the meaning of the upside-down cross. She takes her minuscule cigarette, flips it around, and uses the remaining ground-up tobacco remnants and draws an inverted cross on the padded wall dozens of times, oftentimes retracing the shape for emphasis. She can barely speak, and I feel so sad for this innocent victim.

I turn away from the observation window to meet eyes with my boss when I overhear a couple ER residents talking about the deaf guy who just got admitted. "How in the world does a blind guy have a woodworking shop in his basement?"

I don't need any more data—this is Leonard.

"Where is Mr. Dowdy?" I ask the young doctors.

"Bay 3. You know him?"

"He's a neighbor of mine. He okay?"

"Well, he sliced his fingers, but he'll be okay. You wanna say hi?"

I pull the curtain back, and there is Mr. Dowdy on a gurney, his right hand wrapped in a bloodstained washcloth. His wife is seated next to him, her hand resting on his shoulder. I walk around the gurney, and greet them by squeezing their shoulders, their hands quickly shimmying down my forearms to my hands. Leonard, of course, only uses his left hand, but nonetheless searches for my face.

It only takes a second, and Leonard knows me. A quick flurry of fingerspelling to Betty, and she lets out a "David!" with great relief. She stands up as Leonard breaks into that huge smile that all who know him love.

One day, I will be a surgeon, but sewing up lacerated fingers is not yet my forte. I am not a caregiver here, but a neighbor who stumbled upon a friend in need. I am struck by Leonard's equanimity and bravery—and his bad fortune in suffering an accident despite working so hard to be safe.

Even more important, I am struck by the importance of Leonard's fingers to his well-being. Cutting your fingers is a tragedy; if your eyes and ears don't work and you connect to human beings with your hands, it's a catastrophe.

The timing of Leonard's childhood meningitis dictated that no cure was possible in the 1930s. A similar infection today, if treated promptly with powerful antibiotics, would result in a complete cure. Betty's congenital deafness would obviously have no medicinal treatment, but similar forms of hearing loss are now addressed with cochlear implants. And what does the future hold for vision loss? There can be no doubt that future implantable medical devices will restore lost vision, whether it arises from injury to an eye or to the brain.

But what about the love story between the Dowdys? Of senses lost, companionship, assistance, and guidance? The inspirational camaraderie was born of disability and sustained by mutual dependency and a desire for independence. Will we forfeit resiliency when all our diseases are cured? Do we cheapen triumph when all our tragedies are mitigated?

THE NEXT DECADES

"You are a biased surgeon who is shilling for the medical device industry," you may be thinking.

I don't think so. I have tried to be completely honest about my impressions of the shortcomings of the medical device clearance process. Moreover, I have criticized the appalling lack of implant registries in America, and the embarrassing lack of transparency among some of the device manufacturers in our world.

Recent magazine articles, books, and an exposé documentary highlight the frightening downside of implants. In the 2018 film *The Bleeding Edge*, there is a sickening and cringeworthy scene of an FDA panel debating approval of the Essure implantable birth control device, even joking about what would happen to the reviewers if the device ever proved to be dangerous. This is betrayal of the highest order.

It is actually my belief in the ascendancy of modern medicine in its ability to powerfully and beautifully improve the lives of our fellow man that compels me to simultaneously celebrate the dramatic transformation of medicine over the centuries (and the last seventy-five years in particular) and ferociously criticize our flawed implant evaluation system, the occasional bad actor among device manufacturers, and the infrequent soulless doctor who greedily treats his patients exclusively for his own benefit.

If there is a hell, it will be populated by caregivers who callously, or worse, intentionally, injured the vulnerable in their care. A human is never more susceptible than when they are unconscious on an operating room table. The privilege, and weighty burden of responsibility, of caring for another being when they are defenseless is immense, and to violate that sacred trust is unconscionable. Laura Beil's chilling story of Dr. Christopher Duntsch is gripping because a surgeon's treacherous incompetence is the worst form of betrayal we can imagine. [1]

My hope is that a balanced view of the shortcomings of the medical device industry and an appreciation for the spectacular accomplishments of the scientists, doctors, and businessmen who comprise that industry will provide a sober outlook that will empower our society (and our lawmakers) to sculpt the rules of the implant revolution. But that viewpoint is attained only when an historical comprehension is achieved and an understanding of where we are today is grasped.

Healthcare is expensive because technological breakthroughs are costly; patients who have health insurance can afford pricey solutions, and hospitals are happy to provide therapies that boost their bottom line. As shown in this book, we will soon exceed twenty million implant operations per year in America, at a cost of hundreds of billions of dollars. We love complaining about how "inefficient" the healthcare industry and health insurance are, but none of this is possible without them.

It should be obvious by now that modern healthcare is fantastically expensive because implants, drugs, hospitals, and doctors are expensive. It was far cheaper when doctors were ignorant, hospitals were death houses, drugs were rudimentary, and implants were nonexistent.

To elevate the implant revolution to the next level, the following must occur:

1. Comprehensive, all-inclusive implant registries.
2. Improved FDA regulation of medical devices, perhaps necessitating a complete reinvention of the approval process.
3. Improved price transparencies in health insurance and medical care.
4. Publication of all surgeons' and hospitals' complication rates and outcomes

With these recommendations, I have managed to antagonize both my physician colleagues and everyone in the medical device industry. Given the stakes, the massive costs associated with all four recommendations are justified. Who will pay? Likely, a combination of manufacturers, physicians, insurance companies, and hospitals. There will be pain and sacrifices by all involved, but if America is to maintain our scientific preeminence, a firm commitment to long-term quality is mandatory. More so than in any other industry, the tracking of long-term outcomes is, by far, the most important method of determining the best practices.

THE FUTURE

What will the future look like? The easy answer is *incremental improvements* in joint replacements, pacemakers, catheters, mesh, heart valves, and deep-brain stimulators. Design tweaks and manufacturing modifications will usher in slight improvements, but every manufacturer in every discipline has a pipeline of implants that represent substantial leaps forward, oftentimes opening up whole disease classes that were previously unchallenged.

The device manufacturers are going small.

The initial heart lung machine was the size of an armoire, with multiple whirring drums—pulsating crimson blood through wriggling tubing—and countless moving parts. The mechanical heart was developed several decades later, with the goal of providing cardiopulmonary function to an ambulatory patient. While the first implantable artificial hearts of the 1980s did reside in the thorax of patients, the battery powered units were still the size of a dishwasher. It has been a bumpy ride for artificial hearts, but these devices now run under the control of an external miniature computer that is powered by a battery the size of a ham sandwich. All told, a patient wears a small satchel while walking around, and even exercising. Besides the scars on the chest, modern mechanical heart recipients have one distinct feature: no pulse.

These machines have a continuous internal rotating impeller that drives the blood without valves that open and shut.

Newer machines will become ever smaller, and the power plants of the future will likely resemble an Iron Man–like implantable battery that lasts for years. The artificial hearts of the future will be miniaturized to an unimaginable degree, no longer resembling a machine pump. In fact, while it is impossible to imagine now, the future artificial heart may truly be at nanoscale.

Our kidneys are the size of fists, yet demand about one-fifth of the blood supply of the body. They cleanse the blood of impurities (excreted in the urine) and maintain electrolyte balance. When the kidneys fail, one must undergo kidney transplant of a live organ, or stay on renal dialysis on a regular basis. At present, a hemodialysis machine is about four feet high and resembles a small heart lung machine with its tubing and rotating cylinders. One wonders how small these machines will become in the decades before none are needed. Will the artificial kidney ever be small enough to be implanted? I wouldn't bet against it. But as artificial organs continue to get more minuscule, the march toward disease cure will get more effective, obviating their use.

With the advent of clustered regularly interspaced short palindromic repeats (CRISPR) technology, there can be little doubt that every human will have complete decoding of their DNA, and more importantly, correction of genetic flaws. All diseases that have a genetic basis (genetic susceptibilities or chromosomal damage) will become a thing of the past, including rheumatoid arthritis, heart failure, and skin cancer.

Traumatic injuries (including spleen and liver lacerations, collapsed lungs, fractured bones, and contused brain matter), seem like the one constant that will always attend the human existence. Protective sporting equipment and enhanced vehicular safety systems will likely decrease the severity of injuries, but can trauma be eliminated in a science fiction future? I would guess "no," but could anyone have guessed a few decades ago that tremors could be eliminated with a brain implant?

If life is the most sacred thing in the universe, and if the "right to life is humanity's most fundamental value,"[2] then death is a crime against humanity. The stakes in health care will become even larger once we start to lengthen humankind's days, and if we are fated to live for hundreds of years, will we all become terrified of trauma and a senseless death? It's one thing to be in a car accident and suffer an untimely death in middle age, it's quite another when a forty-year-old is in the same car accident and is expecting to live another several centuries. We could actually end up fearing death more.

Will we ever conquer our bacterial foes? The microbial warriors may be our most formidable challenge, but once we truly harness the power of genetic manipulation, how can we wager that we won't eliminate the threat? Perhaps a bigger threat to humanity will be an extraterrestrial microbe brought back from an interplanetary spaceship that has evolved over billions of years on another planet. Call me crazy.

In a sense, over the next several decades, there will be a three-way war against disease: biological, implant, and genetic. Until the genetic remedy is perfected, biological solutions (pharmaceutical and nutritional) will continue at the cellular level. When these are not effective, implants become the cure. For instance, diabetes, not truly understood until the 1920s with the purification of insulin, is treated with daily insulin injections. These shots have become much less onerous with the advent of insulin pumps. While there is intense research into implanted pancreas tissue with the goal of restoring auto-insulin production, the eventual goal of a genetic cure remains. Thus, the biological, implant, and genetic war on diabetes persists, but it seems unescapable that no one will be injecting insulin a century from now.

One way of testing the certitude of medical futurists is to ask the following question: If you had to invest your entire net worth in a specialized medical company now, and that pot of money would not be accessible for one hundred years, where would you direct that money? A drug company, a joint replacement manufacturer, a genetics company, or a bioelectric company? While I am an orthopedic surgeon, I simply can't imagine that joint replacements will be performed in a century. Similarly, there can be no way that patients will be receiving chemotherapeutic agents. There will be no cancer. Thus, the decision to invest in a bioelectric company versus a genetic company would be based upon what you perceive the role of medicine will be in the future. Will physicians treat disease in the future, or will they exclusively enhance humanoid-cyborgs?

Genetic modifications of humans will certainly be the norm within a decade or two. It is possible (even probable) that everyone in a century will be free from genetic errors. Perhaps this will be forced, where noncompliance is not possible, since those with chromosomal defects would be the "diseased ones." I predict the anti-vaccination rebellion will pale in comparison to "genetic correction" in the 22nd century.

Humans will become the ultimate GMO—genetically modified organisms.

As genetic-based diseases become a thing of the past, and cancer fades from existence, it is likely that future generations won't even need correction of their genomes. Amazingly, it will also be possible that once a generation's communal genome is purified, no further correction would be necessary, since the gametes would share similarly purified genes. Babies will be born with "normal" chromosomes, free from the threat of auto-immune disease, food sensitivities, mental illness, cancer, and heart disease. With illness whipped, attention will turn to enhanced and augmented physical, mental, and social realities. And I believe that that will still necessitate the use of implants—brain-machine interfaces.

This is why I would place my one-hundred-year wager on bioelectric implant companies.

While today's FDA-approved brain implants merely generate an electrical field within a specifically identified location, BMIs will record and stimulate individual neurons. As Edward Evarts pioneered the ability to perform "single unit testing" in lab animals, scientists today are trailblazing the path of recording single units in humans, with the goal of transmitting neurons' signals to a machine.

At present, there are few BMIs. The cochlear implant (with its external sensing device that looks like a beefy hearing aid, and an internal wire deep in the cochlear part of the inner ear) is a BMI, functioning as a receiving device and communicating the sound wave information as electronic signals to the cochlear portion of the auditory nerve. The artificial eye (in early development), will conceptually function like the cochlear implant, with an artificial sensing device that electronically interfaces with the nerves that process the sensory information.

As opposed to the cochlear implant and retinal implants, where there is replacement of a sensory organ, the (even) more advanced implants of the future will be "neural interface" devices, where electronic communication is both afferent (sensory) and efferent (outgoing signals). Efferent communications will one day come from microspecific parts of the brain, and neural interface technology will provide bi-directional information flow.

When you consider that the spinal cord is a conduit of microscopic transmission lines with one billion neurons,[3] is it any wonder that spinal cord injuries have been impossible to repair? As a resident on call for the hand service, I faced the daunting challenge of reconnecting the structures of the wrist following suicide attempts where the desperate patient had slashed her wrist with a knife, severing all nine tendons, two arteries, and the median nerve. This so-called "spaghetti wrist" is challenging because it is difficult to match the tendons that connect the finger muscles (that are surprisingly located in the forearm) to the digits. The surgeon must painstakingly decipher the jumble of tendons and pair them correctly; to cross up even one pair would garble hand function.

If matching those dozen structures is a surgical dilemma, how about one billion neurons of the spinal cord? This is why surgery for displaced vertebral fractures with spinal cord injuries is focused on stabilizing the bones, and never on "repairing the nerves," particulary when those nerve axons are one-quarter the diameter of spider silk.

Alternatively, because the precise location of the motor and sensory strips of the brain cortex are known, it will be likely in the future that spinal cord injuries will be addressed with a bodily house rewiring project (on a cosmic scale). Specialized sensory devices, called BioMEMs (Micro-Electro-Mechanical systems) are being developed that sense our brain's illicitation of movement; the BioMEMs then connect with peripheral nerves, bypassing the spinal cord. These miniaturized neural probes are "integrated with circuitry for amplification, multiplexing, spike detection, and the wireless transmission of power and bidirectional data," and "are facilitating prosthetic devices for many debilitating neurological disorders."[4]

BioMEMs, made possible with advancements in nanotechnology and electronic miniaturization, are transforming medicine. As surgery becomes increasingly minimally invasive, the prospect of treating devastating spinal cord injuries becomes more than a fantasy.

The story of the modernization of medicine usually starts with dread diseases, and then progresses along a path of less severe maladies, until we end up focusing on the inconveniences of life. Witness facial-plastic surgery, initially focused on grotesque and offensive injuries like syphilitic loss of the nose, but in time turning toward asthetic operations based upon fluctuating cultural tastes.

The use of BioMEMs in the brain will surely continue along a path of spinal cord injuries, stroke, brain tumors, cerebral palsy, and seizure, eventually transitioning to less detrimental ailments like early dementia, obsessiveness, moderate depression, and forgetfulness. I can predict with great certitude that over the next generation all these conditions will be treated with some type of brain implants to regulate neurologic function.

But what's next? When it comes to the brain, will we turn from disease to disability, and eventually to super-ability? Unquestionably.

As shown in this book, there are about twenty million implant operations per year in America. Much of our health care dollar revolves around medical devices, but hospitalizations for chronic illnesses represent hundreds of billions of dollars per year. There seems little doubt that with genomic purification, many chronic diseases (and cancer) will be wiped out—leaving mankind focused on achieving the Olympic ideals of *Citius*, *Altius*, and *Fortius* ("Faster, Higher, Stronger").

What are the limitations of our cyborg future? If (when) we cure chronic diseases, the impulse to maximize our physical and mental function will no doubt lead to a destiny where *man is machine*. That is not to say that every human will undergo open brain surgery for implantation of BioMEMs (although that may well happen), but perhaps an unknown process of mechanically embellishing the human mind by altering the physical brain is in our future.

This is a chilling thought to almost all of us. Do we want to live in a world where every human has a partly artificial self?

If you suffered an existential crisis over the relationship of Joaquin Phoenix's character in the motion picture *Her*, in which he falls in love with an operating system representation of a human female "Samantha," (voiced by Scarlett Johansson), you were not alone. It is unsettling to contemplate a reality of interfacing, and falling in love with, super-real computers. Many viewers found it disconcerting (or unbelievable) that someone could feel "love" toward a computer, given the fact that the human subject *knows* it is a faux paramour, but it does highlight the fact that relationships, although typically physical, are grounded in the life of the mind.

Perhaps more unnerving than a computer relationship (which is becoming technically feasible), is the human-robot liaison in *Ex Machina*, where one of the wealthiest men in the world has retreated to an extremely isolated home and research facility, inviting a young computer programmer, Caleb, to his outpost to judge the humanness of his robot creation Ava. Ava has a transparent torso and limbs that reveal her electronic innards, but despite this, the young male protagonist begins to sympathize, and even fall in love with, the convincingly real robot. As the movie progresses, it's tempting to root for the anthropomorphic Ava, even if you are asking yourself, "Would I be fooled in real life?" "The real story in this film is about a machine becoming a girl," says Alex Garland, the writer and director of *Ex Machina*. [5] To this author, Ava becomes a "girl" when we recognize that she has emotions and motives—perhaps most when those human impulses become dangerous.

Robots and Artificial Intelligence (AI) are frightening when they threaten our dominance and sense of control, particularly when that jurisdiction seems capable of being permanently lost with the "rise of machines." But for most of us, the concept of AI is mysterious. Because I cannot write a single line of computer programming code, I am incapable of understanding how the AI is configured and becomes dangerous, but to those who understand and create AI, its future is a certainty.

In some ways, AI is happening all around us—it's not an isolated software program being crafted in Cambridge, Massachusetts. Don't limit your thinking of AI to a computer being smartly programmed to play chess (a very specific type of AI), but consider the future of AI as *Artificial General Intelligence* (AGI), wherein machines will be able to learn, think, and (arguably) have a conscience.

The development of AGI is more important to the cyborg future of man than any implant invention, but there are surprising roadblocks along the way. For instance, a small child can easily discern the difference between a cat and a dog, but computers struggle with this comparison. Computer scientists, therefore, are delving deeply into the way our minds learn and process, expecting that a day will soon come when a collection of computers will finally be able to learn and outfox their programmers. The day this occurs, AGI will become unstoppable.

Today, the elaboration of AGI occurs in real time, all around us. "A company is essentially a cybernetic collective of people and machines . . . there is this collective AI in Google search, where we are all sort of plugged in like nodes in a network, like leaves on a tree. We are all feeding the network with our questions and our search. We are all collectively programming the AI. And Google (and Facebook, Twitter, and Instagram and the social networks), plus all the humans connected to it are all one giant cybernetic collective," claims Elon Musk. [6] The feedback loop reinforces the machines' capabilities, and potentiates all other machines' ability to process information. Remembering is easy—computers run laps around us at the speed of light—but cybernetic cogitation will make us ridiculously superhuman.

In a sense, you are already a cyborg. It is almost a guarantee that you, Dear Reader, have a smart phone, and that it is within reach of you at this very moment. Its memory is infinitely better than yours, particularly when you are connected to the World Wide Web. Forget the capital of Djibouti? Ask your device. (The capital of Djibouti is Djibouti, by the way.) Need a refresher about tying a bow tie, or you can't remember how much butter is in your grandmother's cinnamon roll recipe? You will know in seconds. What if everything in your phone could be placed, not at your fingertips, but inside your mind?

But, you ask, who in the world would be foolish enough to permit the implantation of a computer in their brain? Never gonna happen, you conclude.

The possibility of curing spinal cord injuries, tremors, and seizures with BioMEMs is, pardon the pun, electrifying. In the future, congenital conditions (like cerebral palsy) and acquired diseases (e.g., seizure) will be treated with advanced brain implants. Electroceuticals, implants that "precisely target the medical condition by controlling the neural signals going to a specific organ,"[7] are being tested for conditions that hardly seem to have an electric association, like rheumatoid arthritis and diabetes. Proof that electroceuticals are a real possibility? Pharma powers like GlaxoSmithKline are heavily invested.[8]

As our focus shifts from disease treatment to disease prevention and cure, mankind will no doubt transition to the creation of advanced physical (and mental) *Homo sapiens*. Our future progeny will be the beneficiaries of purified chromosomes and accentuated features. Not only better vision, hearing, dentition, and coronary artery health, but improved physical endurance, memory, and mood. The quantum leap forward, therefore, will be the implantation of neural probes for the dramatic improvement of humans.

If our forefather was *Homo erectus*, our descendant could be *"Homo electrus."*

Can there be any doubt that headphones will continue to go cordless, with the eventual implantation of speakers in our ear canals? A more sophisticated step would be direct wiring to the cochlea, or even the auditory nerve. Covert, even amplified, hearing is possible now.

Conversely, technology is being secretly, and intensely, developed that will sense our brain impulses and one day transmit our thoughts without us moving a muscle. Telepathic speaking, via neural probes to the language parts of our brain, must be possible. Sounds crazy? Plenty of smart people, including Elon Musk, think it is the way of the future.[9]

While movies like *Ex Machina* and *Her* are disquieting, triggering fears of computer control and robot domination, a more salient concern revolves around the future of human interaction in a cyborg future. It's more possible than you think. What will communication be like when everyone in your social sphere is electronically plugged in?

The future rise of the machines will not resemble a bourgeoning rogue robot army, but instead the transformation of humanity into *Homo electrus*. In the Terminator movie series, Skynet is the interconnected artificial intelligence among servers, androids, cyborgs, satellites, and war-machines. The thought of a world where the Terminator and Ava from *Ex Machina* are electronically coordinated is disturbing, as far-fetched as it seems. On our way to a future where *Homo electrus* "speaks" by electronically transmitting a message, there will surely be a day when our offspring will come face-to-face with humanoids who have spare, electronic stares. What will it feel like to coexist with a humankind that has, to paraphrase Timothy Leary, "turned on and tuned in" in a galvanic fashion?

We pine for life before smart phones, where people attended sporting events and concerts *in person*, eyes wide open and biologic senses at their peak. Now, our phones stand in the gap, neither comprehensively capturing the moment, nor allowing us to live *in* the moment. How strange will life be when those electronic gizmos are not in front of our eyes, but implanted into our brains?

I have to say I hope I am wrong. *Homo electrus* mostly freaks me out.

A 13.8-billion-year evolutionary transformation from the beginnings of the earth to the dawn of cyborgs has occurred in three main phases, according to Max Tegmark. Life 1.0 arrived about four billion years ago, and is characterized as simple biological existence, where organisms like bacteria have no ability to communicate and learn. Life 2.0 is modern humankind (appearing a hundred millennia ago), with advanced culture and awareness, and the ability to communicate and improve our "software."[10] Tegmark explains, "Even though the information in our human DNA hasn't evolved dramatically over the past fifty thousand years, the information collectively stored in our brains, books, and computers has exploded. By installing a software module enabling us to communicate through sophisticated spoken languages, we ensured that the most useful information stored in one person's brain could get copied to other brains, potentially surviving even after the original brain died."[11]

At this nascent stage of the implant revolution, it could be argued that we could be categorized as Life 2.1, with artificial joints and pacemakers, but without the dramatic cognitive hardware upgrade that is soon to come. "Many AI researchers think that Life 3.0 may arrive during the coming century, perhaps even during our lifetime, spawned by progress in AI."[12] The key characteristic of Life 3.0 is both a software and hardware upgrade, where cybernetic implants will enable humans to capture, process, communicate, and remember infinitely better than we can today.

The cybernetic organisms (cyborgs) of the near future will usher in the age of *Homo electrus*; the Singularity is near.

Ray Kurzweil, MIT graduate, famed inventor, and futurist, believes that we are only a few decades away from a time when there will be no distinction between human and machine, or between physical and virtual reality.[13] Borrowing a concept from physics and mathematics, *Singularity* is the idea that computer superintelligence will continue its exponential growth unabated, resulting in a future where machine brainpower becomes so powerful and alluring that mankind will have no choice but to partner with computers.

Kurzweil is no amateur futurist with his head in the clouds. This is the man who was the principle inventor of the first CCD flat-bed scanner, the first text-to-speech synthesizer, and the first commercially marketed large-vocabulary speech recognition software, among many other inventions. Using scores of mathematical models and technology assessments, Kurzweil comes to a startling conclusion: by 2045, Singularity will be achieved. "The essence of being human is not our limitations—although we do have many—it's our ability to reach beyond our limitations. We didn't stay on the ground. We didn't even stay of the planet. And we are already not settling for the limitations of our biology."[14]

The three components of the revolution will be "GNR," or Genetics, Nanotechnology, and Robotics. To me, Kurzweil's 2005 book, *The Singularity is Near*, sounds fantastical when he predicts the coming changes in genetic manipulation. Hardly a decade later, with CRISPR technology on the cusp of remedying genetic aberration, it seems that Kurzweil didn't dream fantastical enough. Once we become the ultimate genetically modified organism, augmented humanity with superhuman intelligence is right around the corner. This will occur, claims Kurzweil, with nanotechnology and robotic intelligence.

Biological evolution crafted a species that can think and manipulate its environment. The human species is now "succeeding in accessing—and improving—its own design and is capable of reconsidering and altering these basic tenets of biology."[15] Over the next quarter-century, as machines surpass our general intelligence, and become a billion times more capable of learning and remembering, we will incline toward Life 3.0. As non-biological intelligence transcends biological intelligence, humanity will, no doubt, reverse-engineer our cognitive life, "modeling it, simulating it, reinstating it on more capable substrates, and modifying and extending it . . ."[16] Because biological evolution moves so slowly, building everything from simplistic, folded proteins, all the while processing thought in a still-mysterious manner at a relatively glacial pace, Singularity will usher in an unquenchable thirst to think like a robot.

Utilizing nanotechnology and BioMEMS, first for disease and later for superhuman enhancement, it seems that the march toward a transhuman reality is likely. Lots of smart people think this is our destiny, and it will happen sooner than you think. In fact, the non-biological portion of humanity will ultimately predominate. One hundred years ago, one-third of the Western world was infected with tuberculosis, and it was impossible to imagine a life without those around you hacking and coughing as you plodded along manure-filled dirt roads. But today, I am sure you have never known a single person with TB.

The revolutions that have changed our world, including the industrial, transportation, and energy revolutions, happened more rapidly than anyone could have conceived of at their onset. But they will all pale in comparison to the upcoming phases of the implant revolution.

The final stages of the implant revolution will be the most sophisticated techno-logical revolution in the four-billion-year history of life on earth, because it will be a boutique enterprise of defying evolution and crafting our cyborg future.

The implant revolution had its roots in the founding of the genius societies in the 17th century, the birth of scientific surgery in the 18th century, and the reworking and perfection of the microscope in the early 19th century; the revolution gained steam with the genesis of chemistry, the solidification of germ theory, and the revelation of the organ basis and cellular basis of disease. In the half century between the first elec-tive operations for hernia and the introduction of penicillin in 1941, the first (failed) attempts at implantation occurred.

Two world wars flanked the invention of health insurance, preceding a postwar boom of hospital construction, financed by the US government. In an incredible concentration of developments, polymer sciences, transistors, modern alloys, and antibiotics were industrialized, thus necessitating the feedback loop of insurance and Medicare. *Implant materials, the money to pay for them, gleaming new operating rooms, and the antibiotics to make the surgeries safe launched the revolution.*

After seventy-five years of revolution, there are still glaring flaws in the delivery of medicine in America, and every country. But the dramatic improvement in the treat-ment of arthritis, heart disease, stroke, abdominal organ disease, scoliosis, urinary incontinence, hearing loss, Parkinson's, cancer, and hundreds of other conditions was built upon the foundation of implants. While often dramatically effective, these inter-ventions are exceptionally expensive compared to the old-fashioned "benign neglect" of 18th century medicine, and it is still all too common that implants are not evaluated with enough scientific rigor.

Charles Darwin, in his iconoclastic *On the Origin of Species*, summarized his project in a striking final paragraph:

> Thus, from the war of nature, from famine and death, the most exalted object which we are capable of conceiving, namely, the production of the higher animals, directly follows. There is grandeur in this view of life, with its several powers, having been originally breathed by the Creator into a few forms or into one; and that, whilst this planet has gone cycling on according to the fixed law of gravity, from so simple a beginning end-less forms most beautiful and most wonderful have been, and are being evolved. [17]

Evolution, glacial in its attainment, hid itself from every scientist's view until an extraordinarily perceptive naturalist grasped its grandeur. The inventiveness of Leonardo, the ingenuity of Newton, the imagination of Einstein, and the perception of Darwin could not possibly have conceived of the *"most exalted object"* that will be *Homo electrus*. Millions of years of slight adaptations and, depending upon your philosophical outlook, accidental genetic modifications, can be upended in a matter of hours in a Cambridge laboratory or London operating room. It seems likely that most disease-treating implants will no longer be needed once we conquer degeneration and genetically-dependent disease. Today, the cosmos of implant manufacturers, surgeons, hospitals, insurance companies, governments, and governing bodies are endeavoring together, using a line and race of inventions to alleviate the necessities of life. In the near future, these selfsame federations will collaborate and defy the laws of gravity, shake off the limitations of being *Homo sapiens*, and powerfully embrace a grand new life.

Acknowledgments

There are no eureka moments. At least that's what we are supposed to conclude in this post-modern world, but it simply cannot be true.

When I first starting dating Wendy—for better or worse—she quickly realized that I am an inveterate storyteller, and it didn't take long for her to predict that I would one day be a writer. My three-decade detour in science and surgery was preparation for this book, with countless hours in the world's greatest libraries and sleepless nights responding to editor's critiques when I was still in the business of cranking out research publications in the orthopedic literature. Today, I only rarely conduct scientific experimentation, but maintain a busy orthopedic practice, specializing in shoulder and elbow surgery.

My eureka moment came one day in the shower. I had started this project as an impulse to explain modern surgery to the general public and to medical practitioners alike. I kept stumbling into intriguing stories of the greats of surgery, realizing it *was* possible to link together their lives in the assault against disease. As the warm shower water gushed over my head early one morning, a thought occurred to me: we are in the midst of a revolution.

Sometimes revolutions ignite with the overthrow of a government or the invention of a machine, and there is no denying that you are drowning in upheaval. Other times, it is not possible to detect that you are living in transcendent times without someone pointing it out. I challenge you: name more than a few people you know who have not had surgery. It's almost not possible. And if you've had surgery, you almost certainly have a permanent man-made bit of material inside you—an impossibility back in 1941.

Thus, the implant revolution.

Wendy predicted that I would write a book, but it took the nudging of several friends over the last decade to jump start the project. At a high school basketball game, Al Kileen earnestly looked at me and insisted I get to work. At a Super Bowl party, Michael Mason provided the spark that awakened the long-lost passion to write, and then provided a forum in This Land to kindle the flame. I can't thank Michael enough. My dear sister-in-law Elizabeth Garnsey, a gifted writer and keen editor, gave me early encouragement and genuine, constructive criticism. The renowned Jonathan Cott leant his ear and his affirmation over the last five years. His considerable gift with words is superseded by his loyalty and thoughtfulness, and this book (and any that follow) simply would not exist without Jonathan.

Like every writer, I am the product of my upbringing. Growing up in Manhattan, Kansas I was spoiled to have a few incredible teachers, like my sixth grade English teacher, Mrs. Frazier, and my eighth grade English teacher, Mr. Coleman. Dr. Kremer, my high school British literature teacher was ruthless, and excellent. In college, Mark Williams was an inspiration.

My first medical hero was Jeff Holtgrewe. In 1988, he started being my role model in every way, and anchored what I thought it meant to be an orthopedic surgeon. When medical school and residency were difficult, it was his example that served as an inspirational lifeline, even as he battled serious illness.

I feel so deeply grateful to my medical mentors, particularly H. Clarke Anderson, Vince Pellegrini, and Neal ElAttrache. Fred Reckling, Kevin Black, Spence Reid, Paul Juliano, Sanjiv Naidu, and James Tibone were also monumental in my surgical training.

My orthopedic surgery partners in Colorado have been amazingly supportive of my "hobby" and numerous research trips to Europe and around the United States. In particular, Jared Foran, Michael Ellman, Mitch Robinson, Mark Mills, Ed Rowland, Nimesh Patel, Sameer Lodha, Doug Foulk, Pete Deol, Jim Johnson, Chuck Gottlob, Ron Hugate, Lonnie Loutzenhiser, and John Froelich have been valuable sounding boards and enthusiastic supporters. My friends in the medical device industry have my undying respect, particularly those who reject the status quo. I detest AdvaMed's pedantic falseness regarding the relationship between physicians and corporations, and I assume that thanking world-class entrepreneurs is somehow wrong in their world . . . so I won't. But, for those engineers and business people who are leading the charge: thank you!

Any man is lucky to have a "best friend" throughout life. To start with, my twin brother Daniel has been an incredible wingman for life. He's smart and tough and loyal. Pals like Doug Burton, Rick Kanemasu, and Daniel Wallace are rare. Todd Louis has been my closest friend for thirty years and was the first friend who *knew*, like I knew, that I was meant to be a surgeon. My earnest wish for my own children is that they have (and be) a friend like Todd. Friends like Mark Moulton and Jeff Yanovitch know the meaning of unconditional love. Here in Boulder, my closest friend is Stuart Crespi, a man of integrity and loyalty and my brother from another mother. At the end of *It's a Wonderful Life*, the angel Clarence inscribes his gift copy of *Tom Sawyer* with the phrase, "Remember no man is a failure who has friends." Like George Bailey, I feel like the richest man in town.

You can't write a comprehensive book of history and have a busy surgical practice without an amazing team. Jodi Simcik, Kristy Cooper Neville, Paul Lee, Abby Price, and Ashley Nicholson not only work their fingers to the bone stamping out shoulder and elbow disease, but have always been my test audience for anecdotes and research findings. I am indebted for life!

As a rookie, every step of publishing a book has been a revelation. Profound thanks to Michael Mungiello and Michael Carlisle at InkWell for taking on a greenhorn. Hoping this is the first of several books! To the crew at Pegasus—Claiborne Hancock, Maria Fernandez, Jessica Case—thank you for taking a chance on me! I hope to make you ecstatic! To my editor Drew Wheeler- thank you for honing the manuscript and your kind editorial assistance.

This book is dedicated to my father, J. E. "Gene" Schneider, a warrior, scientist, veterinary surgeon, dedicated father and grandfather, and man of deep faith. He's been gone for more than a decade and I miss him every day. There was no end for his pride in his kids and I'm quite sure he would be peddling this book on Pearl Street in Boulder were he alive today. My mother, Judith Schneider, is still exploring our world at age eighty-six. Her curiosity and refusal to be turned away at closed museums and concert halls were contagious. I am her proud and loving son. My siblings (and in-laws) Mark and Lynne Schneider, Ben and Rochelle Platter, Daniel and Gisella Schneider, Tim and Jenny Brynteson, Elizabeth (and Charlie!) Garnsey, and Herrick and Diane Garnsey have been an enormous source of inspiration and wisdom.

Wendy Garnsey Schneider and I have been together for over thirty years. Our four adult children, Emily, Luke, Jonathan, and Jennifer, are my greatest pride and joy. In the final analysis, their approval and estimation of my job as a father is what I live for. I wish them the best of what life has to offer. Each of our four kids is uniquely talented and fated to accomplish something special. I can't wait to see whom they become.

Finally, to Wendy, my soulmate and best friend. We were college kids when we first met and it didn't take long to fall deeply in love. Often, surgeons marry after all the arduous training is completed, but we married each other a few weeks before medical school started. Year after challenging year, Wendy saw me at my best, and more often, at my worst. Despite the unbelievable demands of med school and residency, our marriage (and budding family) flourished, even as we survived the low times. My life has been vastly improved by your presence and undying support, and I will never be able to describe my adoration and admiration for your wisdom, love, and support. Looks like you're stuck with me forever! Now, on to the next adventure.

—David Schneider

Endnotes

INTRODUCTION

1   Galen, *On the Natural Faculties*, trans. Arthur John Brock (London: William Heinemann, 1928), p. 279.

2   Walter Isaacson, *Leonardo da Vinci* (New York: Simon & Schuster, 2017), p. 9.

3   Stephen Greenblatt, *The Swerve: How the World Became Modern* (New York: W.W. Norton, 2011), p. x.

4   Ibid., p. 186.

5   Ibid., p. x.

6   Galen, *On the Natural Faculties*, trans. Arthur John Brock (London: William Heinemann, 1928), p. x.

7   Owsei Temkin, *Hippocrates in a World of Pagans and Christians* (Baltimore: Johns Hopkins University Press, 1991), p. 10.

8   Siddhartha Mukherjee, *The Emperor of All Maladies* (New York: Scribner, 2010), p. x.

9   Owsei Temkin, *Hippocrates in a World of Pagans and Christians* (Baltimore: Johns Hopkins University Press, 1991), p. x.

10 David Wootton, *Bad Medicine: Doctors Doing Harm Since Hippocrates* (Oxford, UK: Oxford University Press, 2006), p. 42.

11 Ibid., p. 31.

12 Owsei Temkin, *Hippocrates in a World of Pagans and Christians* (Baltimore: Johns Hopkins University Press, 1991), p. 11.

13 Ibid., p. 5.

14 Steven Johnson, *How We Got to Now: Six Innovations That Made the Modern World* (New York: Riverhead Books, 2014), pp. 5–6.

15 Owsei Temkin, *Galenism: Rise and Decline of a Medical Philosophy* (Ithaca, N.Y.: Cornell University Press, 1973), p. 14.

16 Owsei Temkin, *Hippocrates in a World of Pagans and Christians* (Baltimore: Johns Hopkins University Press, 1991), p. 3.

17 Ibid., p. 4.

18 Galen, *On the Natural Faculties*, trans. Arthur John Brock (London: William Heinemann, 1928), p. xix.

ONE: DILEMMA

1   R. I. Harris, "Arthrodesis for Tuberculosis of the Hip," *Journal of Bone and Joint Surgery*, vol. 17, No. 2, 1935

2   E. A. Codman, *The Shoulder* (Boston: Thomas Todd Co., 1934).

3   Charles S. Neer, *Shoulder Reconstruction* (New York: W.B. Saunders, 1990), p. vii.

4   Ibid., p. 146.

5   E. A. Codman, *The Shoulder* (Boston: Thomas Todd Co., 1934), p. 331.

6   Arthur Steindler, *The Traumatic Deformities and Disabilities of the Upper Extremity* (Springfield, Ill.: Charles C. Thomas, 1946), p. 126.

7   A. F. DePalma, *Surgery of the Shoulder* (Philadelphia: J.B. Lippincott, 1950), p. 272.

8   Ibid., p. 423.

9   C. S. Neer, T. H. Brown, H. L. McLaughlin, "Fracture of the neck of the humerus with dislocation of the head fragment," *American Journal of Surgery*, March 1953, pp. 252–58.

TWO: PAPER, PROPHET, AND PRINTING PRESS

1   Sven Beckert, *Empire of Cotton. A Global History* (New York: Knopf, 2014).

2   Matt Ridley, *The Evolution of Everything: How New Ideas Emerge* (New York: Harper Collins, 2015), p. 120.

3   Ibid., p. 125.

4   Kumar Srivastava, "The 'Adjacent Possible' of Big Data: What Evolution Teaches about Insights Generation," *Wired*, Dec. 2014.

5   Steven Johnson, *Where Good Ideas Come From: The Natural History of Innovation* (New York: Riverhead Books, 2010), p. 31.

6   Elizabeth Eisenstein, *The Printing Revolution in Early Modern Europe* (Cambridge, UK: Cambridge University Press, 1983), p. 4.

7   Mark Kurlansky, *Paper: Paging Through History* (New York: W.W. Norton, 2016), p. 13.

8   Craig Kallendorf, "Ancient Book," in *The Book: A Global History*, M. F. Suarez and H. R. Woudhuysen, eds. (Oxford, UK: Oxford University Press, 2013), p. 49.

9   Mark Kurlansky, *Paper: Paging Through History* (New York: W.W. Norton, 2016), p. 14.

10 John Man, *Gutenberg: How One Man Remade the World with Words* (New York: MJF Books, 2002), p. 24.

11 Ibid., p. 48.

12 Ibid., p. 124.

13 Ibid., p. 124.

14 Ibid., p. 164.

15 Ibid., p. 8.

16 Mark Kurlansky, *Paper: Paging Through History* (New York: W.W. Norton, 2016), p. 51.

17 Ibid., p. 26.

18 Ibid., p. 160.

19 Steven Weinberg, *To Explain the World: The Discovery of Modern Science* (New York: HarperCollins, 2015), p. 101.

20 Ibid., p. 104.

21 P. K. Hitti, *History of the Arabs* (London: Macmillan, 1937), p. 315.

22 Hillel Ofek, "Why the Arabic World Turned Away from Science," *The New Atlantis*, Winter 2011, pp. 3–23.

23 Ibid., p. 50.

24 Hillel Ofek, "Why the Arabic World Turned Away from Science," *The New Atlantis*, Winter 2011, p. 7.

25 David Wootton, *Bad Medicine: Doctors Doing Harm Since Hippocrates* (Oxford, UK: Oxford University Press, 2006), p. 50.

26 Michael Flannery, Avicenna entry, *Encyclopedia Britannica* online, quoted August 11, 2016.

27   Sherwin Nuland, *Doctors: The Biography of Medicine* (New York: Vintage Books, 1988), p. 57.

28   Steven Weinberg, *To Explain the World: The Discovery of Modern Science* (New York: HarperCollins, 2015), p. 112.

29   Charles Burnett and Danielle Jacquart, eds., *Constantine the African and Ali ibn al-Abbas al-Magusi; The* Pantegni *and Related Texts* (Leiden, Netherlands: E. J. Brill, 1994), Preface vii–viii.

30   Nicholas Ostler, *Ad Infinitum: A Biography of Latin* (New York: HarperPress, 2009), p. 211.

31   David Osborn, "Constantine the African and Gerard of Cremona," in GreekMedicine .Net, quoted August 20, 2016, http://www.greekmedicine.net/whos_who/Constantine _the_African_Gerard_of_Cremona.html.

32   Christopher de Hamel, "The European Medieval Book," in *The Book: A Global History*, M. F. Suarez and H. R. Woudhuysen, eds. (Oxford, UK: Oxford University Press, 2013), p. 59.

33   John Man, *Gutenberg: How One Man Remade the World with Words* (New York: MJF Books, 2002), p. 88.

THREE: VESALIUS AND *DE HUMANI CORPORIS FABRICA*

1   David Wootton, *The Invention of Science: A New History of the Scientific Revolution* (New York: HarperCollins, 2015), p. 58.

2   Ibid., p. 106.

3   Ibid., p. 75.

4   Ibid., p. 78.

5   Paul Strathern, *The Medici: Power, Money, and Ambition in the Italian Renaissance* (New York: Pegasus Books, 2016), p. 46.

6   Steven Johnson, *How We Got to Now: Six Innovations That Made the Modern World* (New York: Riverhead Books, 2014), p. 17.

7   Ibid., p. 19.

8   Ibid., p. 32.

9   Ibid.

10   Ibid., p. 8.

11   Lewis Mumford, *Technics and Civilization* (Chicago: University of Chicago Press, 2010), p. 129.

12   Steven Johnson, *How We Got to Now: Six Innovations That Made the Modern World* (New York: Riverhead Books, 2014), p. 35.

13   C. D. O'Malley, *Andreas Vesalius of Brussels 1514–1564* (Berkeley: University of California Press, 1964), p. 6.

14   Ibid., p. 10.

15   Ibid., p. 14.

16   Ibid.

17   Ibid., p. 19.

18   Ibid., p. 20.

19   Ibid., p. 44.

20   Ibid., p. 49.

21   Ibid., p. 59.

22   Ibid., p. 64.

23   Ibid., p. 77.
24   Ibid., p. 106.
25   Ibid., p. 113.
26   Ibid., p. 114.
27   Ibid., p. 321.
28   Ibid., p. 317.
29   Ibid., p. 318.
30   Ibid.
31   Ibid., p. 323.
32   S. W. Lambert, W. Wiegand, and W. M. Ivins, *Three Vesalian Essays to Accompany the Icones Anatomicae of 1934* (New York: Macmillan, 1952), p. 27.
33   Ibid., pp. 3–24.
34   C. D. O'Malley, *Andreas Vesalius of Brussels 1514–1564* (Berkeley: University of California Press, 1964), p. 323.

FOUR: THE RISE OF SCIENCE

1   Thomas Sprat, *The History of the Royal Society of London for the Improving of Natural Knowledge* (London: 1667), p. 53.
2   David Wootton, *The Invention of Science: A New History of the Scientific Revolution* (New York: HarperCollins, 2015), p. 24.
3   Ibid., p. 12.
4   Ibid., p. 199.
5   Galilei, Galileo, *Sidereus Nuncius, or the Sidereal Messenger*, Albert van Heiden (trans) (Chicago: University of Chicago Press, 2016), p. 6.
6   David Wootton, *The Invention of Science: A New History of the Scientific Revolution* (New York: HarperCollins, 2015), p. 215.
7   Ibid., p. 39.
8   Perez Zagorin, *Francis Bacon* (Princeton, NJ: Princeton University Press, 1998), p. 122.
9   Ibid., p. 3.
10   John Sutton, *Encyclopedia of the Life Sciences* (New York: Macmillan, 2001), p. 471.
11   David Wootton, *The Invention of Science: A New History of the Scientific Revolution* (New York: HarperCollins, 2015), p. 83.
12   Ibid., p. 75.
13   Perez Zagorin, *Francis Bacon* (Princeton, NJ: Princeton University Press, 1998), p. 79.
14   Ibid., p. 3.
15   David Wootton, *The Invention of Science: A New History of the Scientific Revolution* (New York: HarperCollins, 2015), p. 84.
16   Perez Zagorin, *Francis Bacon* (Princeton, NJ: Princeton University Press, 1998), p. 100.
17   Ibid., p. 123.
18   Francis Bacon, *New Atlantis* (1627), 5:415.
19   Perez Zagorin, *Francis Bacon* (Princeton, NJ: Princeton University Press, 1998), p. 123.
20   Ibid., p. 224.
21   Bill Bryson, ed., *Seeing Further: The Story of Science and the Royal Society* (London, HarperPress, 2010), p. 9.
22   David Wootton, *The Invention of Science: A New History of the Scientific Revolution* (New York: HarperCollins, 2015), p. 35.

23   Bill Bryson, ed., *Seeing Further: The Story of Science and the Royal Society* (London, Harper Press, 2010), p. 3.

24   James Gleick, *Isaac Newton* (New York: Harper Perennial, 2004), p. 3.

25   Edward Dolnick, *The Clockwork Universe: Isaac Newton, the Royal Society, and the Birth of the Modern World* (New York: HarperCollins, 2011), p. 5.

26   Bill Bryson, ed., *Seeing Further: The Story of Science and the Royal Society* (London, HarperPress, 2010), p. 33.

27   Gerek Gjertsen, *The Newton Handbook* (London: Routledge Kegan & Paul, 1987), p. 24.

28   Matthew Green, http://www.telegraph.co.uk/travel/destinations/europe/united -kingdom/england/london/articles/London-cafes-the-surprising-history-of-Londons -lost-coffeehouses/. Accessed October 9, 2019.

29   Ibid.

30   John Maynard Keynes, quoted in James Gleick, *Isaac Newton* (New York: Harper Perennial, 2004), p. 188.

31   Perez Zagorin, *Francis Bacon* (Princeton, NJ: Princeton University Press, 1998), p. 95.

FIVE: HARVEY AND HUNTER

1   Stephen Paget, *John Hunter, Man of Science and Surgeon* (London: Fischer Unwin, 1924), p. 27.

2   Wendy Moore, *The Knife Man: Blood, Body Snatching, and the Birth of Modern Surgery* (New York: Broadway Books, 2005), p. 177.

3   Thomas Wright, *Circulation: William Harvey's Revolutionary Idea* (London: Vintage, 2013), pp. 41–42.

4   Ibid., p. 91.

5   Ibid., p. 119.

6   Ibid., p. 110.

7   Ibid., p. 121.

8   Ibid., p. xiii.

9   Finch, Ernest, "The Influence of the Hunters on Medical Education," *Annals of the Royal College of Surgeons of England*, 1957, vol. 20, pp. 205–48.

10   Wendy Moore, *The Knife Man: Blood, Body Snatching, and the Birth of Modern Surgery* (New York: Broadway Books, 2005), p. 14.

11   Ibid.

12   Ibid., p. 28.

13   Ibid., p. 37.

14   Ibid., p. 39.

15   Ibid., p. 41.

16   Ibid., p. 43.

17   William Hunter, Two Introductory Lectures (London: printed on the order of the trustees for J. Johnson, 1784), p. 73.

18   John Hunter, *The Works of John Hunter*, ed. James Palmer, vol. 4, (London: Longman, Rees, Orme, Brown, Breem, 1835), pp. 81–116.

19   John Hunter, *Essays and Observations on Natural History, Anatomy, Physiology, Psychology and Geology*, ed. Richard Owen, (London: John Van Voorst, 1861), vol. 1, p. 189.

20   Megan Oaten, Richard Stevenson, et al., "Disgust as a Disease-Avoidance Mechanism," *Psychological Bulletin*, vol. 135, No. 2, pp. 303–21.

21   Wendy Moore, *The Knife Man: Blood, Body Snatching, and the Birth of Modern Surgery* (New York: Broadway Books, 2005), p. 62.

22   Ibid., p. 149.

23   Benjamin Franklin, *The Autobiography of Benjamin Franklin* (New York: P.F. Collier, 1909), p. 157.

24   http://www.archives.upenn.edu/people/1700s/shippen_wm.html. Accessed October 9, 2019.

25   Betsy Copping Corner, ed. *William Shippen Jr., Pioneer in American Medical Education, With Notes, and the Original Text of His Edinburgh Dissertation, 1761,* (Philadelphia, PA: American Philosophical Society, 1951), p. 7.

26   Wendy Moore, *The Knife Man: Blood, Body Snatching, and the Birth of Modern Surgery* (New York: Broadway Books, 2005), p. 84.

27   Jessé Foot, *The Life of John Hunter* (London: T. Becket, 1794), pp. 81–2.

28   Wendy Moore, *The Knife Man: Blood, Body Snatching, and the Birth of Modern Surgery* (New York: Broadway Books, 2005), p. 7.

29   Ibid., p. 89.

30   Ibid.

31   Ibid., p. 112.

32   Royal Society Journal Book Copy, vol. 26, 1767–1770, February 5, 1767 (no page numbers).

33   John Hunter, *The Works of John Hunter*, ed. James Palmer, vol. 4, (London: Longman, Rees, Orme, Brown, Breem, 1835), p. 417.

34   Ibid., pp. 417–19.

35   Wendy Moore, *The Knife Man: Blood, Body Snatching, and the Birth of Modern Surgery* (New York: Broadway Books, 2005), p. 6.

36   Ibid., p. 176.

37   Ibid., p. 177.

38   Ibid.

39   Ibid., p. 269.

40   Ibid., p. 170.

41   Ibid., p. 171.

42   Ibid., p. 223.

43   Thomas Wright, *Circulation: William Harvey's Revolutionary Idea* (London: Vintage, 2013), p. 225.

44   Ibid.

SIX: PATHOLOGY

1   Sherwin Nuland, *Doctors: The Biography of Medicine* (New York: Vintage Books, 1995), p. 156.

2   Ibid., p. 157.

3   Ibid., p. 159.

4   Ibid., p. 147.

5   Rudolf Virchow, "Morgagni and the Anatomic Concept," *Bulletin of the History of Medicine*, Oct. 1939; vol. 7, pp. 975-90.

6   Antoni Lewenhoeck, "De Natis'e E Semine Genitali Animalculis," *Philosophical Transactions* (1665–1678). 1753-01-01. 12:1040–1046.

7   Catherine Wilson, *The Invisible World: Early Modern Philosophy and the Invention of the Microscope* (Princeton, New Jersey: Princeton University Press, 1995), p. 36.

8   Ibid., p. 37.

9   Bernard de Fontenelle, p. 9. https://books.google.com/books?id=VOqbtFnjR0C&print
    sec=frontcover&source=gbs_ge_summary_r&cad=0#v=onepage&q&f=false. Accessed
    October 9, 2019.

10 https://www.nationalgallery.org.uk/paintings/vincent-van-gogh-sunflowers. Accessed
    October 9, 2019.

11 http://ursula.chem.yale.edu/~chem220/chem220js/STUDYAIDS/history/chemists
    /perkin.html. Accessed October 9, 2019.

12 "The Top Pharmaceuticals that Changed the World," *Chemical and Engineering News*, vol.
    83, Issue 25, June 2005, https://pubs.acs.org/cen/coverstory/83/8325/8325emergence.html.
    Accessed October 9, 2019.

13 S. I. Hajdu, "Microscopic contributions of pioneer pathologists," *Annals of Clinical &
    Laboratory Science*, vol. 41(2), 2011, p. 201.

14 R. Ali Faisal, et al., "Hematoxylin in History—The Heritage of Histology," *JAMA Der-
    matology*, 2017, 153(3), p. 328.

15 Gary W. Gill, *Cytopreparation: Principles & Practice; Essentials in Cytopathology* (New
    York: Springer, 2012), p. 207.

16 Johannes Steudel, Johannes Müller, German Physiologist, in *Encyclopedia Britannica* online,
    https://www.britannica.com/biography/Johannes-Muller. Accessed October 9, 2019.

17 Sherwin Nuland, *Doctors: The Biography of Medicine* (New York: Vintage Books, 1995),
    p. 310.

18 Ibid., p. 320.

19 Ibid., p. 306.

20 Ibid., p. 307.

21 Ibid., p. 325.

SEVEN: GERMS

1   http://en.muvs.org/topic/the-gate-for-the-secretly-pregnant.pdf. Accessed October 9,
    2019.

2   Sherwin Nuland, *The Doctor's Plague: Germs, Childbed Fever, and the Strange Story of Ignác
    Semmelweis* (New York: Atlas Books, 2003), p. 96.

3   Ibid., p. 96.

4   Ibid., p. 94.

5   Ibid., p. 90.

6   Ibid., pp. 99–100.

7   Ibid., p. 100.

8   Edward Huth and T. J. Murray, *Medicine in Quotations: Views of Health and Disease
    Through the Ages* (Philadelphia: American College of Physicians, 2006), p. 176.

9   Lane, Nick "The Unseen World: Reflections on Leeuwenhoek (1677) 'Concerning Little
    Animals'" *Philosophical Transactions of the Royal Society*, 370: 20140344, 2015, pp. 1–10.

10 Sherwin Nuland, *The Doctor's Plague: Germs, Childbed Fever, and the Strange Story of Ignác
    Semmelweis* (New York: Atlas Books, 2003), p. 156.

11 Ibid., pp. 159–61.

12 Thomas Hodgkin, "On Some Morbid Appearances of the Absorbent Glands and Spleen"
    Medico-Chirurgical Transactions, 1832. 17:68–114.

13 Sherwin Nuland, *Doctors: The Biography of Medicine* (New York: Vintage Books, 1995),
    p. 352.

14   David Wootton, *Bad Medicine: Doctors Doing Harm Since Hippocrates* (Oxford, UK: Oxford University Press, 2006), p. 234.

15   Sherwin Nuland, *Doctors: The Biography of Medicine* (New York: Vintage Books, 1995), p. 354.

16   Ibid., p. 355.

17   Ibid., p. 356.

18   Richard A. Fisher, *Joseph Lister 1827–1912* (New York: Stein and Day, 1977), p. 52.

19   Melvin Santer, *Confronting Contagion: Our Evolving Understanding of Disease* (Oxford, UK: Oxford University Press, 2014), p. 211.

20   Jacob Henle, *On Miasmata and Contagia*, trans. George Rosen (Baltimore: Johns Hopkins Press, 1938), p. 14.

21   Ibid., p. 19.

22   Richard A. Fisher, *Joseph Lister 1827–1912* (New York: Stein and Day, 1977), p. 132.

23   Sherwin Nuland, *Doctors: The Biography of Medicine* (New York: Vintage Books, 1995), p. 362.

24   Ibid., p. 363.

25   Richard A. Fisher, *Joseph Lister 1827–1912* (New York: Stein and Day, 1977), p. 134.

26   Ibid.

27   Edwin S. Gaillard, *The American Medical Weekly*, vols. 8–9, 1878, p. 243.

28   Richard A. Fisher, *Joseph Lister 1827–1912* (New York: Stein and Day, 1977), p. 131–2.

29   Francis Darwin, *The Eugenics Review*, vol. 6:1, 1914, p. 1.

30   Thomas Goetz, *The Remedy: Robert Koch, Arthur Conan Doyle, and the Quest to Cure Tuberculosis* (New York: Gotham Books, 2014), p. 11.

31   Ibid., p. 13.

32   Ibid., p. 6.

33   http://www.merckvetmanual.com/generalized-conditions/anthrax/overview-of-anthrax. Accessed July 23, 2017.

34   Thomas Goetz, *The Remedy: Robert Koch, Arthur Conan Doyle, and the Quest to Cure Tuberculosis* (New York: Gotham Books, 2014), p. 23.

35   Jacob Henle, *On Miasmata and Contagia*, trans. George Rosen (Baltimore: Johns Hopkins Press, 1938), p. 42.

36   Thomas Goetz, *The Remedy: Robert Koch, Arthur Conan Doyle, and the Quest to Cure Tuberculosis* (New York: Gotham Books, 2014), p. 39.

37   Ibid., p. 40.

38   https://www.britannica.com/biography/Robert-Koch. Accessed July 29, 2017.

39   Thomas Goetz, *The Remedy: Robert Koch, Arthur Conan Doyle, and the Quest to Cure Tuberculosis* (New York: Gotham Books, 2014), p. 87.

40   E. Cambau and M. Drancourt, "Steps Towards the Discovery of Mycobacterium Tuberculosis by Robert Koch, 1882" *Clinical Microbiology and Infection*, vol. 20, Issue 3, March 2014, pp. 196–201.

41   Thomas Goetz, *The Remedy: Robert Koch, Arthur Conan Doyle, and the Quest to Cure Tuberculosis* (New York: Gotham Books, 2014), p. 88.

42   Ibid., p. 87.

43   David Wootton, *Bad Medicine: Doctors Doing Harm Since Hippocrates* (Oxford, UK: Oxford University Press, 2006), p. 227.

44   Sherwin Nuland, *Doctors: The Biography of Medicine* (New York: Vintage Books, 1995), p. 379.

EIGHT: ANTIBIOTICS

1   William Rosen, *Miracle Cure: The Creation of Antibiotics and the Birth of Modern Medicine* (New York: Viking, 2017), p. 41.

2   Ibid., p. 39.

3   H. Maruta, "From chemotherapy to signal therapy (1909–2009): A century pioneered by Paul Ehrlich," *Drug Discoveries Therapeutics*, 2009; 3(2): 37–40.

4   William Rosen, *Miracle Cure: The Creation of Antibiotics and the Birth of Modern Medicine* (New York: Viking, 2017), p. 57.

5   Ibid., p. 62.

6   Ibid., p. 63.

7   William Rosen, *Miracle Cure: The Creation of Antibiotics and the Birth of Modern Medicine* (New York: Viking, 2017), p. 53.

8   Ibid., p. 63.

9   Ibid., p. 68.

10   Ibid., p. 107.

11   Ibid., p. 113.

12   Eric Lax, *The Mold in Dr. Florey's Coat: The Story of the Penicillin Miracle* (New York: Henry Holt and Company, 2015), chapter 8.

13   William Rosen, *Miracle Cure: The Creation of Antibiotics and the Birth of Modern Medicine* (New York: Viking, 2017), p. 131.

14   Ibid., p. 135.

15   Robert Bud, *Penicillin: Triumph and Tragedy* (Oxford, UK: Oxford University Press, 2007), p. 36.

16   William Rosen, *Miracle Cure: The Creation of Antibiotics and the Birth of Modern Medicine* (New York: Viking, 2017), p. 135.

17   Paul Starr, *The Social Transformation of American Medicine: The Rise of a Sovereign Profession and the Making of a Vast Industry* (New York: Basic Books, 1982), p. 341.

18   William Rosen, *Miracle Cure: The Creation of Antibiotics and the Birth of Modern Medicine* (New York: Viking, 2017), p. 192.

19   Ibid.

20   Selman Waksman and H. Boyd Woodruff, "Streptothricin, a New Selective Bacteriostatic and Bactericidal Agent, Particularly Active Against Gram-Negative Bacteria" *Proceedings of the Society for Experimental Biology and Medicine*, Feb. 1, 1942, 49(2), pp. 207–10.

21   Albert Schatz, Elizabeth Bugle, and Selman Waksman "Streptomycin, A Substance Exhibiting Antibiotic Activity Against Gram-Positive and Gram-Negative Bacteria," *Proceedings of the Society for Experimental Biology and Medicine*, Jan. 1, 1944, 55(1) pp. 66–9.

22   William Rosen, *Miracle Cure: The Creation of Antibiotics and the Birth of Modern Medicine* (New York: Viking, 2017), p. 211.

23   Ibid., p. 268.

24   Ibid., p. 303.

25   Paul Starr, *The Social Transformation of American Medicine: The Rise of a Sovereign Profession and the Making of a Vast Industry* (New York: Basic Books, 1982), p. 336.

26   William Rosen, *Miracle Cure: The Creation of Antibiotics and the Birth of Modern Medicine* (New York: Viking, 2017), p. 256.

NINE: ANESTHESIA

1   William Mayo, Collected Papers of the Mayo Clinic and the Mayo Foundation, vol. 13, (New York, Saunders, 1922), p. 1274.

2   J. Ashhurst Jr., "Surgery Before the Days of Anesthesia," in J. C. Warren, J. C. White, W. I. Richardson, H. H. Beach, F. C. Shattuck, W. S. Bigelow, eds. *The Semi-Centennial of Anesthesia*, October 16, 1846–October 16, 1896, (Boston: Massachusetts General Hospital, 1897), 27–37.

3   Ann Ellis Hanson, "'Your mother nursed you with bile': anger in babies and small children," in Susanna Braund, and Glenn W. Most, eds., *Ancient Anger, Perspectives from Homer to Galen* (Cambridge, UK: University of Cambridge Press, 2004), p. 185.

4   https://www.greekmyths-greekmythology.com/morpheus-the-god-of-dreams/. Accessed October 4, 2018.

5   M. L. Meldrum, "A capsule history of pain management," *JAMA*, 290(18), Nov. 12, 2003.

6   Ibid., p. 2.

7   Sherwin Nuland, *The Origins of Anesthesia* (Birmingham: Classics of Modern Medicine, 1983), p. 25.

8   https://www.acs.org/content/acs/en/education/whatischemistry/landmarks/joseph priestleyoxygen.html. Accessed October 9, 2019.

9   Ibid.

10 Ibid.

11 Henry Guerlac, "Joseph Black and Fixed Air, a Bicentenary Retrospective, with some New or Little Known Material," *Isis*, vol. 48, No. 2, 1957, p. 125.

12 https://www.acs.org/content/acs/en/education/whatischemistry/landmarks/josephpriest leyoxygen.html. Accessed October 9, 2019.

13 Humphry Davy, *Researches, Chemical and Philosophical: Chiefly Concerning Nitrous Oxide* (Bristol, UK: Biggs and Cottle, 1800), p. 556.

14 Sherwin Nuland, *The Origins of Anesthesia* (Birmingham: Classics of Modern Medicine, 1983), p. 54.

15 Ibid., p. 55.

16 Ibid., p. 63.

17 https://archive.org/stream/101495446.nlm.nih.gov/101495446#page/n1/mode/2up. Accessed October 9, 2019.

18 Sherwin Nuland, *The Origins of Anesthesia* (Birmingham: Classics of Modern Medicine, 1983), p. 65.

19 Ibid., p. 67.

20 Ibid., p. 68.

21 John Collins Warren, "Inhalation of ethereal vapor for the prevention of pain in surgical operations," *Boston Medical and Surgical Journal*, December 9, 1846.

22 Sherwin Nuland, *The Origins of Anesthesia* (Birmingham: Classics of Modern Medicine, 1983), p. 99.

23 Gordon, H. Laing, quoted in Sherwin Nuland, *The Origins of Anesthesia* (Birmingham: Classics of Modern Medicine, 1983), p. 108.

24 http://www.ph.ucla.edu/epi/snow/victoria.html. Accessed October 9, 2019.

TEN: ELECTIVE SURGERY

1   Gerald Imber, *Genius on the Edge: The Bizarre Double Life of Dr. William Stewart Halsted* (New York: Kaplan, 2011), p. 66.

2   James Thomas Flexner and Simon Flexner, "William Henry Welch and the Heroic Age of American Medicine," *New England Journal of Medicine*, 1942; 227: 152–54, July 23, 1942.

3   Gerald Imber, *Genius on the Edge: The Bizarre Double Life of Dr. William Stewart Halsted* (New York: Kaplan, 2011), p. 42.

4   Joshua Berrett, "Doctors Afield: Theodor Billroth," *New England Journal of Medicine*, 264; Jan. 5, 1961, p. 38.

5   A. Cesmebasi, et al., "A Historical Perspective: Bernhard von Langenbeck German Surgeon (1810–1887)," *Clinical Anatomy* 27: 972–75, 2014.

6   Ibid.

7   Siddhartha Mukherjee, *The Emperor of All Maladies* (New York: Scribner, 2010), p. 58.

8   Sherwin Nuland, *Doctors: The Biography of Medicine* (New York: Vintage Books, 1995), p. 391.

9   R. Kazi and R. Peter, "Christian Albert Theodor Billroth: Master of Surgery," *Journal of Postgraduate Medicine*, 2004; 50: 82–3.

10 Sherwin Nuland, *Doctors: The Biography of Medicine* (New York: Vintage Books, 1995), p. 391.

11 M. Goerig, et al., "Carl Koller, Cocaine, and Local Anesthesia. Some less known and forgotten facts," *Regional Anesthesia and Pain Medicine*, 37(3, May-June): 318, 2012.

12 Ibid.

13 G. Gaertner, *Die Entdeckung der Lokalanasthesia* (Vienna: Der neue Tag, 1919), 6.

14 Ibid.

15 H. D. Noyes, "The Ophthalmological Congress in Heidelberg," *Medical Record.* 1884; 26: 417–18.

16 Gerald Imber, *Genius on the Edge: The Bizarre Double Life of Dr. William Stewart Halsted* (New York: Kaplan, 2011), p. 55.

17 Howard Markel, *An Anatomy of Addiction: Sigmund Freud, William Halsted, and the Miracle Drug, Cocaine* (New York: Vintage, 2012), p. 108.

18 Ibid., p. 111.

19 Gerald Imber, *Genius on the Edge: The Bizarre Double Life of Dr. William Stewart Halsted* (New York: Kaplan, 2011), p. 80.

20 Ibid., p. 87.

21 Ibid., p. 98.

22 Ibid., p. 349.

23 Sherwin Nuland, *Doctors: The Biography of Medicine* (New York: Vintage Books, 1995), p. 414.

24 Gerald Imber, *Genius on the Edge: The Bizarre Double Life of Dr. William Stewart Halsted* (New York: Kaplan, 2011), p. 115.

25 Ibid., p. 118.

26 Siddhartha Mukherjee, *The Emperor of All Maladies* (New York: Scribner, 2010), p. 47.

27 Gerald Imber, *Genius on the Edge: The Bizarre Double Life of Dr. William Stewart Halsted* (New York: Kaplan, 2011), p. 228.

28 Ibid., p. 233.

29 Ibid., p. 349.

30 Ibid., p. 296.

31 Ibid., p. 348.

32   Ibid., p. 146.
33   Ibid., p. 350.

ELEVEN: VITALLIUM

1   A. Boire, V. A. Riedel, N. M. Parrish S. Riedel, "Tuberculosis: From an Untreatable
      Disease in Antiquity to an Untreatable Disease in Modern Times?" *Journal of Ancient
      Diseases and Preventable Remedies*, 2013, vol. 1, pp. 1–11.
2   N. J. Eynon-Lewis, D. Ferry, and M. F. Pearse, "Themistocles Gluck, Unrecognized
      Genius," *British Medical Journal*, 1992, vol. 305, pp. 1534–36.
3   R. A. Brand, M. A. Mont, and M. M. Manring, "Biographical Sketch: Themistocles
      Gluck (1853–1942)," *Clinical Orthopedics and Related Research*, 2011, 469, pp. 1525–27.
4   Ibid., p. 1527.
5   M. J. Bankes and R. J. Emery, "Pioneers of Shoulder Replacement: Themistocles Gluck
      and Jules Emile Péan," *Journal of Shoulder and Elbow Surgery*, 1995, vol. 4, pp. 259–62.
6   Ibid., p. 260.
7   M. N. Smith-Petersen, *Journal of Bone and Joint Surgery*, 1953, vol. 35, pp. 1042–44.
8   M. N. Smith-Peterson, "Evolution of mould arthroplasty of the hip joint," *Journal of Bone
      and Joint Surgery*, 1948, vol. 30B, pp. 59–75.
9   C. S. Venable, W. G. Stuck, and A. Beach, "The effects of bone of the presence of metals;
      based upon electrolysis, an experimental study," *Annals of Surgery*, 1937, vol. 105, pp. 917–38.
10   M. N. Smith-Petersen, "Arthroplasty of the hip, a new method," *Journal of Bone and Joint
      Surgery*, 1939, vol. 37, p. 269–88.
11   Ibid., p. 278.
12   E. D. McBride, "A femoral head prosthesis for the hip joint. Four years' experience and
      the results," *Journal of Bone and Joint Surgery*, 1952, vol. 34, pp. 989–96.
13   Ibid., p. 989.
14   M. J. Bankes and R. J. Emery, "Pioneers of Shoulder Replacement: Themistocles Gluck
      and Jules Emile Péan," *Journal of Shoulder and Elbow Surgery*, 1995, vol. 4, p. 262.

TWELVE: OVERSIGHT AND ENTITLEMENT

1   Sue Blevins, *Medicare's Midlife Crisis* (Washington, DC: Cato Institute, 2001), p. 25.
2   Ibid.
3   Ronald L. Numbers, ed., *Compulsory Health Insurance: The Continuing American Debate*
      (Westport, Conn.: Greenwood Press, 1982), p. 6.
4   R. Cunningham and R. M. Cunningham, *The Blues: A History of the Blue Cross and Blue
      Shield System* (Dekalb: Northern Illinois University, 1997), p. ix.
5   Ibid., p. 5.
6   Ibid., p. 4.
7   James E. Stuart, *The Blue Cross Story: An Informal Biography of the Voluntary Nonprofit
      Prepayment Plan for Hospital Care* (self-published), 1952, p. 18.
8   R. Cunningham and R. M. Cunningham, *The Blues: A History of the Blue Cross and Blue
      Shield System* (Dekalb: Northern Illinois University, 1997), p. 35.
9   Ibid., p. 59.
10   Ibid., p. 92.
11   Ibid., p. 118.
12   Oscar Ewing, press statement (Federal Security Agency, Washington, DC, June 25,
      1951).

13   Julian E. Zelizer, "How Medicare Was Made," *New Yorker*, Feb. 15, 2015.

14   Paul Starr, *The Social Transformation of American Medicine: The Rise of a Sovereign Profession and the Making of a Vast Industry* (New York: Basic Books, 1982).

15   Howard, S. Berliner, "The Origins of Health Insurance for the Aged," *International Journal of Health Services* 3, no. 3 (1973): 465.

16   Sue Blevins, *Medicare's Midlife Crisis* (Washington, DC: Cato Institute, 2001), p. 42.

17   Julian Zelizer, "The Contentious Origins of Medicare and Medicaid," in *Medicare and Medicaid at 50, America's Entitlement Programs in the Age of Affordable Care*, Alan B. Cohen, David C. Colby, Keith A. Wailoo, and Julian Zelizer, eds. (Oxford, UK: Oxford University Press, 2015), p. 13.

18   James Morone and Elisabeth Fauquert, "Medicare in American Political History: The Rise and Fall of Social Insurance," in *Medicare and Medicaid at 50, America's Entitlement Programs in the Age of Affordable Care*, Alan B. Cohen, David C. Colby, Keith A. Wailoo, and Julian Zelizer, eds. (Oxford, UK: Oxford University Press, 2015), p. 299.

19   Ibid., p. 299.

20   Ibid.

21   Ibid., p. 300.

22   Ira Katznelson, *Fear Itself, The New Deal and the Origins of Our Time* (New York: Liveright, 2013)

23   Sue Blevins, *Medicare's Midlife Crisis* (Washington, DC: Cato Institute, 2001), p. 46.

24   Ibid.

25   D. B. Smith, "Civil Rights and Medicare, Historical Convergence and Continuing Legacy," in *Medicare and Medicaid at 50, America's Entitlement Programs in the Age of Affordable Care*, Alan B. Cohen, David C. Colby, Keith A. Wailoo, and Julian Zelizer, eds. (Oxford, UK: Oxford University Press, 2015), p. 35.

26   Ibid.

27   Nathaniel Wesley, *Black Hospitals in America: History, Contributions, and Demise* (Tallahassee, Fla., NRW Associates Publications, 2010).

28   Cited in Rick Mayes, "The Origins, Development, and Passage of Medicare's Revolutionary Prospective Payment System," *Journal of the History of Medicine* 62, Jan. 2007, p. 25.

29   Uwe Reinhardt, "Medicare Innovations in the War Over the Key to the US Treasury," in *Medicare and Medicaid at 50, America's Entitlement Programs in the Age of Affordable Care*, Alan B. Cohen, David C. Colby, Keith A. Wailoo, and Julian Zelizer, eds. (Oxford, UK: Oxford University Press, 2015), p. 172.

30   Ibid.

31   Ibid., p. 173.

32   Ibid.

33   Ibid., p. 174.

34   Ibid., p. 175.

35   American Medical Association, "History of the RBRVS," http://www.ama-assn.org //ama/pub/physician-resources/solutions-managing-your-practice/coding-billing-insurance/medicare/the-resource-based-relative-value-scale/history-of-rbrvs.page. Accessed October 9, 2019.

36   Uwe Reinhardt, "Medicare Innovations in the War Over the Key to the US Treasury," in *Medicare and Medicaid at 50, America's Entitlement Programs in the Age of Affordable Care*, Alan B. Cohen, David C. Colby, Keith A. Wailoo, and Julian Zelizer, eds. (Oxford, UK: Oxford University Press, 2015), p. 178.

37   Uwe Reinhardt, "Medicare Innovations in the War Over the Key to the US Treasury," in
      Medicare and Medicaid at 50, America's Entitlement Programs in the Age of Affordable Care,
      Alan B. Cohen, David C. Colby, Keith A. Wailoo, and Julian Zelizer, eds. (Oxford, UK:
      Oxford University Press, 2015), p. 179.
38   Kaiser Family Foundation, "10 Essential Facts About Medicare's Finan-
      cial Outlook," Feb. 2, 2017, http://kff.org/medicare/issue-brief/10-essential-facts
      -about-medicares-financial-outlook/. Accessed October 9, 2019.
39   Uwe Reinhardt, "Medicare Innovations in the War Over the Key to the US Treasury," in
      Medicare and Medicaid at 50, America's Entitlement Programs in the Age of Affordable Care,
      Alan B. Cohen, David C. Colby, Keith A. Wailoo, and Julian Zelizer, eds. (Oxford, UK:
      Oxford University Press, 2015), p. 182.

THIRTEEN: DEVICE CLEARANCE

1   Philip J. Hilts, *Protecting America's Health: The FDA, Business, and One Hundred Years of
      Regulation* (Chapel Hill: University of North Carolina Press, 2004), p. ix.
2   Ibid., p. 3.
3   Ibid., p. x.
4   Ibid., p. xi.
5   David Greenberg, "How Teddy Roosevelt Invented Spin," Atlantic, Jan. 24, 2016,
      https://www.theatlantic.com/politics/archive/2016/01/how-teddy-roosevelt
      -invented-spin/426699/. Accessed October 9, 2019.
6   Ibid.
7   Philip J. Hilts, *Protecting America's Health: The FDA, Business, and One Hundred Years of
      Regulation* (Chapel Hill: University of North Carolina Press, 2004), p. 24.
8   Ibid., p. 55.
9   Ibid.
10   Ibid., p. 93.
11   Ibid.
12   Carol Rados, "Medical Device and Radiological Health Regulations Come of Age,"
      FDA Consumer Magazine, Jan.–Feb., 2006, https://www.fda.gov/aboutfda/whatwedo
      /history/productregulation/medicaldeviceandradiologicalhealthregulationscomeofage
      /default.htm. Accessed October 9, 2019.
13   Meryl Gordon, "A Cash Settlement, but No Apology", *New York Times*, Feb. 20, 1999,
      http://www.nytimes.com/1999/02/20/opinion/a-cash-settlement-but-no-apology.html.
      Accessed October 9, 2019.
14   I. D. Learmonth, C. Young, C. Rorabeck, "The Operation of the Century," *The Lancet*,
      2007, 1508–19.
15   G. K. McKee, J. Watson-Farrar, "Replacement of arthritic hips by the McKee-Farrar
      prosthesis," *Journal of Bone and Joint Surgery*, 1966, 48 B:245, 59.
16   D. Cohen, "Out of joint: The Story of the ASR," *British Medical Journal*, May 14, 2011,
      342:d2905.
17   https://www.depuysynthes.com/asrrecall/depuy-asr-recall-usen.html. Accessed October
      9, 2019.
18   C. Delaunay, "Registries in Orthopaedics," *Orthopaedics & Traumatology: Surgery &
      Research*, 101 (2015), S69–S75.
19   P. Slatis and B. Veraart, "Goran Carl Harald Bauer: 1923–1994," *Acta Orthopaedica
      Scandinavica*, 65: 5, 491–8, 1994.

20   Barry Meier, "A Call for a Warning System on Artificial Joints," *New York Times*, July 29, 2008.

21   Australian Orthopedic Association. National Joint Replacement Registry, annual report 2007. AOA, 2008.

22   D. Cohen, "Out of joint: The Story of the ASR," *British Medical Journal*, May 14, 2011, 342:d2905.

23   Ibid.

24   Barry Meier, "A Call for a Warning System on Artificial Joints," *New York Times*, July 29, 2008.

25   Barry Meier, "House Bill Would Create Artificial Joints Registry," *New York Times*, June 10, 2009.

26   Barry Meier, "Concerns Over Metal on Metal Hip Implants," *New York Times*, March 3, 2010.

27   Ibid.

28   http://www.mcminncentre.co.uk/research-lectures-debate.html. Accessed October 9, 2019.

29   Barry Meier, "Doctors Who Don't Speak Out," *New York Times*, Feb. 15, 2013.

30   DePuy Orthopedics Inc. 2010. "DePuy Orthopedics Voluntarily Recalls Hip System," https://www.depuysynthes.com/about/news-press/qs/depuy-orthopaedics -voluntarily-recalls-asr-hip-system---depuy. Accessed October 9, 2019.

31   http://www.annualreports.com/HostedData/AnnualReportArchive/j/NYSE_JNJ_2013.pdf. Accessed October 9, 2019.

32   Barry Meier, "Frustrations from a Deal on Flawed Hip Implants," *New York Times*, Nov. 25, 2013.

33   Matthias Wienroth, et al., "Precaution, governance and the failure of medical implants: The ASR hip in the UK," *Life Sciences, Society and Policy*, 2014, 10:19.

34   Andrew Barry, *Political Machines: Governing a Technological Society* (London: Athlone Press, 2001).

35   Matthias Wienroth, et al., "Precaution, governance and the failure of medical implants: The ASR hip in the UK," *Life Sciences, Society and Policy*, 2014, 10:19.

36   D. Cohen, "How Safe are Metal-on-Metal Hip Implants?" *British Medical Journal*, Feb. 28, 2012, 344: e1410.

FOURTEEN: MEDICAL INDUSTRIAL COMPLEX AND MEDICAL DEVICES

1   William Henry Kellar, *Enduring Legacy: The M.D. Anderson Foundation and The Texas Medical Center* (College Station: Texas A&M University Press, 2014), p. 37.

2   Ibid., p. 41.

3   Ibid., p. xxi.

4   Ibid., p. 182.

5   Ibid., p. 196.

6   Society for Assisted Reproductive Technology, http://www.sart.org/globalassets/__sart /infographics/number-of-clinics-treatments-births.png. Accessed October 9, 2019.

7   G. S. Dawe, et al., "Cell Migration from Baby to Mother," *Cell Adhesion & Migration*, 1(1): 2007, pp. 19–27.

8   M. F. Maitz, "Applications of synthetic polymers in clinical medicine," *Biosurface and Biotribology*, vol. 1, 2015, pp. 161–76.

9   http://education.seattlepi.com/can-minerals-form-deep-within-earth-6008.html. Accessed October 9, 2019.

10   https://www.britannica.com/technology/chromium-processing. Accessed October 9, 2019.

11   http://www.mining.com/web/global-cobalt/. Accessed October 9, 2019.

12   https://www.theatlantic.com/health/archive/2017/11/placebo-effect-of-the-heart /545012/. Accessed October 9, 2019.

13   https://globenewswire.com/news-release/2016/12/15/897773/0/en/Global-Cardiac -Pacemaker-Market-will-exceed-USD-12-00-billion-by-2021-Zion-Market-Research. html. Accessed October 9, 2019.

FIFTEEN: SURGERY OF THE HEART

1   Stephen Paget, *The Surgery of the Chest* (Bristol, England: John Wright, 1896), p. 121.

2   http://www.timesleader.com/news/local/455923/dr-victor-greco-operated-heart-lung -machine-during-first-successful-open-heart-surgery. Accessed October 9, 2019.

3   William Stoney, *Evolution of Cardiopulmonary Bypass*, vol. 119, pp. 2844–53, 2009.

4   Vincent Gott, Lewis Lillehei, and Owen Wangensteen, "The Right Mix for Giant Achievement in Cardiac Surgery," *Annals of Thoracic Surgery*, vol. 79, 2005, pp. S2210–13.

5   Ibid., p. S2211.

6   Ibid.

7   https://medicine.wright.edu/about/news-and-events/vital-signs/article/a-real-life -macgyver-builds-a-medical-school. Accessed October 9, 2019.

8   Earl Bakken, *A Full Life, The Autobiography of Earl Bakken*. Self published, p. 32.

9   Ibid.

10   Ibid.

11   http://www.pbs.org/transistor/album1/. Accessed October 9, 2019.

12   http://www.pbs.org/transistor/background1/corgs/bellabs.html. Accessed October 9, 2019.

13   Earl Bakken, *A Full Life, The Autobiography of Earl Bakken*. Self published, p. 38.

14   http://www.medtronic.com/us-en/about/facts-stats.html. Accessed October 9, 2019.

15   Henry Ellis, and John W. Kirklin, "Aortic Stenosis," *Surgical Clinics of North America*, Aug. 1955, p. 1033.

16   W. Bruce Fye, *Caring for the Heart: Mayo Clinic and the Rise of Specialization* (Oxford, UK: Oxford University Press, 2015), p. 250.

17   Ibid., p. 253.

18   A. F. Crocetti, "Cardiac Diagnostic and Surgical Facilities in the United States," Public Health Rep., 1965, 80: 1035–53.

19   W. Bruce Fye, *Caring for the Heart: Mayo Clinic and the Rise of Specialization* (Oxford, UK: Oxford University Press, 2015), p. 323.

20   W. Bruce Fye, *Caring for the Heart: Mayo Clinic and the Rise of Specialization* (Oxford, UK: Oxford University Press, 2015), p. 323, quoting Hurst, "History of Cardiac Catheterization," in S. B. King III and J. S. Douglas, eds. *Coronary Arteriography and Angioplasty* (New York: McGraw-Hill, 1985), pp. 5–6.

21   W. Bruce Fye, *Caring for the Heart: Mayo Clinic and the Rise of Specialization* (Oxford, UK: Oxford University Press, 2015), p. 326, quoting D. B. Effler to F. A. LeFevre, Nov. 8, 1960, Effler Papers, CCA.

22   A. Roguin, *Cardiovascular Interventions*. Circulation: 2011;4:206–209.

23   https://www.theatlantic.com/technology/archive/2013/12/no-old-maps-actually -say-here-be-dragons/282267/. Accessed October 9, 2019.

24   R. P. Hudson, "Eisenhower's heart attack: How Ike beat heart disease and held onto the presidency," (review). *Bulletin of the History of Medicine*, vol. 72 (1), p. 161–62.

25   https://www.azquotes.com/quote/1267465. Accessed October 9, 2019.

SIXTEEN: SPECIALIZATION IN SURGERY

1   William Osler, "Why is it so? Is it so?" *Journal of the Tennessee State Medical Association*, 1919, 12: 222.

2   Paul Starr, *The Social Transformation of American Medicine: The Rise of a Sovereign Profession and the Making of a Vast Industry* (New York: Basic Books, 1982), p. 38.

3   Ibid.

4   US Bureau of the Census, *Historical Statistics of the United States: Colonial Times to 1970* (Washington, DC: US Department of Commerce, 1975), p. 78.

5   W. Bruce Fye, *Caring for the Heart: Mayo Clinic and the Rise of Specialization* (Oxford, UK: Oxford University Press, 2015), p. 9.

6   https://www.thoughtco.com/how-skyscrapers-became-possible-1991649. Accessed October 9, 2019.

7   https://www.spc.noaa.gov/faq/tornado/f-scale.html. Accessed October 9, 2019.

8   S. H. Severson, *Rochester: Mecca for Millions* (Rochester, MN: Marquette Bank & Trust, 1979).

9   W. W. Mayo, "Address," in Memorial of St. Mary's Hospital, (Rochester, Minn.: St. Mary's Hospital, 1894), pp. 7–8.

10   W. J. Mayo, "John[s] Hopkins, May 1895," handwritten notebook, MCA.

11   W. Bruce Fye, *Caring for the Heart: Mayo Clinic and the Rise of Specialization* (Oxford, UK: Oxford University Press, 2015), p. 16.

12   Ibid., p. 17.

13   Ibid., p. 19.

14   Ibid., p. 23.

15   Ibid., p. 29.

16   W. J. Mayo, "Commencement Address," in *Collected Papers of the Staff of St. Mary's Hospital, Mayo Clinic* (Philadelphia: W.B. Saunders, 1911), pp. 557–66.

17   Rosemary Stevens, *American Medicine and the Public Interest*, rev. ed., (Berkeley: University of California Press, 1998), p. ix.

18   David B. Levine, *Anatomy of a Hospital: Hospital for Special Surgery, 1863–2013* (New York: Hospital for Special Surgery, 2013), p. xi.

19   Ibid., p. 4.

20   M. M. Manning and J. H. Calhoun, "Royal Whitman, 1857–1946" *Journal of Bone and Joint Surgery American Vol.*, 1946, vol. 28, pp. 890–92.

21   Rosemary Stevens, *American Medicine and the Public Interest*, rev. ed., (Berkeley: University of California Press, 1998), p. ix.

22   David B. Levine, *Anatomy of a Hospital: Hospital for Special Surgery, 1863–2013* (New York: Hospital for Special Surgery, 2013), p. 185.

23   Ibid., p. 215.

SEVENTEEN: IMPLANT REVOLUTION

1   H. P. Platt, Sir John Charnley in *Some Manchester Doctors* W. J. Elwood, A. F. Tuxford, eds. (Manchester, UK: Manchester University Press, 1985).

2   C. S. Neer, T. H. Brown, and H. L. McLaughlin, "Fracture of the neck of the humerus with dislocation of the head fragment," *American Journal of Surgery*, March 1953, pp. 252–58.

3   C. S. Neer, "Articular replacement for the humeral head," *Journal of Bone and Joint Surgery*, 1955, 37-A, pp. 215–28.

4   C. S. Neer, "Replacement arthroplasty for glenohumeral osteoarthritis," *Journal of Bone and Joint Surgery*, 1974, 56-A, pp. 1–13.

5   William Waugh, *John Charnley: The Man and the Hip* (Berlin: Springer-Verlag, 1990), p. 114.

6   John Charnley, "Arthroplasty of the hip—a new operation," 1961, *The Lancet* I:1129–32.

7   William Waugh, *John Charnley: The Man and the Hip* (Berlin: Springer-Verlag, 1990), p. 122.

8   Ibid.

9   C. S. Neer, "Recent experience in total shoulder replacement," *Journal of Bone and Joint Surgery*, 1982, 64-A, pp. 319–37.

EIGHTEEN: THE BIRTH OF SPORTS MEDICINE

1   https://www.forbes.com/forbes-400/list/3/#version:static. Accessed October 9, 2019.

2   Reinhold Schmieding, "Helping Surgeons Treat their Patients Better: A history of Arthrex's contribution to Arthroscopic Surgery" Arthrex publication, 2006, p. 12.

3   Reinhold Schmieding, Personal communication, June 2, 2017.

4   The *New York Tribune* the following day would declare, ". . . the sun smiled cheerfully, now and then dodging behind clouds as if he had got a black eye at football . . . all that was wanted was a little warmth for there were thousands of 'tiger' men and pretty girls who shivered in the chill November air."

5   Cowan, a Presbyterian minister, would go on to become the football coach at the University of North Carolina and the University of Kansas (where he coached John Outland). Cowan resigned as head football coach after three years, but continued as a physical culture professor for an additional two years, before being replaced by a new professor . . . James Naismith.

6   http://drs.library.yale.edu:8083/HLTransformer/HLTransServlet?stylename=yul.ead2002.xhtml.xsl&pid=mssa:ms.0125&clear-stylesheet-cache=yes. Accessed October 9, 2019.

7   Ibid.

8   Zezima, Katie, May 29, 2014 Washington Post https://www.washingtonpost.com/news/the-fix/wp/2014/05/29/teddy-roosevelt-helped-save-football-with-a-white-housemeeting-in-1905. Accessed October 9, 2019.

9   Almost never done anymore. We typically use the *palmaris longus* from the *ipsilateral* (same side) arm, or one of the smaller hamstring tendons from the *ipsilateral* leg.

10   Dr. Frank Jobe has received special recognition at the Baseball Hall of Fame. It is hard to name another figure who has had such a profound impact on the game. Contemplate the difference between Sandy Koufax walking away from the game at thirty, and Mariano Rivera having elbow surgery from Dr. Jobe before ever having played a single big league game.

NINETEEN: CALCULATING THE IMPACT

1   https://www.cbo.gov/about/products/budget-economic-data#2. Accessed October 9, 2019.

2   D. P. Rice, B. S. Cooper, National Health Expenditures, 1950–67, Bulletin, Jan. 1969, https://www.ssa.gov/policy/docs/ssb/v32n1/v32n1p3.pdf. Accessed October 9, 2019.

3   https://data.bls.gov/cgi-bin/cpicalc.pl?cost1=5500&year1=196712&year2=201712. Accessed October 9, 2019.

4   https://www.hcup-us.ahrq.gov/db/nation/nis/nisdbdocumentation.jsp. Accessed October 9, 2019.

5   Mark Coventry, "The History of Joint Replacement Arthroplasty,' *Joint Replacement Arthroplasty* (Philadelphia: Churchill Livingstone, 2003), p. 6.

6   S. Kurtz, et al., "Projections of Primary and Revision Hip and Knee Arthroplasty in the United States from 2005 to 2030," *Journal of Bone and Joint Surgery*, 2007, pp. 780–85.

7   K. McDermott, et al., "Overview of Operating Room Procedures During Inpatient Stays in U.S. Hospitals," 2014. Statistical Brief #233. December, 2017. Agency for Healthcare Research and Quality, Rockville, Md. HCUP-Operating-Room-Procedures-United-States-2014 (1).pdf. Accessed Jan. 13, 2018.

8   Ibid.

9   Ibid.

10   Ibid.

11   Ibid.

12   S. Kurtz, et al., "Projections of Primary and Revision Hip and Knee Arthroplasty in the United States from 2005 to 2030," *Journal of Bone and Joint Surgery*, 2007, pp. 780–85.

13   R. Westermann, "Reverse shoulder arthroplasty in the United States: A comparison of national volume, patient demographics, complications, and surgical indications," *Iowa Orthopedic Journal*, (35), 2015, pp. 1–7.

14   E. Melamed, et al., "Trends in the Utilization of Total Wrist Arthroplasty versus Wrist Fusion for Treatment of Advanced Arthritis," *Journal of Wrist Surgery*, 5 (3), 2016, pp. 211–16.

15   SmartTRAK, 2018 Orthopedic Industry report.

16   S. Raikin, "Trends in Treatment of Advanced Ankle Arthroplasty by Total Ankle Replacement or Ankle Fusion," *Foot Ankle International*, March, 35(3); 2014, pp. 216–24.

17   SmartTRAK, 2018 Orthopedic Industry report.

18   K. McDermott, et al., "Overview of Operating Room Procedures During Inpatient Stays in U.S. Hospitals," 2014. Statistical Brief #233. December, 2017. Agency for Healthcare Research and Quality, Rockville, Md. HCUP-Operating-Room-Procedures-United-States-2014 (1).pdf. Accessed Jan. 13, 2018.

19   Ibid.

20   SmartTRAK, 2018 Orthopedic Industry report.

21   K. McDermott, et al., "Overview of Operating Room Procedures During Inpatient Stays in U.S. Hospitals," 2014. Statistical Brief #233. December, 2017. Agency for Healthcare Research and Quality, Rockville, Md. HCUP-Operating-Room-Procedures-United-States-2014 (1).pdf. Accessed Jan. 13, 2018.

22   SmartTRAK, 2018 Orthopedic Industry report.

23   Ibid.

24   L. T. Buller, et al., "Trends in Anterior Cruciate Ligament Reconstruction in the United States," *Orthopedic Journal of Sports Medicine*, 3(1), 2015, pp. 1–8.

25   M. P. Leathers, "Trends and demographics in anterior cruciate ligament reconstruction in the United States," *Journal of Knee Surgery*, Oct. 28(5); pp. 390–94.

26   L. T. Buller, et al., "Trends in Anterior Cruciate Ligament Reconstruction in the United States," *Orthopedic Journal of Sports Medicine*, 3(1), 2015, pp. 1–8.

27   https://www.census.gov/popclock/. Accessed Feb. 22, 2018.

28   SmartTRAK, 2018 Orthopedic Industry report.

29   Ibid.

30   A. Chiang Colvin, et al., "National Trends in Rotator Cuff Repair" *Journal of Bone and Joint Surgery*, Feb. 94(3), 2012, pp. 227–33.

31   N. Bonazza, et al., "Trends in surgical management of shoulder instability," *Orthopedic Journal of Sports Medicine*, June, 5(6), 2017, pp. 1–7.

32   SmartTRAK, 2018 Orthopedic Industry report.

33   Ibid.

34   R. Lee, et al., "Fifteen-year outcome trends for valve surgery in North America," *Annals of Thoracic Surgery*, 91, 2011, pp. 677–84.

35   Ibid.

36   F. Algahtani, et al., "Contemporary trends in the use and outcomes of surgical treatment of tricuspid regurgitation," *Journal of the American Heart Association*, Dec., 6(12): e007597, pp. 1–10.

37   J. S. Gammie, et al., "Trends in mitral valve surgery in the United States: Results from the Society of Thoracic Surgeons Adult Cardiac Database," *Annals of Thoracic Surgery*, 87, 2009, pp. 1431–9.

38   A. R. Opotowsky, et al., "A shifting approach to management of the thoracic aorta in bicuspid aortic valve," *Journal of Thoracic and Cardiovascular Surgery*, Aug., 146(2), 2013, pp. 339–46.

39   K. McDermott, et al., "Overview of Operating Room Procedures During Inpatient Stays in U.S. Hospitals," 2014. Statistical Brief #233. December, 2017. Agency for Healthcare Research and Quality, Rockville, Md. HCUP-Operating-Room-Procedures-United-States-2014 (1).pdf. Accessed Jan. 13, 2018.

40   R. Lee, et al., "Fifteen-year outcome trends for valve surgery in North America," *Annals of Thoracic Surgery*, 91, 2011, pp. 677–84.

41   K. McDermott, et al., "Overview of Operating Room Procedures During Inpatient Stays in U.S. Hospitals," 2014. Statistical Brief #233. December, 2017. Agency for Healthcare Research and Quality, Rockville, Md. HCUP-Operating-Room-Procedures-United-States-2014 (1).pdf. Accessed Jan. 13, 2018.

42   Ibid.

43   S. Kurtz, et al., "Implantation trends and patient profiles for pacemakers and implantable cardioverter defibrillators in the United States: 1993–2006," *Pacing and Clinical Electrophysiology*, June 1, 2010.

44   M.J.P. Raatikainen, et al.; "Statistics on the use of cardiac electronic devices and electrophysiological procedures in the European Society of Cardiology countries: 2014 report from the European Heart Rhythm Association," *Europace* 17, 2015, i1–i75.

45   Anahad O'Connor, "Heart Stents Still Overused, Experts Say," *New York Times*, Aug. 15, 2013, https://well.blogs.nytimes.com/2013/08/15/heart-stents-continue-to-be-overused/. Accessed March 8, 2018.

46   L. Szabo, "Stents open clogged arteries of 1M Americans annually," Aug. 6, 2013. https://www.usatoday.com/story/news/politics/2013/08/06/bush-stent-heart-surgery/2623111/. Accessed March 8, 2018.

47   Ilene McDonald, "Half of cardiac stent procedures overused, unnecessary," *Fierce Healthcare*. https://www.fiercehealthcare.com/healthcare/half-cardiac-stent-procedures-overused-unnecessary. Accessed March 8, 2018.

48   R. Riley, et al., "Trends in coronary revascularization in the United States from 2001 to 2009, recent declines in percutaneous coronary intervention volumes," *Circulation: Cardiovascular Quality and Outcomes*, March 1; 4(2); 2011, pp. 193–97.

49   A. Epstein, "Coronary revascularization trends in the United States, 2001–2008," *JAMA*, May 4, vol. 305,(17), 2011, pp. 1769–776.

50   https://blog.mediligence.com/2009/05/05/drug-eluting-bare-metal-and-absorbable-stents-segment-growth-2009-and-2017/. Accessed March 8, 2018.

51   D. Buck, et al., "The Impact of endovascular treatment on isolated iliac artery aneurysm treatment and mortality," *Journal of Vascular Surgery*, Aug., 62(2), 2015, pp. 331–335.

52   https://www.healio.com/cardiac-vascular-intervention/aneurysm-repair/news/online/%7B51a14891-cdd4-439e-9dd5-368cc492e92a%7D/total-number-of-aaa-repairs-in-us-declining-annually-since-2005. Accessed March 6, 2018.

53   L. Mureebe, et al., "National trends in the repair of ruptured abdominal aortic aneurysms," *Journal of Vascular Surgery*, vol. 48 (5), Nov. 2008, pp. 1101–07.

54   R. Parwardhan, "Implanted ventricular shunts in the United States: the billion-dollar-a-year cost of hydrocephalus treatment," *Neurosurgery*, 56; 2005, pp. 139–45.

55   F. Khan, et al., "Factors affecting ventriculoperitoneal shunt survival in adult patients," *Surgical Neurology International* (6), 2015, p. 25.

56   J. Jalbert, "Clipping and coiling of unruptured intracranial aneurysms among Medicare beneficiaries, 2000 to 2010," *Stroke* (46); 2015, pp. 2452–457.

57   https://www.cms.gov/Research-Statistics-Data-and-Systems/Statistics-Trends-and-Reports/MedicareMedicaidStatSupp/Downloads/2011_Section2.pdf#Table2.1. Accessed March 11, 2018.

58   A. A. Brinjikji, et al., "Better outcomes with treatment by coiling relative to clipping of unruptured intracranial aneurysms in the United States, 2001–2008." *American Journal of Neuroradiology*, June 2011 vol. 32 (6), pp. 1071–75.

59   https://www.medicalalley.org/media/22695/neuromod_pages.pdf. Accessed March 11, 2018.

60   B. Youngerman, et al., "A decade of emerging indications: deep-brain stimulation in the United States," *Journal of Neurosurgery*, vol. 125 (2), 2016, pp. 461–71.

61   J. Prager, "Estimates of annual spinal cord stimulator implant rises in the United States," *Neuromodulation*, vol. 13 (1), 2010, pp. 68–9.

62   https://www.grandviewresearch.com/industry-analysis/neurostimulation-devices-industry. Accessed March 25, 2018.

63   https://www.nidcd.nih.gov/health/statistics/quick-statistics-hearing. Accessed March 11, 2018.

64   http://www.medel.com/cochlear-implants-facts/. Accessed March 9, 2018.

65   https://unos.org/data/transplant-trends/#transplants_by_organ_type+year+2014. Accessed March 9, 2018.

66   https://unos.org/data/transplant-trends/#transplants_by_donor_type+organ+All Organs. Accessed March 9, 2018.

67   https://www.cdc.gov/art/pdf/2015-report/ART-2015-National-Summary-Report.pdf#page=65. Accessed March 18, 2018.

68   Ibid.

69   https://www.forbes.com/sites/davidsable/2014/04/24/ivf-and-infertility-by-the-numbers/. Accessed March 18, 2018.

70   https://www.medpagetoday.com/urology/erectiledysfunction/52233. Accessed October 9, 2019.

71   S. MacDonald, "Waves of change: national trends in surgical management of male stress incontinence," *Urology*, vol. 108, October, 2017, pp. 175–79.

72   C. Steiner, et al., "Surgeries in Hospital-Based Ambulatory Surgery and Hospital Inpatient Settings, 2014," Statistical Brief #223. May, 2017. Agency for Healthcare

Research and Quality, Rockville, MD. HCUP-Ambulatory-Inpatient-Surgeries-2014.pdf, Accessed Jan. 13, 2018.

73 W. Stark, et al., "Trends in Intraocular Lens Implantation in the United States," *Archives of Opthalmology*, vol. 104, Dec., 1986, pp. 1769–70.

74 https://www.healio.com/ophthalmology/cataract-surgery/news/print/premier-surgeon/%7B6c74b954-0386-4638-957e-9f58eff91c3f%7D/refractive-surgery-and-iols--future-trends. Accessed March 25, 2018.

75 https://www.aao.org/eyenet/article/simultaneous-bilateral-cataract-surgery-debate-con. Accessed March 25, 2018.

76 https://www.sciencedaily.com/releases/2016/10/161018094928.htm. Accessed March 21, 2018.

77 K. Baylon, et al., "Past, present and future of surgical meshes: a review," *Membranes*, vol. 7(3), pp. 1–23.

78 https://emedicine.medscape.com/article/1534321-overview. Accessed March 21, 2018.

79 https://www.goremedical.com/conditions/hernia. Accessed March 21, 2018.

80 https://asmbs.org/resources/estimate-of-bariatric-surgery-numbers. Accessed March 18, 2018.

81 Ibid.

82 http://obgyn.ucla.edu/mesh-related-complications. Accessed March 21, 2018.

83 Ibid.

84 Ibid.

85 Michele Jonsson Funk, et al., "Trends in the surgical management of stress urinary incontinence," *Obstetrics & Gynecology*, April; 119(4), 2012, pp. 845–51.

86 Ibid.

87 https://d2wirczt3b6wjm.cloudfront.net/News/Statistics/2014/plastic-surgery-statistics-full-report-2014.pdf. Accessed March 21, 2018.

88 http://breastimplantinfo.org/fda-breast-implants/. Accessed March 21, 2018.

89 L. Gaviria, et al., "Current trends in dental implants," *Journal of the Korean Association of Oral and Maxillofacial Surgeons*, 40(2), 2014, pp. 50–60.

TWENTY: BRAIN IMPLANTS

1 J. W. Langston, "The MPTP Story," *Journal of Parkinson's Disease*, (7), pp. S11–S19, 2017.

2 R. Lewin, "Trail of Ironies to Parkinson's Disease," *Science*, (224), pp. 1083–5, 1984.

3 J. M. Harlow, "Recovery from the passage of an iron bar through the head." *Publications of the Massachusetts Medical Society*. 2(3), pp. 327–47, 1868.

4 Maria Konnikova, *Scientific American*, Feb. 8, 2013, https://blogs.scientificamerican.com/literally-psyched/the-man-who-couldnt-speakand-how-he-revolutionized-psychology/. Accessed July 15, 2018.

5 Aubertin, 1861, quoted by L. L. LaPointe, *Paul Broca and the Origins of Language in the Brain* (San Diego: Plural Publishing, 2012), p. 129.

6 A. P. Wickens, *A History of the Brain: From Stone Age Surgery to Modern Neuroscience* (London: Psychology Press, 2014), p. 171.

7 Bahar Gholipour, "A visual history of neurons," *Brain Decoder*, April 13, 2015. http://behdad.org/mirror/www.braindecoder.com/a-visual-history-of-neurons-1089282606.html. Accessed July 19, 2018.

8 A. B. Keener, "The first neuron drawings, 1870s," *The Scientist*. https://www.the-scientist.com/foundations/the-first-neuron-drawings-1870s-34751. Accessed July 19, 2018.

9   Stanley Finger, "Santiago Ramón y Cajal: From Nerve Nets to Neuron Doctrine," *Minds Behind the Brain: A History of the Pioneers and their Discoveries* (New York: Oxford University Press, 2000), pp. 197–216.

10   E. A. Newman, A. Araque, and J. M. Dubinsky, eds., *The Beautiful Brain: The Drawings of Santiago Ramón y Cajal* (New York: Abrams, 2018), p. 12.

11   L. Swanson, in E. A. Newman, A. Araque, and J. M. Dubinsky, eds., *The Beautiful Brain: The Drawings of Santiago Ramón y Cajal* (New York: Abrams, 2018), p. 12.

12   M. Fessenden, Smithsonian.com, https://www.smithsonianmag.com/arts-culture/revel-these-wondrous-drawings-father-neuroscience-180961881/ Jan. 23, 2017, Accessed July 27, 2018.

13   Ibid.

14   E. V. Evarts, "Activity of neurons in visual cortex of the cat during sleep with low voltage fast EEG activity," *Journal of Neurophysiology* 25: 812–6, 1962.

15   E. V. Evarts, "Temporal patterns of discharge of pyramidal tract neurons during sleep and waking in the monkey," *Journal of Neurophysiology*, 27: 152–71, 1964.

16   E. V. Evarts, "Pyramidal tract activity associated with a conditioned hand movement in the monkey," *Journal of Neurophysiology*, 29: 1011–27, 1966.

17   W. T. Thach, *Edward Vaughan Evarts 1926–1985, A biographical memoir*. National Academy of Sciences. Biographical Memoirs, 2000, vol. 78, pp. 1–15.

18   Ibid., p. 6.

19   A. Mehta, Mahlon DeLong profile part 1. The Dana Foundation. http://www.dana.org /News/Details.aspx?id=42940. Accessed July 29, 2018.

20   Ibid.

21   M. R. DeLong, "Activity of pallidal neurons during movement," *Journal of Neurophysiology*,. 34: 414–27. 1971.

22   A. Mehta, Mahlon DeLong profile part 1. The Dana Foundation. http://www.dana.org /News/Details.aspx?id=42940. Accessed July 29, 2018.

23   https://med.emory.edu/gamechangers/researchers/delong/bio.html. Accessed July 29, 2018.

24   A. Mehta, Mahlon DeLong profile part 1. The Dana Foundation. http://www.dana.org /News/Details.aspx?id=42940. Accessed July 29, 2018.

25   Ibid.

26   http://www.dana.org/News/Details.aspx?id=42940. Accessed July 29, 2018.

27   H. Bergman, T. Wichmann, M. R. DeLong, "Reversal of experimental parkinsonism by lesions of the subthalamic nucleus," *Science*, vol. 249, Issue 4975, pp. 1436–1438, Sept. 1990.

28   Ibid.

29   J. L. Vitek, et al., "Randomized trial of pallidotomy versus medical therapy for Parkinson's disease," *Annals of Neurology*, vol. 53, 2003, pp. 558–569.

30   R. Williams, "Alim-Louis Benabid: Stimulation and Serendipity," *The Lancet Neurology*, vol. 9, Issue 12, Dec. 2010, p. 1152.

31   Ibid.

32   A. L. Benabid, P. Pollak, A. Louveau, S. Henry, and J. de Rougemont, "Combined (thalamotomy and stimulation) stereotactic surgery of the VIM thalamic nucleus for bilateral Parkinson disease," *Applied Neurophysiology*, vol. 50, 344–46, 1987.

33   G. E. Alexander, M. R. DeLong, P. L. Strick, "Parallel organization of functionally segregated circuits linking basal ganglia and cortex," *Annual Review of Neuroscience*, vol. 9, pp. 357–81, 1986.

34   https://www.epo.org/learning-events/european-inventor/finalists/2016/benabid.html. Accessed August 4, 2018.

35   Michael S. Okun, "Deep-Brain Stimulation—Entering the Era of Human Neural-Network Modulation." *New England Journal of Medicine*, vol. 371, Oct. 9, pp. 1369–73, 2014.

TWENTY-ONE: CYBORG FUTURE AND *HOMO ELECTRUS*

1   https://wondery.com/shows/dr-death/. Accessed October 9, 2019.

2   Yuval Noah Harari, *Homo Deus: A Brief History of Tomorrow* (New York: Harper Perennial, 2018), p. 21.

3   https://faculty.washington.edu/chudler/facts.html. Accessed Aug. 15, 2018.

4   T. James, et al., "BioMEMs—Advancing the Frontiers of Medicine," *Sensors*, 8(9): pp. 6077–107, 2008.

5   *Ex Machina* 2015—Behind the Scenes https://www.youtube.com/watch?v=nZcHPhGsNi0

6   Joe Rogan Experience #1169- Elon Musk https://www.stitcher.com/podcast/the-joe-rogan-experience/e/56151455. Accessed October 9, 2019.

7   M. Rozenfeld, "The future of medicine might be bioelectronic implants," http://theinstitute.ieee.org/technology-topics/life-sciences/the-future-of-medicine-might-be-bioelectronic-implants. Accessed October 9, 2019.

8   Ibid.

9   https://www.nextbigfuture.com/2017/03/elon-musk-has-gone-public-with-his.html. Accessed October 9, 2019.

10   Max Tegmark, *Life 3.0: Being Human in the Age of Artificial Intelligence* (New York: Vintage Books, 2017).

11   Ibid., p. 28.

12   Ibid., p. 29.

13   Ray Kurzweil, *The Singularity is Near: When Humans Transcend Biology* (New York: Penguin Books, 2005) p. 9.

14   Ibid., p. 311.

15   Ibid., p. 310.

16   Ibid., p. 298.

17   Charles Darwin, *On the Origin of Species*, (London: John Murray, 1885), p. 429.

Index